FUNDAMENTALS OF CALCULUS

FUNDAMENTALS OF CALCULUS

CARLA C. MORRIS
University of Delaware

ROBERT M. STARK
University of Delaware

WILEY

Published by John Wiley & Sons, Inc., Hoboken, New Jersey
Published simultaneously in Canada

Library of Congress Cataloging-in-Publication Data:

Morris, Carla C.
Fundamentals of Calculus / Carla C. Morris, Robert M. Stark.
pages cm
Includes bibliographical references and index.
ISBN 978-1-119-01526-0 (cloth)
1. Calculus–Textbooks. I. Stark, Robert M., 1930- II. Title.
QA303.2.M67 2015
515–dc23

2014042182

Printed in the United States of America

10 9 8 7 6 5 4 3 2 1

CONTENTS

Preface

A fundamental calculus course is a staple for students in Business, Economics, and the Natural, Social, and Environmental Sciences, among others. Most topics within this book parallel conventional texts, and they appear here directly, unimpeded by lengthy examples, explanations, data, irrelevances, and redundancies. Examples are the primary means to illustrate concepts and techniques as students readily respond to them. While there are ample Exercises and Supplementary Exercises, distracting abundance is avoided. Arrow symbols interspersed in the text, UP (↑) and DOWN (↓), are used to signal student tips, insights, and general information much as an instructor might to students in class to help their understanding. UP arrows modestly increase text depth while DOWN arrows expand aid to students.

Students often question the importance and usefulness of calculus, and some find math courses confusing and difficult. To address such issues, one goal of the text is for students to understand that calculus techniques involve basic rules used in combinations to solve complex problems. The challenge to students is to disassemble problems into manageable components. For example, a derivative of $[(f(x)g(x))/h(x)]^r$. The text encourages students to use power, quotient, and product rules for solutions. Another goal is to encourage students to understand calculus as the mathematics of change. To help, text examples guide students in modeling skills.

The elements of finite calculus lacking in most texts is a feature in this book, which serves multiple purposes. First, it offers an easier introduction by focusing on "change" and enables students to compare corresponding topics with differential calculus. Many may argue that it is the more natural calculus for social and other sciences. Besides, and equally important, finite calculus lends itself to modeling and spreadsheets. In Chapters 2 and 5, finite calculus is applied to marginal economic analysis, finance, growth, and decay.

Each chapter begins with an outline of sections, main topics of discussion, and examples. The outline displays topical chapter coverage and aids in finding items of particular interest. The Historical Notes sketch some of the rich 4000-year history of mathematics and its people.

Some students may skip Chapter 1, **Linear Equations and Functions** while others find it a useful review.

In Chapter 2, **The Derivative** finite differences are introduced naturally in forming derivatives and is a topic in its own right. Usually absent from applied calculus texts, finite calculus emphasizes understanding calculus as the "mathematics of change" (not simply rote techniques) and is an aid to popular spreadsheet modeling.

In Chapter 3, **Using the Derivative** students' newly acquired knowledge of a derivative appears in everyday contexts including marginal economic analysis. Early on, it shows students an application of calculus.

Chapter 4, **Exponential and Logarithmic Functions** delineates their principles. This chapter appears earlier than in other texts for two reasons. One, it allows for more complex derivatives to be discussed in Chapter 5 to include exponentials and logarithms. Two, it allows for the discussion on finite differences in Chapter 5 to be a lead in for integration in Chapter 6, **Integral Calculus**.

Chapter 5, **Techniques of Differentiation** treats the derivatives of products and quotients, maxima, and minima. Finite differences (or finite calculus) appear again in a brief section of this chapter. The anti-differences introduced in Chapter 5 anticipate the basics of integration in Chapter 6.

Chapters 7 and 8, **Integration Techniques** and **Functions of Several Variables** respectively, typify most texts. An exception is our inclusion of partial fractions.

Chapter 9, **Series and Summations** includes important insight to applications.

Chapter 10, **Applications to Probability** links calculus and probability.

The table below suggests sample topic choices for a basic calculus course:

Chapter	Traditional Course	Enhanced Course	Two-Semester Course
1 Linear Equations and Functions	Selections	Selections	Selections
2 The Derivative	✓	✓	✓
3 Using the Derivative	✓	✓	✓
4 Exponential and Logarithmic Functions	✓	✓	✓
5 Differentiation Techniques	Selections	Selections	✓
6 Integral Calculus	✓	✓	✓
7 Integration Techniques	Selections	Selections	✓
8 Functions of Several Variables	Selections	Selections	✓
9 Series and Summations	Optional	Selections	✓
10 Applications to Probability	Optional	Selections	✓

SUPPLEMENTS

A modestly priced Student Solutions Manual contains complete solutions.

SUGGESTIONS

Suggestions for improvements are welcome.

ACKNOWLEDGEMENTS

We have benefitted from advice and discussions with Professors Louise Amick, Washington College; Nancy Hall, University of Delaware Associate in Arts Program-Georgetown; Richard Schnackenberg, Florida Gulf Coast University; Robert H. Mayer, Jr, US Naval Academy, Dr Wiseley Wong, University of Maryland; and Carolyn Krause, Delaware Technical and Community College-Terry Campus. Amanda Seiwell assisted with a PowerPoint® supplement. We acknowledge the University of Delaware's Morris Library for use of its resources during the preparation of this text.

ABOUT THE COMPANION WEBSITE

This book is accompanied by a companion website:

http://www.wiley.com/go/morris/calculus

The website includes:

- Instructors' Solutions Manual
- PowerPoint® slides by chapter
- Test banks by chapter
- Teacher Commentary

About The Authors

Carla C. Morris has taught courses ranging from college algebra to calculus and statistics since 1987 at the Dover Campus of the University of Delaware. Her B.S. (Mathematics) and M.S. (Operations Research & Statistics) are from Rensselaer Polytechnic Institute, and her Ph.D. (Operations Research) is from the University of Delaware.

Robert M. Stark is Professor Emeritus of Mathematical Sciences at the University of Delaware. His undergraduate and doctoral degrees were awarded by, respectively, Johns Hopkins University and the University of Delaware in physics, mathematics, and operations research. Among his publications is the 2004 Dover Edition of Mathematical Foundations for Design with R. L. Nicholls.

1 Linear Equations and Functions

Fundamentals of Calculus, First Edition. Carla C. Morris and Robert M. Stark.

Companion Website: http://www.wiley.com/go/morris/calculus

1.1 SOLVING LINEAR EQUATIONS

Mathematical descriptions, often as **algebraic expressions**, usually consist of alphanumeric characters and special symbols.

↑ The name "algebra" has fascinating origins in early Arabic language (Historical Notes).

For example, physicists describe the distance, s, that an object falls under gravity in a time, t, by $s = (1/2)gt^2$. Here, the letters s and t represent **variables** since their values may change while, g, the acceleration of gravity, is considered as constant. While any letters can represent variables, typically, later letters of the alphabet are customary. Use of x and y is generic. Sometimes, it is convenient to use a letter that is descriptive of a variable, as t for time.

Earlier letters of the alphabet are customary for fixed values or **constants**. However, exceptions are common. The equal sign, a special symbol, is used to form an **equation**. An equation equates algebraic expressions. Numerical values for variables that preserve equality are called **solutions** to the equations.

For example, $5x + 1 = 11$ is an equation in a single variable, x. It is a **conditional** equation since it is only true when $x = 2$. Equations that hold for all values of the variable are called **identities**. For example, $(x + 1)^2 = x^2 + 2x + 1$ is an identity. By solving an equation, values of the variables that satisfy the equation are determined.

An equation in which only the first powers of variables appear is a **linear equation**. Every linear equation in a single variable can be solved using some or all of these properties:

Substitution – Substituting one expression for an equivalent one does not alter the original equation. For example, $2(x - 3) + 3(x - 1) = 21$ is equivalent to $2x - 6 + 3x - 3 = 21$ or $5x - 9 = 21$.

Addition – Adding (or subtracting) a quantity to each side of an equation leaves it unchanged. For example, $5x - 9 = 21$ is equivalent to $5x - 9 + 9 = 21 + 9$ or $5x = 30$.

Multiplication – Multiplying (or dividing) each side of an equation by a non-zero quantity leaves it unchanged. For example, $5x = 30$ is equivalent to $(5x)(1/5) = (30)(1/5)$ or $x = 6$.

↓ Here are examples of linear equations: $5x - 3 = 11$, $y = 3x + 5$, $3x + 5y + 6z = 4$. They are linear in one, two, or three variables, respectively. It is the unit exponent on the variables that identifies them as linear.

↓ By "solving an equation" we generally intend the numerical values of its variables.

To Solve Single Variable Linear Equations

1. **Resolve fractions.**
2. **Remove grouping symbols.**
3. **Use addition (and/or subtraction) to have variable terms on one side of the equation.**
4. **Divide the equation by the variable's coefficient.**
5. **As a check, verify the solution in the original equation.**

Example 1.1.1 Solving a Linear Equation

Solve $(3x/2) - 8 = (2/3)(x - 2)$.

Solution:
To remove fractions, multiply both sides of the equation by 6, the least common denominator of 2 and 3. (Step 1 above)
The revised equation becomes

$$9x - 48 = 4(x - 2).$$

Next, remove grouping symbols (Step 2). That leaves

$$9x - 48 = 4x - 8.$$

Now, subtract 4x and add 48 to both sides (Step 3). Now,

$$9x - 4x - 48 + 48 = 4x - 4x - 8 + 48 \quad or \quad 5x = 40.$$

Finally, divide both sides by the coefficient 5 (Step 4). One obtains $x = 8$.
The result, $x = 8$*, is checked by substitution in the original equation (Step 5):*

$$3(8)/2 - 8 = (2/3)(8 - 2)$$

$$4 = 4 \quad checks!$$

$$\text{The solution } x = 8 \text{ is correct!}$$

Equations often have more than one variable. To solve linear equations in several variables simply bring a variable of interest to one side. Proceed as for a single variable regarding the other variables as constants for the moment.

↓ If y is the variable of interest in $3x + 5y + 6z = 2$, it can be written as $y = (2 - 3x - 6z)/5$ regarding x and z as constants for now.

Example 1.1.2 Solving for y

Solve for y: $5x + 4y = 20$.

Solution:

Move terms with y to one side of the equation and any remaining terms to the opposite side. Here, $4y = 20 - 5x$. Next, divide both sides by 4 to yield $y = 5 - (5/4)x$.

Example 1.1.3 Simple Interest

"Interest equals Principal times Rate times Time" expresses the well-known Simple Interest Formula, $I = PRT$. Solve for the time, T.

Solution:

Grouping, $I = (PR)T$ so PR becomes a coefficient of T. Dividing by PR gives $T = I/PR$.

Mathematics is often called "the language of science" or "the universal language". To study phenomena or situations of interest, mathematical expressions and equations are used to create **mathematical models**. Extracting information from the mathematical model provides solutions and insights. Mathematical modeling ideas appear throughout the text. These suggestions may aid your modeling skills.

To Solve Word Problems

1. **Read problems carefully.**
2. **Identify the quantity of interest (and possibly useful formulas).**
3. **A diagram may be helpful.**
4. **Assign symbols to variables and other unknown quantities.**
5. **Use symbols as variables and unknowns to translate words into an equation(s).**
6. **Solve for the quantity of interest.**
7. **Check your solution and whether you have answered the proper question.**

Example 1.1.4 Investment

Ms. Brown invests \$5000 at 6% annual interest. Model her resulting capital for one year.

Solution:
Here the principal (original investment) is \$5000. The interest rate is 0.06 (expressed as a decimal) and the time is 1 year.
Using the simple interest formula, $I = PRT$, Ms. Brown's interest is

$$I = (\$5000)(0.06)(1) = \$300.$$

After one year a model for her capital is $P + PRT = \$5000 + \$300 = \$5300$.

Example 1.1.5 Gasoline Prices

Recently East Coast regular grade gasoline was priced about \$3.50 per gallon. West Coast prices were about \$0.50/gallon higher.

a) What was the average regular grade gasoline price on the East Coast for 10 gallons?
b) What was the average regular grade gasoline price on the West Coast for 15 gallons?

Solution:

a) On average, a model for the East Coast cost of ten gallons was $(10)(3.50) = \$35.00$.
b) On average, a model for the West Coast of fifteen gallons was $(15)(\$4.00) = \$60.00$.

◆ Consumption as a function of disposable income can be expressed by the linear relation $C = mx + b$, where C is consumption (in \$); x, disposable income (in \$); m, marginal propensity to consume and b, a scaling constant. This consumption model arose in Keynesian economic studies popular during The Great Depression of the 1930s.

EXERCISES 1.1

In Exercises 1–6 identify equations as an identity, a conditional equation, or a contradiction.

1. $3x + 1 = 4x - 5$
2. $2(x + 1) = x + x + 2$
3. $5(x + 1) + 2(x - 1) = 7x + 6$
4. $4x + 3(x + 2) = x + 6$
5. $4(x + 3) = 2(2x + 5)$
6. $3x + 7 = 2x + 4$

In Exercises 7–27 solve the equations.

7. $5x - 3 = 17$
8. $3x + 2 = 2x + 7$
9. $2x = 4x - 10$
10. $x/3 = 10$
11. $4x - 5 = 6x - 7$
12. $5x + (1/3) = 7$
13. $0.6x = 30$
14. $(3x/5) - 1 = 2 - (1/5)(x - 5)$
15. $2/3 = (4/5)x - (1/3)$
16. $4(x - 3) = 2(x - 1)$
17. $5(x - 4) = 2x + 3(x - 7)$
18. $3x + 5(x - 2) = 2(x + 7)$
19. $3s - 4 = 2s + 6$
20. $5(z - 3) + 3(z + 1) = 12$
21. $7t + 2 = 4t + 11$
22. $(1/3)x + (1/2)x = 5$
23. $4(x + 1) + 2(x - 3) = 7(x - 1)$
24. $1/3 = (3/5)x - (1/2)$
25. $\dfrac{x + 8}{2x - 5} = 2$
26. $\dfrac{3x - 1}{7} = x - 3$
27. $8 - \{4[x - (3x - 4) - x] + 4\}$
$= 3(x + 2)$

In Exercises 28–35 solve for the indicated variable.

28. Solve: $5x - 2y + 18 = 0$ for y.
29. Solve: $6x - 3y = 9$ for x.
30. Solve: $y = mx + b$ for x.
31. Solve: $3x + 5y = 15$ for y.
32. Solve: $A = P + PRT$ for P.
33. Solve: $V = LWH$ for W.
34. Solve: $C = 2\pi r$ for r.
35. Solve: $Z = \dfrac{x - \mu}{\sigma}$ for x.

Exercises 36–45 feature mathematical models.

36. The sum of three consecutive positive integers is 81. Determine the largest integer.
37. Sally purchased a used car for \$1300 and paid \$300 down. If she plans to pay the balance in five equal monthly installments, what is the monthly payment?
38. A suit, marked down 20%, sold for \$120. What was the original price?
39. If the marginal propensity to consume is $m = 0.75$ and consumption, C, is \$11 when disposable income is \$2, develop the consumption function.
40. A new addition to a fire station costs \$100,000. The annual maintenance cost increases by \$2500 with each fire engine housed. If \$115,000 has been allocated for the addition and maintenance next year, how many additional fire engines can be housed?

41. Lightning is seen before thunder is heard as the speed of light is much greater than the speed of sound. The flash's distance from an observer can be calculated from the time between the flash and the sound of thunder.
    The distance, d (in miles), from the storm can be modeled as $d = 4.5t$ where time, t, is in seconds.
    a) If thunder is heard two seconds after lightning is seen, how far is the storm?
    b) If a storm is 18 miles distant, how long before thunder is heard?
42. A worker has forty hours to produce two types of items, A and B. Each unit of A takes three hours to produce and each item of B takes two hours. The worker made eight items of B and with the remaining time produced items of A. How many of item A were produced?
43. An employee's Social Security Payroll Tax was 6.2% for the first $87,000 of earnings and was matched by the employer. Develop a linear model for an employee's portion of the Social Security Tax.
44. An employee works 37.5 hours at a $10 hourly wage. If Federal tax deductions are 6.2% for Social Security, 1.45% for Medicare Part A, and 15% for Federal taxes, what is the take-home pay?
45. The body surface area (BSA) and weight (Wt) in infants and children weighing between 3 kg and 30 kg has been modeled by the linear relationship
    $\text{BSA} = 1321 + 0.3433\text{Wt}$ (where BSA is in square centimeters and weight in grams)
    a) Determine the BSA for a child weighing 20 kg.
    b) A child's BSA is 10,320 cm^2. Estimate its weight in kilograms.

    Current, J.D.,"A Linear Equation for Estimating the Body Surface Area in Infants and Children.", The Internet Journal of Anesthesiology 1998:Vol2N2.

1.2 LINEAR EQUATIONS AND THEIR GRAPHS

Mathematical models express features of interest. In the managerial, social, and natural sciences and engineering, linear equations often relate quantities of interest. Therefore, a thorough understanding of linear equations is important.

The standard form of a linear equation is $ax + by = c$ where a, b, and c are real valued constants. It is characterized by the first power of the exponents.

Standard Form of a Linear Equation

$$ax + by = c$$

a, b, c are real numbered constants; a and b, not both zero

Example 1.2.1 Ordered Pair Solutions

Do the points (3, 5) and (1, 7) satisfy the linear equation $2x + y = 9$?

Solution:
*A point satisfies an equation if equality is preserved. The point (3, 5) yields: $2(3) + 5 \neq 9$. Therefore, the ordered pair (3, 5) is not a solution to the equation $2x + y = 9$.
For (1, 7), the substitution yields $2(1) + 7 = 9$. Therefore, (1, 7) is a point on the line $2x + y = 9$.*

↓ An ordered (coordinate) pair, (x, y) describes a (graphical) point in the x, y plane. By convention, the x value always appears first.

A **graph** is a pictorial representation of a function. It consists of points that satisfy the function. **Cartesian Coordinates** are used to represent the relative positions of points in a plane or in space. In a plane, a point P is specified by the coordinates or ordered pair (x, y) representing its distance from two perpendicular intersecting straight lines, called the x-axis and the y-axis, respectively (figure).

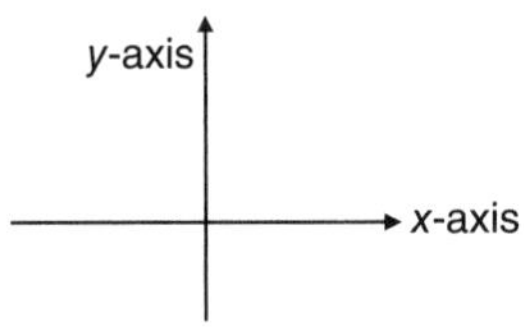

Cartesian coordinates are so named to honor the mathematician **René Descartes** (Historical Notes).

The **graph of a linear equation is a line**. It is uniquely determined by two distinct points. Any additional points can be checked as the points must be **collinear** (i.e., lie on the same line). The coordinate axes may be differently scaled. To determine the ***x*-intercept** of a line (its intersection with the x-axis), set $y = 0$ and solve for x. Likewise, for the ***y*-intercept** set $x = 0$ and solve for y.

↓ For the linear equation $2x + y = 9$, set $y = 0$ for the x-intercept $(x = 4.5)$ and $x = 0$ for the y-intercept $(y = 9)$. As noted, intercepts are intersections of the line with the respective axes.

Example 1.2.2 Intercepts and Graph of a Line

Locate the x and y-intercepts of the line $2x + 3y = 6$ and graph its equation.

Solution:
When $x = 0$, $3y = 6$ so the y-intercept is $y = 2$. When $y = 0$, $2x = 6$ so the x-intercept is $x = 3$. The two intercepts, (3, 0) and (0, 2), as two points, uniquely determine the line. As

⟶

a check, arbitrarily choose a value for x, say $x = -1$*. Then,* $2(-1) + 3y = 6$ *or* $3y = 8$ *so* $y = 8/3$*. Therefore* $(-1,\ 8/3)$ *is another point on the line. Check that these three points lie on the same line.*

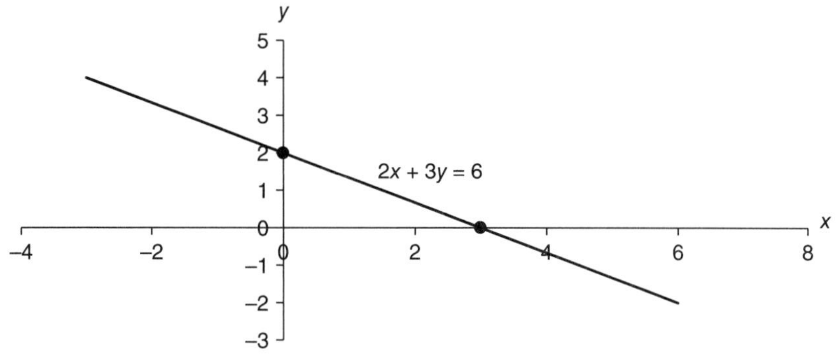

↑ Besides the algebraic representation of linear equations used here, many applications use elegant matrix representations. So $2x + 3y = 6$ (algebraic) can also be expressed as $(2\quad 3)\begin{pmatrix} x \\ y \end{pmatrix} = (6)$ in matrix format.

Price and quantity often arise in economic models. For instance, the demand $D(p)$ for an item is related to its unit price, p, by the equation $D(p) = 240 - 3p$. In graphs of economic models price appears on the (vertical) y-axis and quantity on the (horizontal) x-axis.

Example 1.2.3 Intercepts of a Demand Function

Given $D(p) = 240 - 3p$*.*

a) Determine demand when price is 10.
b) What is demand when the goods are free?
c) At what price will consumers no longer purchase the goods?
d) For what values of price is D(p) meaningful?

Solution:

a) Substituting $p = 10$ *yields a demand of 210 units.*
b) When the goods are free $p = 0$ *and* $D(p) = 240$*. Note that this is an intercept.*

⟶

c) Here, $D(p) = 0$ and $p = 80$ is the price that is too high and results in no demand for the goods. Note, this is an intercept.

d) Since price is at least zero, and the same for demand, therefore, $0 \leq p \leq 80$ and $0 \leq D(p) \leq 240$.

When either a or b in $ax + by = c$ is zero, the standard equation reduces to a single value for the remaining variable. If $y = 0$, $ax = c$ so $x = c/a$; a vertical line. If $x = 0$, $by = c$ so $y = c/b$; a horizontal line.

↓ Remember: horizontal lines have zero slopes while vertical lines have infinite slopes.

Vertical and Horizontal Lines

For

$$ax + by = c$$

When $b = 0$ the graph of $ax = c$ (or $x =$ constant) is a vertical line.

When $a = 0$ the graph of $by = c$ (or $y =$ constant) is a horizontal line.

It is often useful to express equations of lines in different (and equivalent) algebraic formats. The **slope**, m, of a line can be described in several ways; "the rise divided by the run", or "the change in y, denoted by Δy, divided by the change in x, Δx". From left to right, positive sloped lines "rise" (/) while negative sloped lines "fall" (\).

♦ In usage here, Δ denotes "a small change" or "differential." Later, the same symbol is used for a "finite difference." Unfortunately, the dual usage, being nearly universal, compels its usage. However, usage is usually clear from the context.

Slope

$$m = \frac{\text{rise}}{\text{run}} = \frac{\text{change in } y}{\text{change in } x} = \frac{\Delta y}{\Delta x} = \frac{y_2 - y_1}{x_2 - x_1}$$

The equation of a line can be expressed in different, but equivalent, ways. The **slope-intercept form** of a line is $y = mx + b$, where m is the slope and b its y-intercept. A horizontal line has zero slope. A vertical line has an infinite (undefined) slope, as there is no change in x for any value of y.

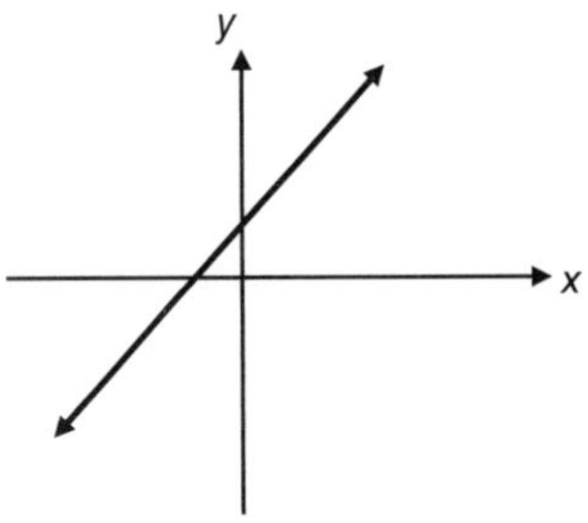

Positive slope

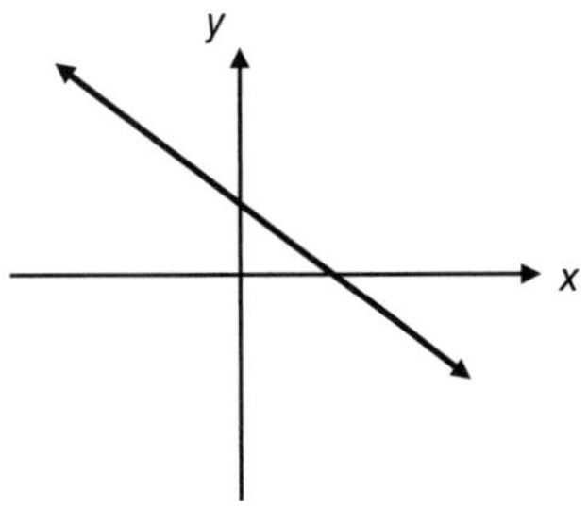

Negative slope

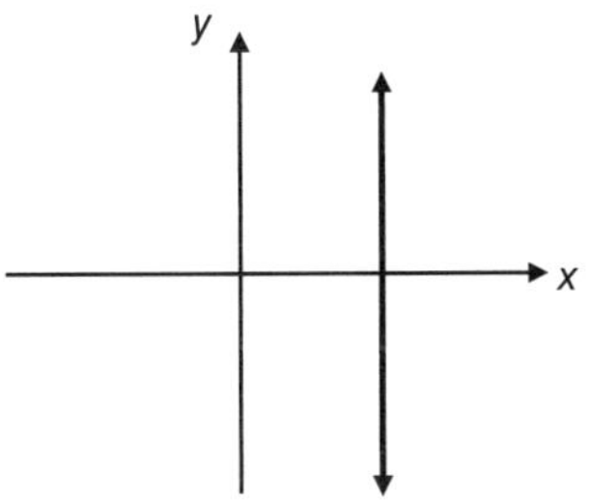

Undefined slope

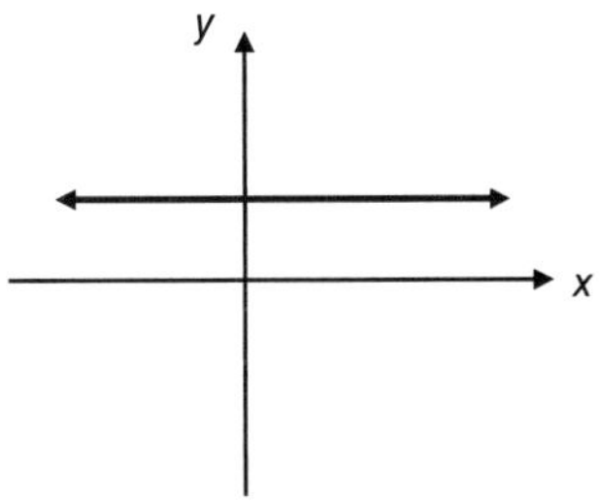

Zero slope

Slope-Intercept Form

$$y = mx + b$$

where m is the slope and b the y-intercept

A linear equation in standard form, $ax + by = c$, is written in slope-intercept form by solving for y.

Example 1.2.4 Slope-Intercept Form

Write $2x + 3y = 6$ in slope-intercept form and identify the slope and y-intercept.

Solution:
Solving, $y = (-2/3)x + 2$. By inspection, the slope is $-2/3$ ("line falls") and the y-intercept is (0, 2); in agreement with the previous Example.

Linear equations are also written in **point-slope form:** $y - y_1 = m(x - x_1)$. Here, (x_1, y_1) is a given point on the line and m is the slope.

Point-Slope Form

$$y - y_1 = m(x - x_1)$$

where m is the slope and (x_1, y_1) a point on the line.

Example 1.2.5 Point-Slope Form

Determine the equation of a line in point-slope form passing through (2, 4) and (5, 13).

Solution:

First, the slope $m = \frac{13-4}{5-2} = \frac{9}{3} = 3$. *Now, using (2, 4) in the point-slope form we have* $y - 4 = 3(x - 2)$. *[Using the point (5, 13) yields the equivalent* $y - 13 = 3(x - 5)$*]. For the slope-intercept form, solving for y yields* $y = 3x - 2$. *The standard form is* $3x - y = 2$.

↓ Various representations of linear equations are equivalent but can seem confusing. Usage depends on the manner in which information is provided. If the slope and y-intercept are known, use the slope-intercept form. If coordinates of a point through which the line passes is known, use the point-slope form. Using a bit of algebraic manipulation, you can simply remember to use $y = mx + b$.

♦ Incidentally, as noted earlier, while the generic symbols x and y are most common for variables, other letters are also used to denote variables. The equation $I = PRT$ was introduced earlier to express accrued interest. Economists use q and p for quantity and price, respectively, and scientists often use F and C for Fahrenheit and Celsius temperatures, respectively, and so on.

Example 1.2.6 Temperature Conversion

Water boils at 212°F(100°C) and freezes at 32°F (0°C). What linear equation relates Celsius and Fahrenheit temperatures?

Solution:

Denote Celsius temperatures by C and Fahrenheit temperatures by F. The ordered pairs are (100, 212) and (0, 32). Using the slope-intercept form of a line, $m = \frac{32-212}{0-100} = \frac{-180}{-100} = \frac{9}{5}$. *The second ordered pair, (0, 32) is its y-intercept. Therefore,* $F = 9/5C + 32$ *is the widely used relation to enable conversion of Celsius to Fahrenheit temperatures. An Exercise seeks the Fahrenheit to Celsius relation.*

♦ A common mathematical model for depreciation of equipment or buildings is to relate current value, y (dollars) to age, x (years). Straight Line Depreciation (SLD) is a common choice. In an SLD model, annual depreciation, d, is the same each year of useful life. Any remaining value is the "salvage value," s. Therefore, $y = dx + s$ is the desired model.

Example 1.2.7 Salvage Value

Equipment value at time t is $V(t) = -10{,}000t + 80{,}000$ and its useful life expectancy is 6 years. Develop a model for the original value, salvage value, and annual depreciation.

Solution:
The original value, at $t = 0$, is \$80,000. The salvage value, at $t = 6$, the end of useful life, is \$20,000 = $(-10{,}000(6) + 80{,}000)$. The slope, which is the annual depreciation, is \$10,000.

Lines having the same slope are **parallel**. Two lines are **perpendicular** if their slopes are negative reciprocals.

Parallel and Perpendicular Lines

$$y = m_1x + b_1 \qquad y = m_2x + b_2$$

Two lines are parallel if their slopes are equal

$$m_1 = m_2$$

Two lines are perpendicular if their slopes are negative reciprocals

$$m_1 = -\frac{1}{m_2}$$

(regardless of their y-intercepts b_1 and b_2)

Example 1.2.8 Parallel or Perpendicular Lines

Are these pairs of lines parallel, perpendicular, or neither?

a) $3x - y = 1$ *and* $y = (1/3)x - 4$
b) $y = 2x + 3$ *and* $y = (-1/2)x + 5$
c) $y = 7x + 1$ *and* $y = 7x + 3$

⟶

Solution:

a) *The two slopes are required. By inspection, the slope of the second line is 1/3. The line $3x - y = 1$ in slope-intercept form is $y = 3x - 1$ so its slope is 3. Since the slopes are neither equal nor negative reciprocals the lines intersect.*
b) *The slopes are 2 and $(-1/2)$. Since these are negative reciprocals the two lines are perpendicular.*
c) *The lines have the same slope (and different intercepts) so they are parallel.*

♦ "What Makes an Equation Beautiful", once the title of a New York Times article isn't likely to excite widespread interest; especially in linear equations.

However, linear equations are main building blocks for more advanced – and more interesting equations.

Some physicists were asked, "Which equations are the greatest?" According to the article some were nominated for the breadth of knowledge they capture, for their historical importance, and some for reshaping our perception of the universe.

EXERCISES 1.2

1. Determine the x and y-intercepts for the following:
   a) $5x - 3y = 15$
   b) $y = 4x - 5$
   c) $2x + 3y = 24$
   d) $9x - y = 18$
   e) $x = 4$
   f) $y = -2$
2. Determine slopes and y-intercepts for the following:
   a) $y = (2/3)x + 8$
   b) $3x + 4y = 12$
   c) $2x - 3y - 6 = 0$
   d) $6y = 4x + 3$
   e) $5x = 2y + 10$
   f) $y = 7$
3. Determine the slopes of lines defined by the points:
   a) (3, 6) and (−1, 4)
   b) (1, 6) and (2, 11)
   c) (6, 3) and (12, 7)
   d) (2, 3) and (2, 7)
   e) (2, 6) and (5, 6)
   f) (5/3, 2/3) and (10/3, 1)
4. Determine the equation for the line
   a) with slope 4 that passes through (1, 7).
   b) passing through (2, 7) and (5, 13).
   c) with undefined slope passing through (2, 5/2).
   d) with x-intercept 6 and y-intercept −2.
   e) with slope 5 and passing through (0, −7).
   f) passing through (4, 9) and (7, 18).

5. Plot graphs of:
   a) $y = 2x - 5$
   b) $x = 4$
   c) $3x + 5y = 15$
   d) $2x + 7y = 14$

6. Plot graphs of:
   a) $2x - 3y = 6$
   b) $y = -3$
   c) $y = (-2/3)x + 2$
   d) $y = 4x - 7$

7. Are the pairs of lines parallel, perpendicular, or neither?
   a) $y = (5/3)x + 2$ and $5x - 3y = 10$
   b) $6x + 2y = 4$ and $y = (1/3)x + 1$
   c) $2x - 3y = 6$ and $4x - 6y = 15$
   d) $y = 5x - 4$ and $3x - y = 4$
   e) $y = 5$ and $x = 3$

8. Determine equation for the line
   a) through (2, 3) and parallel to $y = 5x - 1$.
   b) through (1, 4) and perpendicular to $2x + 3y = 6$.
   c) through (5, 7) and perpendicular to $x = 6$.
   d) through (4, 1) and parallel to $x = 1$.
   e) through (2, 3) and parallel to $2y = 5x + 4$.

9. When does a linear equation lack an x-intercept? Have more than one x-intercept? Lack a y-intercept? Have more than one y-intercept?

10. Model the conversion of Fahrenheit to Celsius temperatures. Hint: Use the freezing and boiling temperatures for water.

11. A new machine cost $75,000 and has a salvage value of $21,000 after nine years. Model its straight line depreciation.

12. A new car cost $28,000 and has a trade-in value of $3000 after 5 years. Use straight line depreciation.

13. The distance a car travels depends on the quantity of gasoline available. A car requires seven gallons to travel 245 miles and 12 gallons to travel 420 miles. What linear relationship expresses distance (miles) as a function of gasoline usage (gallons)?

14. Office equipment is purchased for $50,000 and after ten years has a salvage value of $5000. Model its depreciation with a linear equation.

15. Monthly rent on a building is $1100 (a fixed cost). Each unit of the firm's product costs $5 (a variable cost). Form a linear model for the total monthly cost to produce x items.

16. A skateboard sells for $24. Determine the revenue function for selling x skateboards.

17. A car rental company charges \$50 per day for a medium sized car and 28 cents per mile driven.
    a) Model the cost for renting a medium sized car for a single day.
    b) How many miles can be driven that day for \$92?

18. At the ocean's surface water and air pressures are equal (15 lb / in^2). Below the surface water pressure increases by 4.43 lb / in^2 for every 10 feet of depth.
    a) Express water pressure as a function of ocean depth.
    b) At what depth is the water pressure 80 lb / in^2?

19. A man's suit sells for \$84. The cost to the store is \$70. A woman's dress sells for \$48 and costs the store \$40. If the store's markup policy is linear, and is reflected in the price of these items, model the relationship between retail price, R, and store cost, C.

20. Demand for a product is linearly related to its price. If the product is priced at \$1.50 each, 40 items can be sold. Priced at \$6, only 22 items are sold. Let x be the price and y the number of items sold. Model a linear relationship between price and items sold.

1.3 FACTORING AND THE QUADRATIC FORMULA

Factoring is a most useful mathematical skill. A first step in factoring expressions is to seek the **greatest common factor** (GCF) among terms of an algebraic expression. For example, for $5x + 10$, the GCF is 5. So, to factor this expression one writes $5(x + 2)$.

Often one is unsure whether an expression is factorable. One aid is to note that a linear (first power) expression is only factorable for a common constant factor. The expression $5(x + 2)$ is completely factored because the term $(x + 2)$ is linear and does not have any constant common factor. When higher power (order) terms are involved, it is more difficult to know whether additional factoring is possible. We develop a few guidelines in this Section and with practice you should have more success with this essential skill.

The GCF may not be a single expression. When factoring $10x^3y^2 + 30x^2y^3$, 10 is a common factor. However, the variables x and y must be considered. The GCF is $10x^2y^2$ and leads to $10x^2y^2(x + 3y)$. The expression $(x + 3y)$ cannot be factored.

Example 1.3.1 Finding the GCF

Find the greatest common factor for these expressions.

a) $5x^4(a + b)^3 + 10x^2y(a + b)^2$

b) $2x^3 + 10x^2 - 48x$

c) $(x + y)^{10}(x^2 + 4x + 7)^3 + (x + y)^8(x^2 + 4x + 7)^4$

Solution:

a) Here, 5, x^2*, and* $(a + b)^2$ *are common factors to each term. The GCF is* $5x^2(a + b)^2$ *and results in* $5x^2(a + b)^2[x^2(a + b) + 2y]$.

⟶

b) Since 2 and x are factors, the GCF is 2x and $2x(x^2 + 5x - 24)$ *is the first stage of factoring.*

c) Here, $(x + y)^8$ *and* $(x^2 + 4x + 7)^3$ *are common factors. The GCF is* $(x + y)^8 (x^2 + 4x + 7)^3$ *and results in* $(x + y)^8(x^2 + 4x + 7)^3[(x + y)^2 + (x^2 + 4x + 7)]$.

The number of terms in an expression is a hint as some factoring rules depend upon the number of terms involved.

For two terms, one seeks a sum or difference of squares or of cubes. With three terms, factoring is usually by trial and error as there are fewer dependable rules and mastery comes from experience (and luck!). With four terms, it may be possible to group terms in "compatible" pairs for continued factoring. In other instances, grouping three terms may be useful.

Sum and Differences of Squares and Cubes

1. $x^2 - a^2 = (x + a)(x - a)$
2. $x^2 + a^2$ **is not factorable**
3. $x^3 - a^3 = (x - a)(x^2 + ax + a^2)$
4. $x^3 + a^3 = (x + a)(x^2 - ax + a^2)$

↓ Here are a couple of illustrations:

$$x^3 - 16a^2x = x(x^2 - 16a^2) = x(x - 4a)(x + 4a)$$

$$x^4y^4 + xy = xy(x^3y^3 + 1) = xy(xy + 1)(x^2y^2 - xy + 1)$$

Example 1.3.2 Sum and Difference of Squares

Factor these expressions of sums or differences of squares.

a) $x^2 - 100$ *b)* $2x^4y - 50y$ *c)* $(x + y)^2(a + b) - 9(a + b)$.

Solution:

a) This is a difference of squares. Therefore, the expression factors as $(x + 10)(x - 10)$.

b) First, the common factor 2y yields $(2y)[x^4 - 25]$. *Within the brackets is a difference of two squares,* $(x^2)^2$ *and* $(5)^2$. *Continuing,* $(2y)(x^2-5)(x^2 + 5)$ *is the factorization.*

⟶

c) A common factor $(a + b)$, yields $(a + b)[(x + y)^2 - 9]$. Within the brackets is a difference of two squares $(x + y)^2$ and $(3)^2$. Continuing, yields a completely factored expression $(a + b)[(x + y) + 3][x + y) - 3]$.

Example 1.3.3 Sum and Difference of Cubes

Factor these expressions of sums or differences of cubes.

a) $x^3 + 1000$ *b)* $5x^3 - 625$ *c)* $2x^3y^2(a + b)^3 + 2x^3y^2$

Solution:

a) Factoring of a sum of two cubes $(x)^3$ and$(10)^3$ yields $(x + 10)(x^2 - 10x + 100)$.
b) First, the GCF of 5 yields $5(x^3 - 125)$. Next, for a difference of cubes, the factoring yields $5(x - 5)(x^2 + 5x + 25)$. (Note that the trinomial factor, $x^2 + 5x + 25$, cannot be simplified).
c) Here, the GCF is $2x^3y^2$ so $2x^3y^2[(a + b)^3 + 1]$ is the first stage result. Next, noting the difference of cubes yields the completely factored
$2x^3y^2[(a + b) + 1][(a + b)^2 - (a + b) + 1]$.

Factoring trinomials is more complicated; however, some ideas may be useful. One approach is to use trial and error that is somewhat the reverse of the FOIL technique used to multiply two binomial factors to yield a trinomial. For instance, to factor $x^2 - 10x + 21$ the lead term, x^2, suggests two linear factors that multiply x by itself. Next, focus on the constant and its factors. To multiply two numbers to yield +21 they must have the same sign. The coefficient of the linear term being negative, each of the two factors must be negative. We seek two negative factors of 21. Possible choices are −1 and −21, or −3 and −7. The latter is the correct option since −3 − 7 sums to the linear coefficient −10. The factored trinomial is $(x-3)(x-7)$.

Quadratic trinomials $x^2 + bx + c$, with unit coefficient of x^2 are fairly simple to factor. Always be alert for a GCF before attempting to factor. Expressions that cannot be factored are known as being **prime** or **irreducible**.

↓ Here are a couple of illustrations:
For $x^2 + 11x + 28$, note that 4 and 7 are factors of 28 that sum to 11. Therefore, $(x + 7)(x + 4)$ is the factorization. For $2x^2 + 7x - 4$ the coefficient of x^2 immediately suggests a start with $(2x+?)(x+?)$.
The −4 suggests trial and error with numbers +4 and −1 or −4 and +1. The result is $(2x-1)(x + 4)$.

↑ You may know the quadratic formula and wonder why it has not been used so far. It can always be used to factor second order trinomials and appears later in this Section.

Example 1.3.4 Factoring Trinomials

Factor these trinomials:

a) $x^2 + 20x + 36$

b) $2x^3 - 24x^2 + 54x$

c) $x^2 - 3x - 10$

d) $x^2 + 9x + 12$

Solution:

a) As 36 is positive, the two factors have like positive sign since 20x is positive. Therefore, all positive pairs of factors of 36 are prospects. Here, 2 and 18 are the factors which sum to 20. The trinomial is factored as $(x + 2)(x + 18)$.

b) First the GCF 2x is factored to yield $2x(x^2 - 12x + 27)$. *The constant term and the −12 indicate the need of two negative factors of 27 which sum to 12. The two factors are −9 and −3 so the expression factors as* $(2x)(x - 3)(x - 9)$.

c) Here, the constant term indicates that the factors are of opposite sign. We seek two factors of −10, one positive and one negative that sum to −3. These are −5 and +2 so the trinomial factors as $(x - 5)(x + 2)$.

d) Here we seek two positive factors of 12 that add to 9. This is not possible. Therefore, this trinomial is not factorable.

When the lead term coefficient is not unity, trial and error factoring seems more difficult. For instance, to factor $2x^2 + 5x - 12$ the lead term can only result from multiplying $2x$ by x. The constant is negative so one positive factor and one negative factor are needed here as

$$(2x \pm \quad)(x \pm \quad).$$

The possibilities are (−1, 12), (1, −12), (−2, 6), (2, −6), (−3, 4), (3, −4), (−4, 3), (4, −3), (−6, 2), (6, −2), (−12, 1), (12, −1). Some of these twelve options are quickly eliminated. The first term with $2x$, cannot have a factor that is an even number as it would contain a multiple of 2 to factor at the start. This leaves the four possibilities (−1, 12), (1, −12), (−3, 4), and (3, −4). Investigating these possibilities yields (−3, 4) as the correct pair. The trinomial factors as $(2x - 3)(x + 4)$.

Example 1.3.5 Factoring Trinomials (revisited)

Factor these trinomials:

a) $3x^2 + 13x + 10$ b) $6x^2 - 11x - 10$ c) $60x^3 + 74x^2 - 168x$.

Solution:

a) *First, use* $(3x + \quad)(x + \quad)$ *for the first term. Next, since 10 and 13 are both positive only two positive factors of ten need to be investigated. There are four possibilities (1, 10), (2, 5), (5, 2), and (10, 1). Investigating the (2, 5) and (5, 2) possibilities first, determine that neither fits. Finally, looking at the remaining two pairs the trinomial is factored as* $(3x + 10)(x + 1)$.

b) *Here, there are two possibilities for* $6x^2$. *They are* $(2x + \quad)(3x + \quad)$ *or* $(6x + \quad)(x + \quad)$. *Sometimes it is helpful to use factors closer together unless the middle term coefficient is large. Here,* $(2x + \quad)(3x + \quad)$ *is first choice. The* -10 *indicates positive and negative factors to try to obtain the* -11 *coefficient of the middle term. Therefore* $(-1, 10)$, $(1, -10)$, $(-5, 2)$ *and* $(5, -2)$ *are the possibilities to evaluate since none of them have a multiple of 2 for a factor. Checking further, the trinomial factors* $(2x - 5)(3x + 2)$ *emerge.*

c) *Here, the GCF is 2x so* $(2x)[30x^2 + 37x - 84]$. *The* $30x^2$ *can come from multiplying* $(x + \quad)(30x + \quad)$, $(2x + \quad)(15x + \quad)$, $(3x + \quad)(10x + \quad)$, *or* $(5x + \quad)(6x + \quad)$. *Quite a list of options! Again, as a rule of thumb, first try the factors closer together. If that fails, proceed to the next closest pair and so on. We try* $(2x)(5x + \quad)(6x + \quad)$ *first. The* -84 *can be found in many ways. However, a multiple of 2 or 3 cannot be the second factor since the 6x term would have a common factor then. This narrows the options to* $(-12, 7)$, $(12, -7)$, $(-84, 1)$, *and* $(84, -1)$. *If none of these are correct, investigate the 3x and 10x as factors. The 5x and 6x option is correct to yield* $(2x)(5x + 12)(6x - 7)$ *as the completely factored trinomial.*

Factoring trinomials when the lead coefficient is not unity becomes easier with practice. One learns to eliminate possibilities by the value of the second coefficient or by noticing that certain factors cannot be a common factor in the terms. To factor expressions with four terms, they are often grouped for common factors.

For instance, suppose one wants to factor $ax + 5x + ay + 5y$. The four terms of this expression lack a common factor (GCF). However, pairs of factors do share a common factor. Group the front pair and back pair as $(ax + 5x) + (ay + 5y)$ to begin the factoring. Now, factoring the pairs yields $(x)(a + 5) + (y)(a + 5)$. These terms have $(a + 5)$ as a GCF to yield $(a + 5)(x + y)$ to complete the factoring. An equivalent alternative groups the terms as $(ax + ay) + (5x + 5y)$. Factoring the GCF of the pairs yields
$(a)(x + y) + (5)(x + y) = (x + y)(a + 5)$.

Example 1.3.6 Factoring by Grouping

Completely factor these expressions by grouping.

a) $ax - 7x + ay - 7y$

b) $x^2 + 9x - a^2 - 9a$

c) $2x^3 - 10x + 3x^2 - 15$

d) $x^2 + 10x + 25 - y^2$

Solution:

a) *Grouping as* $(ax - 7x) + (ay - 7y)$ *yields* $x(a - 7) + y(a - 7)$. *The common factor is* $(a - 7)$. *Therefore, the expression factors as* $(a - 7)(x + y)$.

b) *Grouping as* $(x^2 + 9x) - (a^2 + 9a)$ *yields* $x(x + 9) - a(a + 9)$; *not helpful. Try regrouping by powers, as* $(x^2 - a^2) + (9x - 9a)$ *which yields* $(x + a)(x - a) + 9(x - a)$. *This has a common factor* $(x - a)$. *The expression factors as* $(x - a)[(x + a) + 9]$.

c) *Grouping as* $(2x^3 - 10x) + (3x^2 - 15)$ *yields* $2x(x^2 - 5) + 3(x^2 - 5)$. *It has a common factor* $(x^2 - 5)$. *Therefore, the expression factors as* $(x^2 - 5)(2x + 3)$.

d) *Grouping in pairs makes no sense here since no two pairs have a common factor. Grouping as* $(x^2 + 10x + 25) - y^2$, *the first three terms of the expression are the perfect square trinomial* $(x + 5)^2$. *Now,* $(x + 5)^2 - y^2$ *is a difference of squares and factors as* $[(x + 5) + y][(x + 5) - y]$.

The Quadratic Formula

The **quadratic equation**, whose largest exponent is two, is an important trinomial. It has the form: $y = ax^2 + bx + c$, where a, b, c are real constants. The graph of this quadratic is a **parabola**.

To determine any x-intercepts (called "roots" or zeros of the quadratic) set $y = 0$ and solve $0 = ax^2 + bx + c$. Sometimes this can be accomplished by factoring the trinomial and then setting its factors to zero. However, often it may not possible to factor a trinomial. Fortunately, it may factor using the **quadratic formula**.

↓ That is, for $ax^2 + bx + c = 0$, $x = \dfrac{-b \pm \sqrt{b^2 - 4ac}}{2a}$ *(Quadratic Formula).*

For instance, to find the factors of $x^2 - 5x + 4$ using the quadratic formula, substitute $a = 1,\ b = -5$, and $c = 4$ as $x = \dfrac{-(-5) \pm \sqrt{(-5)^2 - 4(1)(4)}}{2(1)} = \dfrac{5 \pm 3}{2}$. So $x = 4$ and $x = 1$ and the factors are $(x - 4)$ and $(x - 1)$. Given the equation $x^2 - 5x + 4 = 0$, $x = 4$ and $x = 1$ are its roots.

Example 1.3.7 The Quadratic Formula

Derive the Quadratic Formula.
Hint: Complete the square of $ax^2 + bx + c = 0$ *to solve for its roots.*

Solution:
First, rewrite the equation as $ax^2 + bx = -c$.
To complete the square, set the coefficient of x^2 *to unity ("1") by dividing by "a" as*

$$\frac{ax^2 + bx}{a} = \frac{-c}{a} = x^2 + \frac{b}{a}x = \frac{-c}{a}$$

Next, square 1/2 the coefficient of the linear term, x. Next, add the result to both sides. That is, square
$\frac{1}{2}\left(\frac{b}{a}\right)$ *to become* $\left[\left(\frac{1}{2}\right)\left(\frac{b}{a}\right)\right]^2 = \left[\frac{b}{2a}\right]^2 = \frac{b^2}{4a^2}$ *and add to each side as*

$$x^2 + \frac{b}{a}x + \frac{b^2}{4a^2} = \frac{-c}{a} + \frac{b^2}{4a^2}$$
$$x^2 + \frac{b}{a}x + \frac{b^2}{4a^2} = \frac{-4ac + b^2}{4a^2}$$

Now, the left side is the perfect square,

$$\left(x + \frac{b}{2a}\right)^2 = \frac{b^2 - 4ac}{4a^2}$$

Next, take the square root of each side

$$\sqrt{\left(x + \frac{b}{2a}\right)^2} = \sqrt{\frac{b^2 - 4ac}{4a^2}}$$

$$\left(x + \frac{b}{2a}\right) = \pm\frac{\sqrt{b^2 - 4ac}}{2a}$$

$$x = -\frac{b}{2a} \pm \frac{\sqrt{b^2 - 4ac}}{2a} = \frac{-b \pm \sqrt{b^2 - 4ac}}{2a}$$

the familiar quadratic formula!

The Quadratic Formula

The roots (solutions) of

$$\mathbf{ax^2 + bx + c = 0}$$

are $\mathbf{x = \frac{-b \pm \sqrt{b^2 - 4ac}}{2a}} \quad \mathbf{(a \neq 0)}$

Example 1.3.8 Zeros of Quadratics

Use the quadratic formula to determine the roots of:

a) $x^2 + 20x + 50 = 0$

b) $x^2 + 12 = 13x$

c) $x^2 - 3x = -5$

d) $6x^2 - 13x - 5 = 0.$

Solution:

a) *Note the corresponding values of a, b, and c are 1, 20 and 50, respectively. The quadratic formula yields*

$$x = \frac{-(20) \pm \sqrt{(20)^2 - 4(1)(50)}}{2(1)} = \frac{-20 \pm \sqrt{200}}{2} = \frac{20 \pm 10\sqrt{2}}{2}$$
$$= 10 \pm 5\sqrt{2}$$

The two irrational roots $10 + 5\sqrt{2}$*and* $10 - 5\sqrt{2}$ *indicate the quadratic was not factorable.*

b) *First, rewrite the equation as* $x^2 - 13x + 12 = 0$*. The values of a, b, and c are 1, −13, and 12, respectively. The quadratic formula yields*

$$x = \frac{-(-13) \pm \sqrt{(-13)^2 - 4(1)(12)}}{2(1)} = \frac{13 \pm \sqrt{121}}{2} = \frac{13 \pm 11}{2} \quad or$$
$$x = \frac{13 + 11}{2} = 12, \quad \frac{13 - 11}{2} = 1.$$

Note in this example that the radicand is a perfect square. The two rational solutions indicate that the equation is factorable $[(x - 12)(x - 1)]$.

c) *First, rewrite the equation as* $x^2 - 3x + 5 = 0$*. The values of a, b, and c are 1, −3, and 5, respectively. The quadratic formula yields* $x = \frac{-(-3) \pm \sqrt{(-3)^2 - 4(1)(5)}}{2(1)} = \frac{3 \pm \sqrt{-11}}{2}$*. The negative radicand indicates that there are no real valued solutions(the solutions are complex numbers).*

d) *The equation is already set to zero so the values of a, b, and c are 6, −13, and −5, respectively. The quadratic formula yields* $x = \frac{-(-13) \pm \sqrt{(-13)^2 - 4(6)(-5)}}{2(6)} = \frac{13 \pm \sqrt{289}}{12} = \frac{13 \pm 17}{12}$. $x = \frac{13 + 17}{12} = \frac{5}{2}$, $\frac{13 - 17}{12} = \frac{-1}{3}$*. The two rational roots here indicate the original equation was factorable as*

$$\left[\left(x - \frac{5}{2}\right)\left(x + \frac{1}{3}\right)\right] \text{ or as } [(2x - 5)(3x + 1)].$$

Example 1.3.9 A Quadratic Supply Function

Use the quadratic formula to determine the price below which producers would not manufacture a product whose ***supply function*** *is* $S(p) = p^2 + 10p - 200$. *The supply function,* $S(p)$, *relates manufactured supply and unit selling price,* p.

Solution:
First, note the values of a, b, and c are 1, 10 and −200, respectively. Using the quadratic formula,

$$p = \frac{-(10) \pm \sqrt{(10)^2 - 4(1)(-200)}}{2(1)} = \frac{-10 \pm \sqrt{900}}{2} = \frac{-10 \pm 30}{2}.$$

The two rational roots are −20 and 10. However, price cannot be negative. For prices above $p = 10$ *the supply function is positive. Therefore, producers will not manufacture the product when the price is below 10.*

EXERCISES 1.3

In Exercises 1–8 factor out the greatest common factor.

1. $8x - 24$
2. $20c^2 - 10c$
3. $5x^3 - 10x^2 + 15x$
4. $36y^4 - 12y^3$
5. $5a^3b^2c^4 + 10a^3bc^3$
6. $x(a + b) + 2y(a + b)$
7. $20x^3y^5z^6 + 15x^4y^3z^7 + 20x^2y^4z^5$
8. $4(x + y)^3(a + b)^5 + 8(x + y)^2(a + b)^5$

In Exercises 9–16 factor completely using the rules for sums or differences of squares or cubes.

9. $x^2 - 25$
10. $2x^3 - 8x$
11. $3x^2 + 27$
12. $x^4 - 81$
13. $2x^3 - 16$
14. $x^3 + 125$
15. $7x^2(a + b) - 28(a + b)$
16. $3x^3(x + y)^4 + 24(x + y)^4$

In Exercises 17–32 factor each trinomial completely or indicate primes.

17. $x^2 + 5x + 4$
18. $x^2 - 37x + 36$
19. $x^2 + 3x + 1$
20. $x^2 - 10x + 9$
21. $x^2 - 6x - 16$
22. $x^2 - 7x + 6$
23. $2x^2 + 12x + 16$
24. $x^3 - x^2 - 12x$

25. $a^2b^2 + 9ab + 20$
26. $x^2(a + b) - 15x(a + b) + 36(a + b)$
27. $2x^2y^2 + 28xy + 90$
28. $x^2 - 20x + 36$
29. $x^2 + 7x + 5$
30. $x^2 - 19x + 48$
31. $x^4 - 5x^2 + 4$
32. $x^6 - 7x^3 - 8$

In Exercises 33–36 factor by grouping.

33. $x^2 - a^2 + 5x - 5a$
34. $x^2 - y^2 + 3x - 3y$
35. $4ab - 8ax + 6by - 12xy$
36. $x^2 + 6x + 9 - y^2$

In Exercises 37–46 solve using the quadratic formula.

37. $x^2 + 9x + 8 = 0$
38. $x^2 + 5x + 3 = 0$
39. $x^2 + 17x + 72 = 0$
40. $x^2 + 3x + 5 = 0$
41. $x^2 + 4x + 7 = 0$
42. $x^2 + 11x = -24$
43. $x^2 + 18 = 9x$
44. $x^2 - 4x = 45$
45. $2x^2 - 3x + 1 = 0$
46. $5x^2 - 16x + 3 = 0$

1.4 FUNCTIONS AND THEIR GRAPHS

Recall that any real number can be displayed on a number line. Often, solutions are displayed as intervals, especially inequalities. We assume real numbers here.

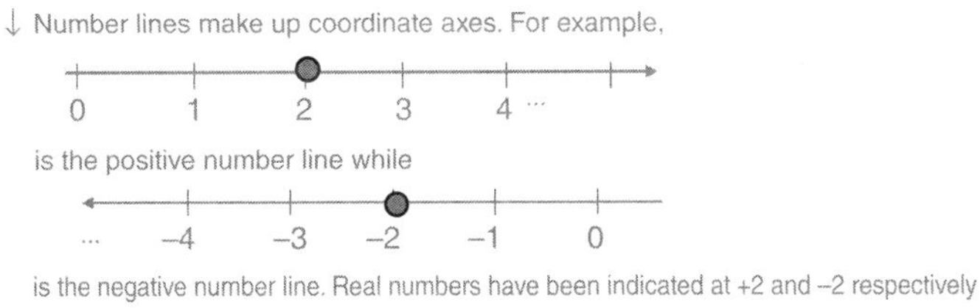

An **open interval** does not include endpoints. Thus, $x > 5$ is an example of an open interval since $x = 5$ is excluded. In a **closed interval** endpoints are included. Thus, $-4 \leq x \leq -2$ is a closed interval as the endpoints are included. An example of a **half-open** interval is $3 < x \leq 7$.

The table displays intervals on a number line and in **interval notation.** In interval notation, **parentheses**, ( ), represent open endpoints and **brackets**, [ ], closed endpoints. Parentheses are used for $\pm\infty$. An open endpoint represents an absent point while a closed one is an actual point.

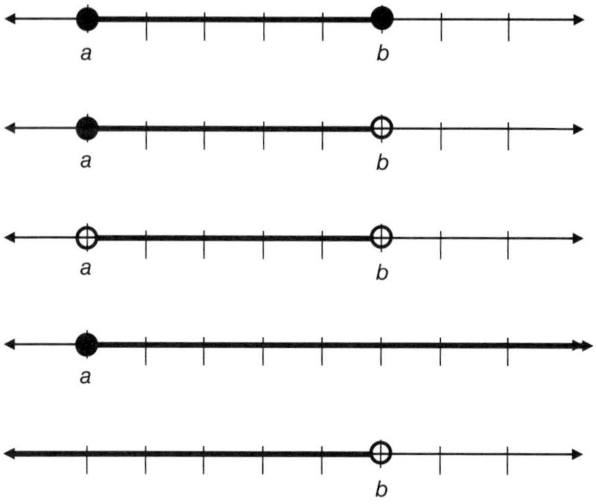

Interval notation	Inequality
$[a, b]$	$a \leq x \leq b$
$[a, b)$	$a \leq x < b$
(a, b)	$a < x < b$
$[a, \infty)$	$x \geq a$
$(-\infty, b)$	$x < b$

Example 1.4.1 Interval Notation

Display these inequalities on a number line and express them in interval notation.

a) $x > 5$ *b)* $-2 \leq x \leq 3$

Solution:

a)

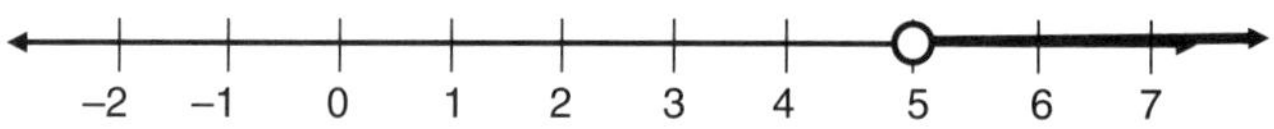

Notice the open endpoint at $x = 5$*. The interval notation for the inequality is* $(5, \infty)$.

b)

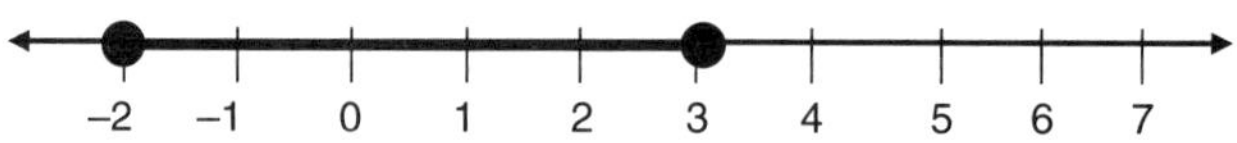

Notice the closed endpoints are used here because of the indicated equality. The interval notation for the inequality is $[-2, 3]$.

Interval notation requires some care. For example, a bracket with $\pm\infty$ implies closure at infinity which is not sensible!

Functions

You may have encountered several ways to define **functions**, (usually denoted as $y = f(x)$). A function is a rule that assigns to each value of x a unique value of y. Stated differently, each element of the **domain** is assigned one element of the **range**. Another, drawing on the language of computers or economics, is that to each "input" (x) there is a corresponding "output" (y).

The domain can be any (sensible) input. For example, the domain for polynomials is any real number. Remember, a denominator cannot be zero. If there are radicals, the *radicand* (expression under the radical sign) and the root determine the domain. The *index* of a radical is its "root." That is, the index is 2 for a square root, 3 for a cube root, and 4 for a fourth root, and so on. There is no sign restriction on domains of radicals with an odd index. If the index is even, as in square roots or fourth roots, the radicand cannot be negative.

Example 1.4.2 Finding Domains

Find the domain for each of the following functions.

a) $x^3 + 5x^2 + 7x + 3$

b) $\sqrt[3]{4x - 7}$

c) $\sqrt[4]{4x - 7}$

d) $\dfrac{5x + 1}{(x - 1)(x + 2)}$.

Solution:

a) *Recall, a polynomial is an expression whose exponents are whole numbers* $\{0,\ 1,\ 2,\ \dots\}$. *Therefore,* $x^3 + 5x^2 + 7x + 3$ *is a polynomial and, as such, its domain is all real numbers; or in interval notation* $(-\infty,\ \infty)$.

b) *The domain for* $\sqrt[3]{4x - 7}$ *is the real numbers since the index* ($n = 3$) *is odd.*

c) *Not all real numbers can be the domain for* $\sqrt[4]{4x - 7}$ *since its index* ($n = 4$) *is even. The radicand cannot be negative so* $4x - 7 \geq 0$ *yields* $x \geq 7/4$. *In interval notation, the domain is* $[7/4,\ \infty)$.

d) *The real numbers are the domain for* $\dfrac{5x + 1}{(x - 1)(x + 2)}$ *except for* $x = 1$ *or* $x = -2$, *In interval notation the domain is* $(-\infty,\ -2) \cup (-2,\ 1) \cup (1,\ \infty)$.

Example 1.4.3 Function Values

If $f(x) = x^2 - 3$ *determine the following:*

a) $f(0)$

b) $f(2)$

c) $f(a)$

d) $f(a + 2)$.

Solution:

a) *Replace x by 0 to yield* $0^2 - 3 = -3$. *So,* $f(0) = -3$.

b) *Replace x by 2 to yield* $2^2 - 3 = 1$. *So,* $f(2) = 1$.

c) *Replace x by a to yield* $a^2 - 3$. *So,* $f(a) = a^2 - 3$.

⟶

d) Replace x by a + 2 to yield $(a+2)^2 - 3 = (a^2 + 4a + 4) - 3$. *So,* $f(a+2) = a^2 + 4a + 1$.

Example 1.4.4 Function Notation and Piecewise Intervals

Suppose f(x) is the piecewise function

$$f(x) = \begin{cases} x^2 + 1 & x > 4 \\ 5x + 3 & -1 \le x \le 4 \\ 7 & x < -1 \end{cases}$$

determine the following: a) $f(-3)$ b) $f(0)$ c) $f(2)$ d) $f(4)$ e) $f(7)$

Solution:

a) *The value* $x = -3$ *belongs in the interval* $x < -1$. *Substituting* $x = -3$ *into the third portion of the function above doesn't alter the constant value of 7.*

b) *The value* $x = 0$ *lies in the interval* $-1 \le x \le 4$. *Substituting* $x = 0$ *into* $5x + 3$ *yields* $f(0) = 3$.

c) *The value 2 also lies in the interval* $-1 \le x \le 4$. *Substituting* $x = 2$ *into* $5x + 3$ *yields* $f(2) = 13$.

d) *The value* $x = 4$ *also lies in the interval* $-1 \le x \le 4$ *(note the equality for* $x = 4$*). Substituting* $x = 4$ *into* $5x + 3$ *yields* $f(4) = 23$.

e) *The value 7 lies in the interval* $x > 4$. *Substituting* $x = 7$ *into* $x^2 + 1$ *yields* $f(7) = 50$.

To determine whether or not a graph actually represents a function, it must pass the **vertical line test**: *if an intersection with any vertical line is not unique, the graph cannot be that of a function.*

The graph below cannot be that of a function as it fails the vertical line test at $x = a$.

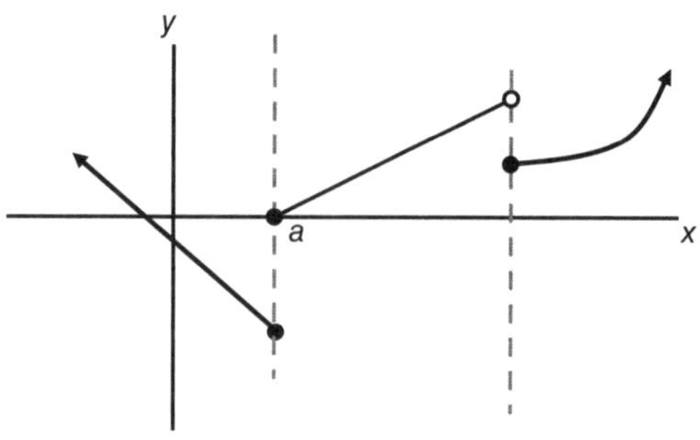

↑ The closed endpoint indicates that a point is included in the graph while an open endpoint is not an actual point of the graph. Therefore, one open endpoint and one closed circle at a value of x does not violate the conditions for the vertical line test.

Example 1.4.5 Determining Functions

Determine whether or not the graph shown below is that of a function.

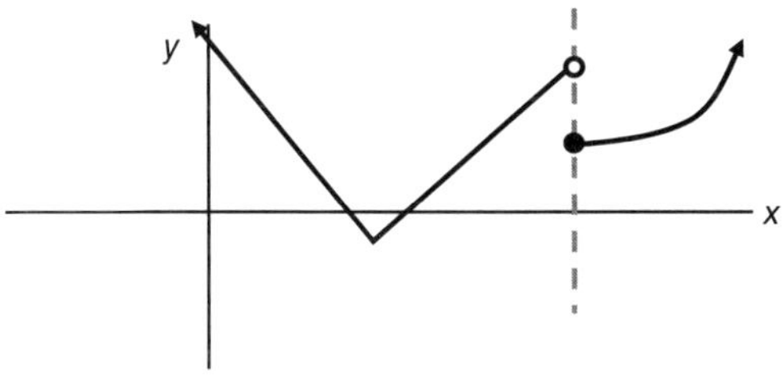

Solution:
It is a function that passes the vertical line test (vertical line passes through one open and one closed circle). There is a unique value of y for every value of x.

Graphs of Functions

In Section 1.2 you learned to graph linear functions using intercepts and, as a check, to plot at least three points. In general, the graph of quadratics (2nd degree) or higher degree polynomials should have at least five plotted points to be reasonable representations. To graph such functions, the y-intercept, x-intercepts, and key points (such as the vertex for a parabola) are useful.

Example 1.4.6 Graph of a Parabola

Sketch $y = x^2 - 2x - 8$.

Solution:
The axis of symmetry is $x = -b/2a$. *Here,* $a = 1$ *and* $b = -2$ *so the axis of symmetry is at* $x = 1$. *Recall, that when* $a > 0$ *the parabola opens upward (concave upward). The y-intercept is the constant term,* -8, *here. The vertex, or turning point, of the parabola occurs at the axis of symmetry. Substituting* $x = 1$ *yields* $y = -9$ *so the vertex occurs at the ordered pair* $(1, -9)$. *To determine the x-intercepts, y is set to zero. Here,* $0 = x^2 - 2x - 8$. *Solving yields* $0 = (x - 4)(x + 2)$ *so the x -intercepts are at* $x = 4$ *and* $x = -2$. *Now, four points of the parabola are identified. A fifth point, using symmetry, is* $(2, -8)$ *and substituting* $x = 2$ *into the original equation correctly yields* -8.

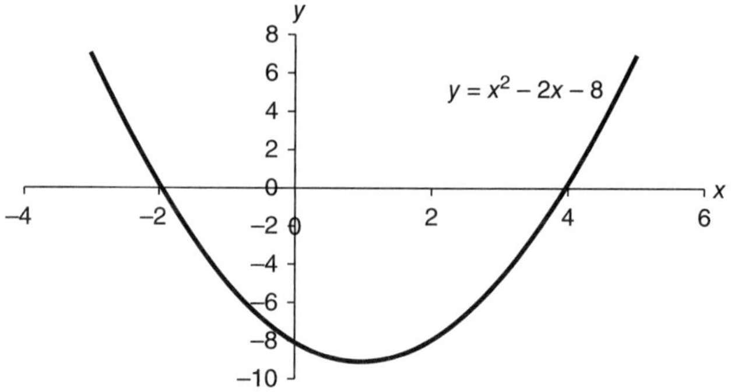

↑ When there are two x-intercepts a useful check is to verify that their average is the x value for the axis of symmetry. In the previous Example $(-2+4)/2 = 1$ and $x = 1$ was the axis of symmetry.

Other graphs useful to recognize are $y = |x|$, $y = x^3$, $y = 1/x$, and $x^2 + y^2 = r^2$ (shown below).

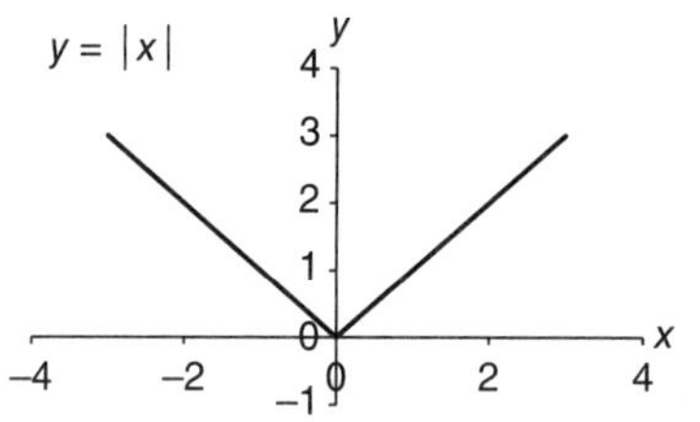

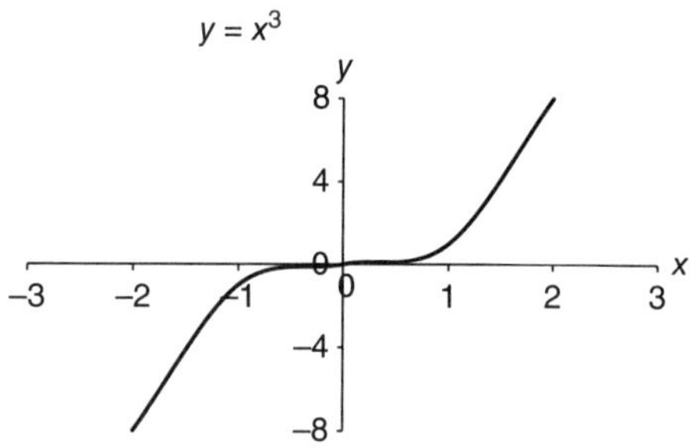

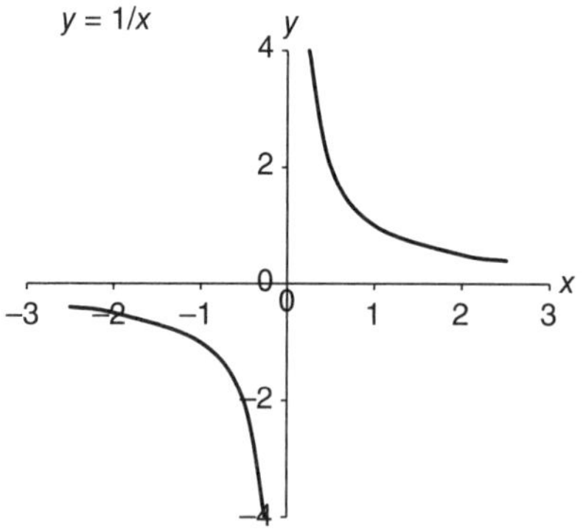

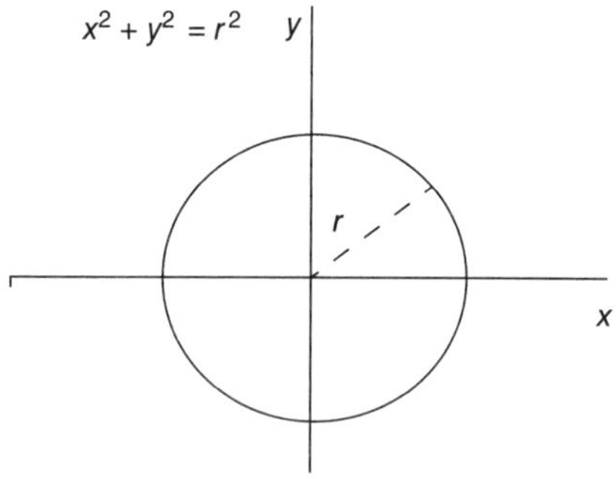

Sometimes one uses the graph of a **piecewise function**. You can use portions of the familiar graphs with the domains restricted for the employed pieces.

Example 1.4.7 A Piecewise (segmented) Graph

$$\text{Graph the function} f(x) = \begin{cases} 2x & x < -1 \\ x^2 & -1 \le x \le 1 \\ 4 & x > 1 \end{cases}$$

Solution:
The graph is the line $y = 2x$ as long as $x < -1$. The restriction on x means that an open circle is used at $(-1, \ -2)$. The second segment of the graph is parabolic with closed circles at $(-1, \ 1)$ and $(1, 1)$. The third and final segment of the graph is a portion of the horizontal line $y = 4$ with an open circle at $(1, 4)$.

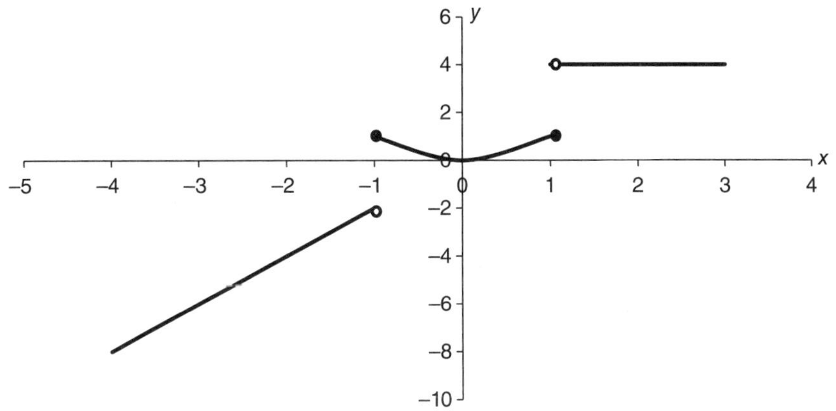

The Algebra of Functions

Many functions are composed of other functions. For example, for sales volume x, the profit function, $P(x)$, is the revenue $R(x)$ minus the cost $C(x)$ functions, $P(x) = R(x) - C(x)$.

Functions may be added, subtracted, multiplied or divided. For instance, $f(x) + g(x)$ is the addition of $g(x)$ to $f(x)$ and can be written $(f + g)(x)$. Likewise, other operations are denoted by

$$f(x) - g(x) = (f - g)(x), \ f(x)g(x) = (f \cdot g)(x), \text{ and } \frac{f(x)}{g(x)} = \left(\frac{f}{g}\right)(x).$$

Algebra of Functions

The notation for addition, subtraction, multiplication, and division of two functions f(x) and g(x) are, respectively,

1. $(f + g)(x) = f(x) + g(x)$ **3.** $(f \cdot g)(x) = f(x) \cdot g(x)$

2. $(f - g)(x) = f(x) - g(x)$ **4.** $\left(\frac{f}{g}\right)(x) = \frac{f(x)}{g(x)}$

Example 1.4.8 Algebra of Functions

Suppose $f(x) = x^3 - 7x^2 + 6x + 3$ *and* $g(x) = 2x^2 - 3x$. *Determine*

a) $(f + g)(x)$

b) $(f - g)(x)$

c) $(f \cdot g)(x)$

d) $\left(\frac{f}{g}\right)(x)$

Solution:

a) $(f + g)(x) = (x^3 - 7x^2 + 6x + 3) + (2x^2 - 3x) = x^3 - 5x^2 + 3x + 3.$

b) $(f - g)(x) = (x^3 - 7x^2 + 6x + 3) - (2x^2 - 3x) = x^3 - 9x^2 + 9x + 3.$

c) $(f \cdot g)(x) = (x^3 - 7x^2 + 6x + 3)(2x^2 - 3x) = 2x^5 - 17x^4 + 33x^3 - 12x^2 - 9x.$

d) $\left(\frac{f}{g}\right)(x) = \dfrac{x^3 - 7x^2 + 6x + 3}{2x^2 - 3x}.$

Composite functions result from one function being the variable for another. A composite function is denoted by $(f \circ g)(x) = f(g(x))$. In other words, the notation means that $g(x)$ is substituted for x in the function $f(x)$. Likewise, $(g \circ f)(x) = g(f(x))$. In other words, substitute $f(x)$ for x in the function $g(x)$. Take care not to confuse the composite operation with the multiplication operation.

Example 1.4.9 Composite Functions

Suppose $f(x) = x^5 - 3x^3 + 4x + 1$ *and* $g(x) = 5x^3 - 7x + 8$. *Determine* $(f \circ g)(x)$ *and* $(g \circ f)(x)$

Solution:

$$\begin{aligned}(f \circ g)(x) &= f(5x^3 - 7x + 8)\\ &= (5x^3 - 7x + 8)^5 - 3(5x^3 - 7x + 8)^3 + 4(5x^3 - 7x + 8) + 1.\\ (g \circ f)(x) &= g(x^5 - 3x^3 + 4x + 1)\\ &= 5(x^5 - 3x^3 + 4x + 1)^3 - 7(x^5 - 3x^3 + 4x + 1) + 8.\end{aligned}$$

EXERCISES 1.4

In Exercises 1–6 identify intervals on a number line.

1. $(3, 7)$ 2. $[-3, 6]$ 3. $[5, \infty)$ 4. $(-\infty, 10]$ 5. $(-2, 1]$ 6. $[-5, 9)$

In Exercises 7–11 use interval notation to describe the inequalities.

7. $x > 4$ 8. $x \leq -3$ 9. $-3 < x < 7$ 10. $x \geq -5$ 11. $1 \leq x < 8$

In Exercise 12–14 use interval notation to describe the inequalities depicted on the number line.

12.
13.
14.

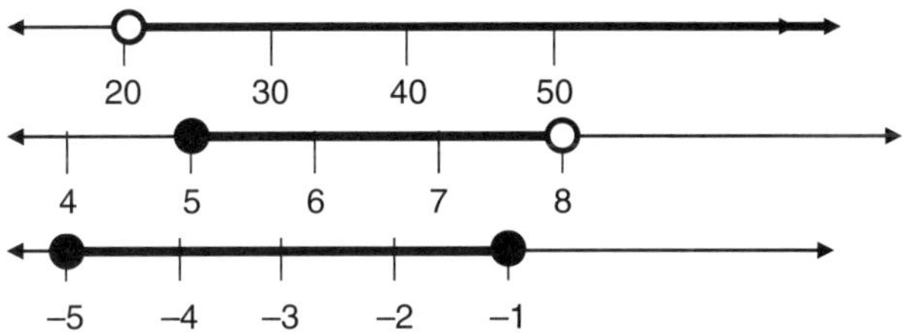

In Exercises 15–20 determine domains for the given functions.

15. $f(x) = 2x^4 - 3x^2 + 7$
16. $f(x) = \sqrt[5]{x - 3}$
17. $f(x) = \sqrt{2x - 5}$
18. $f(x) = 5x^3 - 2x + 1$
19. $f(x) = \dfrac{3x + 5}{(x - 1)(x + 3)}$
20. $f(x) = \dfrac{5}{\sqrt[3]{x - 4}}$

21. If $f(x) = 7x^3 + 5x + 3$ determine $f(0), f(1)$, and $f(x + 3)$.
22. If $f(x) = 4x^2 + 3x + 2$ determine $f(-1)$, $f(2)$, and f(a).
23. If $f(x) = x^5 + 11$ determine $f(-1), f(a^2)$, and $f(x + h)$.
24. If $f(x) = x^2 + 6x + 8$ determine $f(a), f(x + h)$, and $f(x) + h$.

In Exercises 25–27 determine whether or not the graph is that of a function.

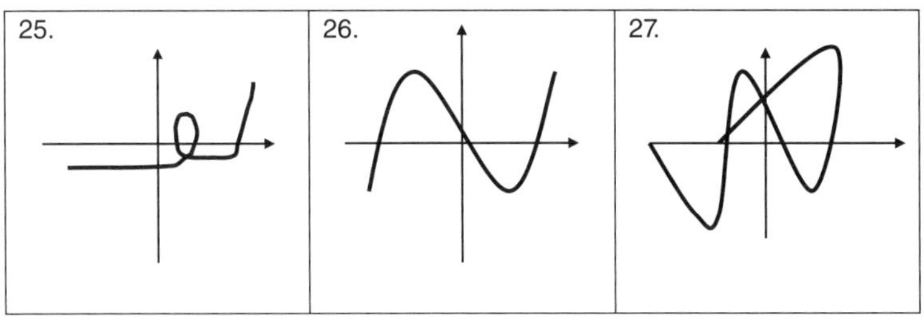

In Exercises 28–33 graph the indicated functions

28. $f(x) = 3x - 9$
29. $f(x) = x^2 - 4$
30. $f(x) = x^2 - 3x + 2$
31. $f(x) = x^3 - 8$
32. $f(x) = \frac{2}{x}$
33. $f(x) = \begin{cases} 2x & x < -1 \\ |x| & -1 \leq x < 3 \\ 5 & x \geq 3 \end{cases}$

34. Let $f(x) = 5x^4 - 3x^2 + 2$ and $g(x) = 3x^3 + 7x^2 + 2x$. Determine the following:

a) $(f + g)(x)$ b) $(f - g)(x)$ c) $\left(\frac{f}{g}\right)(x)$ d) $(f \circ g)(x)$

35. Let $f(x) = 4x^5 - 2x^3 + 2x$ and $g(x) = 3x^5 + 7x^3 + 8$. Determine the following:

a) $(g - f)(x)$ b) $(f - g)(x)$ c) $(f \cdot g)(x)$ d) $(g \circ f)(x)$

36. Let $f(x) = \frac{x}{x+1}$ and $g(x) = \frac{1}{x-3}$. Determine the following:

a) $(f + g)(x)$ b) $(f \cdot g)(x)$ c) $(f \circ g)(x)$ d) $(g \circ f)(x)$

37. Let $f(x) = 2x^5$ and $g(x) = x^2 + 4$. Determine the following:

a) $f(x) + h$ b) $g(x + h)$ c) $f(a)g(a)$ d) $f(x + 1)g(x + 2)$

1.5 LAWS OF EXPONENTS

For any b ($\neq 0$), and any positive integer, x, we define

$$b^x = b \cdot b \cdots b$$

$$x \text{ times} \quad x = 1, 2, \ \ldots$$

That is, b^x is b multiplied by itself x times. For example, $2^6 = (2)(2)(2)(2)(2)(2) = 64$. Some laws of exponents are summarized below and most are most likely to be familiar.

Laws of Exponents

1. $b^0 = 1$

2. $b^{-x} = \frac{1}{b^x}$

3. $b^x b^y = b^{x+y}$

4. $\frac{b^x}{b^y} = b^{x-y}$

5. $(b^x)^y = b^{xy}$

6. $(ab)^x = a^x b^x$

7. $\left(\frac{a}{b}\right)^x = \frac{a^x}{b^x}$

8. $\left(\frac{a}{b}\right)^{-x} = \left(\frac{b}{a}\right)^x = \frac{b^x}{a^x}$

In general, positive exponents are preferred to simplify expressions.

Example 1.5.1 Using Exponent Laws

Use laws of exponents to calculate

a) $3^2 3^3$

b) 7^0

c) $\frac{5^4}{5^3}$

d) $\frac{1}{2^{-5}}$.

Solution:

a) *Here, exponents are added to yield* $3^5 = 243$.

b) *Any nonzero number with an exponent of zero is 1.*

c) *Here, the exponents are subtracted to yield* $5^{4-3} = 5^1 = 5$.

d) *The expression is rewritten as* $2^5 = 32$ *to use a positive exponent.*

Example 1.5.2 Using Exponent Laws (revisited)

Use laws of exponents to calculate

a) x^3x^7

b) x^{-2}

c) $(x^3)^4$

d) $\frac{x^5x^2}{x^{-4}}$

e) $\left(\frac{x^3}{y^5}\right)^{-3}$.

Solution:

a) *Here, add exponents to yield* x^{10}.

b) *Using positive exponents this expression is rewritten as* $1/x^2$.

c) *Here, multiply exponents to yield* x^{12}.

d) *Here, too laws are used.* $\frac{x^5x^2}{x^{-4}} = \frac{x^7}{x^{-4}} = x^7x^4 = x^{11}$.

e) *Here it is easier to invert and separate the fraction as*

$$\left(\frac{y^5}{x^3}\right)^3 = \frac{(y^5)^3}{(x^3)^3} = \frac{y^{15}}{x^9}.$$

Fractional exponents are sometimes written in radical format. For instance, $4^{1/2} = \sqrt{4} = 2$ and $125^{1/3} = \sqrt[3]{125} = 5$. The previous laws of exponents apply to any real numbered exponent; not only integer valued ones.

Additional Laws of Exponents

$$\sqrt[x]{b} = b^{1/x} \qquad \sqrt[x]{b^y} = b^{y/x}$$

In numerical expressions it is often more useful to write $\sqrt[x]{b^y}$ *as* $(b^{1/x})^y$. In other words, first the xth root reduces the number b, then raises it to the yth power. For instance, to evaluate $16^{3/4}$, the fourth root of 16 is 2. Next, cube 2 to get 8. This is easier (without a calculator) than cubing 16 and seeking its fourth root.

In evaluating $-16^{3/2}$ first take the square root of 16, which is 4, and its cube is 64. Finally, use the negative sign for the result, -64. This contrasts with evaluating $(-16)^{3/2}$ where the square root of -16 is a complex number and beyond our scope.

The next Example continues rules of exponents with fractional exponents and numerical coefficients.

Example 1.5.3 Using Fractional Exponents

Simplify the following expressions:

a) $\left(\dfrac{27x^6}{8y^3}\right)^{2/3}$ *b)* $\sqrt[3]{x^2}(5x^{1/3} + x^{7/3})$ *c)* $\left(\dfrac{25x^{5/2}y^{9/2}}{x^{-7/2}y^{1/2}}\right)^{3/2}$

Solution:

a) First, evaluate the numerical coefficients and then the individual variables. Here,

$$\left(\frac{27}{8}\right)^{2/3} = \left(\frac{3}{2}\right)^2 = \frac{9}{4},\ (x^6)^{2/3} = x^4,\ \textit{and}\ (y^3)^{2/3} = y^2.$$

Therefore, the expression simplifies to $\dfrac{9x^4}{4y^2}$.

b) First, rewrite $\sqrt[3]{x^2}$ *as* $x^{2/3}$ *to yield* $x^{2/3}(5x^{1/3} + x^{7/3})$. *Simplify using the distributive law as* $5x + x^3$.

c) First, simplify the expression within the parenthesis. Here, the expression becomes $(25x^6y^4)^{3/2} = 125x^9y^6$.

EXERCISES 1.5

In Exercises 1–12 compute numerical values.

1. $1^{3/7}$
2. $(-1)^{4/5}$
3. $25^{3/2}$
4. $-36^{1/2}$
5. $64^{5/6}$
6. $(-27)^{4/3}$
7. $(2/3)^{-2}$
8. $(1.75)^0$
9. $(0.008)^{1/3}$
10. $7^{2/3}7^{4/3}$
11. $\dfrac{15^3}{5^3}$
12. $(25^{3/4})^{2/3}$

In Exercises 13–30 use the laws of exponents to simplify the expressions.

13. x^3x^5
14. $x^{2/5}x^{13/5}$
15. $(2xy)^3$
16. $(-4xy)^2$
17. $\dfrac{x^3x^5}{x^{-4}}$
18. $\dfrac{x^8x^{-5}}{x^4}$
19. $\dfrac{x^4y^5}{x^2y^{-2}}$
20. $\left(\dfrac{x^2x^5}{y^{-3}}\right)^2$
21. $\left(\dfrac{2x^3}{y^2}\right)^2$
22. $\left(\dfrac{3x^4}{y^3}\right)^4$
23. $\sqrt[3]{x^5}\,\sqrt[3]{x^4}$
24. $\sqrt[4]{x^6y^3}\,\sqrt[4]{x^2y^9}$
25. $(81x^4y^8)^{1/4}$
26. $(-27x^{-3}y^{-6})^{-1/3}$
27. $\dfrac{(16x^4y^5)^{3/2}}{\sqrt{y}}$
28. $\dfrac{(-8x^5y^6)^{2/3}}{\sqrt[3]{x}}$
29. $\dfrac{(8x^5y^7)^{2/3}}{\sqrt[3]{xy^2}}$
30. $\dfrac{(64x^2y^3)^{-2/3}}{x^{5/3}y^{-6}}$

1.6 SLOPES AND RELATIVE CHANGE

A **secant line** joins two points on the curve $y = f(x)$. Its slope, "the rise over the run" is

$$\frac{f(x+h)-f(x)}{(x+h)-x} = \frac{f(x+h)-f(x)}{h}, \; h \neq 0$$

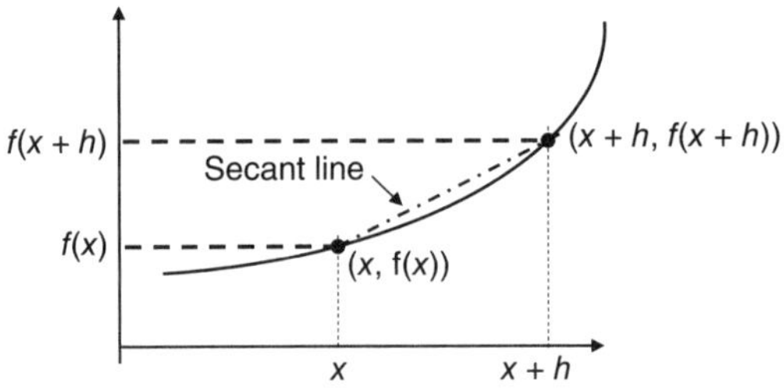

This relative change in f(x) with a change, h, in x is a **difference quotient**. It has a basic role in differential calculus.

Example 1.6.1 Another Difference Quotient

Express the slope of the secant line as a difference quotient.

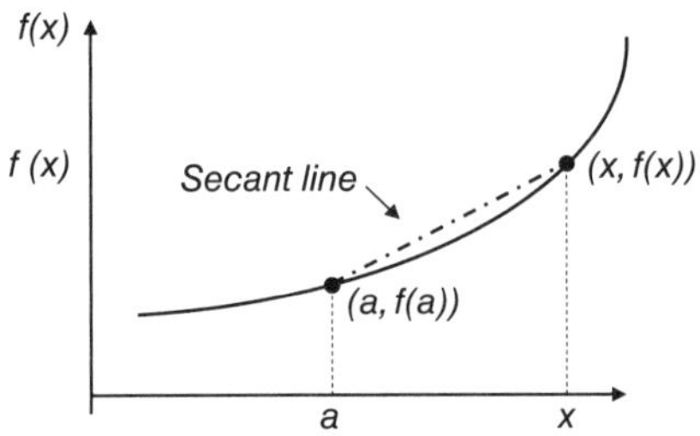

Solution:

The slope of this secant line is:

$$\frac{f(x) - f(a)}{x - a}.$$

Example 1.6.2 Difference Quotients

When $f(x) = x^2 + 2x + 6$ *determine*

a) $\dfrac{f(x+h) - f(x)}{h}$ *b)* $\dfrac{f(x) - f(a)}{x - a}.$

Solution:

a) First, write $f(x + h)$ *as* $(x + h)^2 + 2(x + h) + 6 = x^2 + 2xh + h^2 + 2x + 2h + 6.$ *Note that* $(x + h)$ *has been substituted for* x*.*

⟶

Next, the numerator is obtained by subtracting f(x), as

$$f(x+h)-f(x) = (x^2+2xh+h^2+2x+2h+6)-(x^2+2x+6)$$
$$= 2xh+h^2+2h$$

Forming the difference quotient:

$$\frac{f(x+h)-f(x)}{h} = \frac{h(2x+h+2)}{h} = 2x+h+2.$$

(The denominator, h, is a factor of the numerator).

b) Again, in sequence,

$$f(a) = a^2+2a+6 \text{ so } f(x)-f(a) = (x^2+2x+6)-(a^2+2a+6)$$
$$= x^2+2x-a^2-2a.$$

Since the numerator has $(x-a)$ as a factor, regroup by exponent as

$$\frac{f(x)-f(a)}{x-a} = \frac{(x^2-a^2)+(2x-2a)}{x-a} = \frac{(x-a)(x+a)+2(x-a)}{x-a}$$
$$= \frac{(x-a)[(x+a)+2]}{x-a}$$
$$= x+a+2.$$

↓ Do not confuse $f(x+h)$ and $f(x)+h$. The latter adds h to $f(x)$ while the former replaces x by $x+h$.

↑ The denominator of a difference quotient is a factor of the numerator so the difference quotient can be simplified.

EXERCISES 1.6

1. If $f(x) = 6x+11$ determine

   a) $\dfrac{f(x+h)-f(x)}{h}$ b) $\dfrac{f(x)-f(a)}{x-a}$.

2. If $f(x) = 9x+3$ determine

   a) $\dfrac{f(x+h)-f(x)}{h}$ b) $\dfrac{f(x)-f(a)}{x-a}$.

3. If $f(x) = 7x - 4$ determine

a) $\dfrac{f(x+h) - f(x)}{h}$ b) $\dfrac{f(x) - f(a)}{x - a}$.

4. If $f(x) = 10x + 1$ determine

a) $\dfrac{f(x+h) - f(x)}{h}$ b) $\dfrac{f(x) - f(a)}{x - a}$.

5. If $\mathrm{f(x) = x^2 - 7x + 4}$ determine

a) $\dfrac{f(x+h) - f(x)}{h}$ b) $\dfrac{f(x) - f(a)}{x - a}$.

6. If $f(x) = x^2 + 3x + 9$ determine

a) $\dfrac{f(x+h) - f(x)}{h}$ b) $\dfrac{f(x) - f(a)}{x - a}$.

7. If $f(x) = x^2 + 6x - 8$ determine

a) $\dfrac{f(x+h) - f(x)}{h}$ b) $\dfrac{f(x) - f(a)}{x - a}$.

8. If $f(x) = 3x^2 - 5x + 2$ determine

a) $\dfrac{f(x+h) - f(x)}{h}$ b) $\dfrac{f(x) - f(a)}{x - a}$.

9. If $f(x) = 5x^2 - 2x - 3$ determine

a) $\dfrac{f(x+h) - f(x)}{h}$ b) $\dfrac{f(x) - f(a)}{x - a}$.

10. If $f(x) = x^3 - 1$ determine

a) $\dfrac{f(x+h) - f(x)}{h}$ b) $\dfrac{f(x) - f(a)}{x - a}$.

11. If $f(x) = x^3 - 4x + 5$ determine

a) $\dfrac{f(x+h) - f(x)}{h}$ b) $\dfrac{f(x) - f(a)}{x - a}$.

12. If $f(x) = x^3 + 3x + 9$ determine

a) $\dfrac{f(x+h) - f(x)}{h}$ b) $\dfrac{f(x) - f(a)}{x - a}$.

13. If $f(x) = 2x^3 - 7x + 3$ determine

a) $\dfrac{f(x+h) - f(x)}{h}$ b) $\dfrac{f(x) - f(a)}{x - a}$.

14. If $f(x) = 1/x^2$ determine

a) $\dfrac{f(x+h) - f(x)}{h}$ b) $\dfrac{f(x) - f(a)}{x - a}$.

15. If $f(x) = \dfrac{3}{x^3}$ determine

a) $\dfrac{f(x+h) - f(x)}{h}$ b) $\dfrac{f(x) - f(a)}{x - a}$.

HISTORICAL NOTES

According to the Oxford Dictionary of the English Language (OED), the word **Algebra** became an accepted form among many similar ones in the seventeenth century. It is believed to have origins in early Arabic language from root words as "reunion of broken parts" and "to calculate." While the word "algebra" has contexts other than this chapter as, for example, in the surgical treatment of fractures and bone setting, and as the branch of mathematics which investigates the relations and properties of numbers by means of general symbols; and, in a more abstract sense, a calculus of symbols combined according to certain defined laws. Algebra is a principal branch of mathematics.

René Descartes (1596–1650) – Descartes was born in France in 1596, the son of an aristocrat. He traveled in Europe studying a variety of subjects including mathematics, science, law, medicine, religion, and philosophy. He was influenced by other thinkers of The Enlightenment.

Descartes ranks as an important and influential thinker. Some refer to him as a founder of modern philosophy as well as an outstanding mathematician. Descartes founded analytic geometry and sought simple universal laws governing all physical change. He originated Cartesian coordinates and contributed to the treatment of negative roots and the convention of exponent notation.

CHAPTER 1 SUPPLEMENTARY EXERCISES

1. Solve $9(x - 3) + 2x = 3(x + 1) - 2$.
2. Solve $\dfrac{3x+1}{4} + \dfrac{1}{2} = x$.
3. Solve $Z = \dfrac{x - \mu}{\sigma}$ for μ.
4. Determine three consecutive odd integers whose sum is 51.
5. Graph $3x + 5y = 15$ by using the intercepts as two of the points.
6. Graph $2x - 3y = 6$.

7. Determine the equation of the line through (5, 7) and (2, 1).
8. Are the lines $5x + 2y = 7$ and $y = (2/5)x + 1$ parallel, perpendicular, or neither?
9. Determine the equation of the line through (2, 5) parallel to $4x - 3y = 12$.
10. Factor $6x^3y^5 - 216xy^3$ completely.
11. Factor $2x^3 - 18x^2 - 20x$ completely.
12. Solve $6x^2 - 7x - 5 = 0$ by using a) factoring b) quadratic formula.
13. Factor $2ax - 2ay - by + bx$.
14. Solve $3x^2 + 10 = 13x$ by using the quadratic formula.
15. Solve $2/3(x - 1) < x - 2$ and write the solution in interval notation.
16. Solve $2(x - 3) + 5(x + 3) \geq 2x + 19$.
17. Determine the domain of $\dfrac{2x + 5}{x^3 + 9x^2 + 8x}$.
18. Graph $y = x^2 + 4x - 12$ and make sure to determine the intercepts as well.
19. If $f(x) = x^2 + 3x + 1$ and $g(x) = x^3 - 3x^2 - 4$ find

    a) (f − g)(x) b) (f · g)(x) c) $(f \circ g)(x)$.

20. If $f(x) = 5x^3 + 7x + 4$ and $g(x) = 2x^3 - 3x^2 - 4$ find

    a) $(f + g)(x)$ b) $\left(\dfrac{f}{g}\right)(x)$ c) $(g \circ f)(x)$.

21. Simplify $\left(\dfrac{-2x^5}{3x^2}\right) \cdot \left(\dfrac{3x^{-2}}{4x^5}\right)$.
22. Simplify $\left(\dfrac{625x^{7/2}y^{13}}{x^{-9/2}y^{-3}}\right)^{1/4}$.
23. Simplify $\left(\dfrac{2x^3}{3y^{-2}z^4}\right)^{-3}$.
24. $f(x) = x^2 + 11x + 2$ find

    a) $\dfrac{f(x+h) - f(x)}{h}$ b) $\dfrac{f(x) - f(a)}{x - a}$

25. If $f(x) = x^3 + 3x + 1$ find

    a) $\dfrac{f(x+h) - f(x)}{h}$ b) $\dfrac{f(x) - f(a)}{x - a}$

2 The Derivative

Fundamentals of Calculus, First Edition. Carla C. Morris and Robert M. Stark.

Companion Website: http://www.wiley.com/go/morris/calculus

2.1 SLOPES OF CURVES

The equation of a line, $y = mx + b$, and its slope, m, were featured in the previous chapter. For a line having positive slope, y increases ("rises") as x increases, while for a negative slope, y decreases ("falls") as x increases. A horizontal line has a zero slope, and a vertical line an infinite slope.

The concept of slope extends to nonlinear functions. The **tangent to a curve** generally changes from one point to another. This contrasts with a line whose tangent is the line itself. The slope of a tangent line to a curve at a point P, a circle in this case, is the slope of the curve at that point.

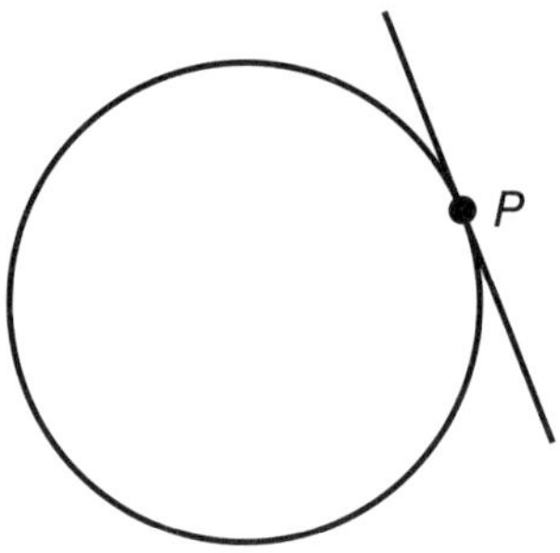

↓ A moment's reflection suggests a conceptual problem. How can a point be a slope? The answer is that it cannot. However, we can imagine a tiny section of the curve, having much the same appearance of a tiny line segment, whose slope is "very close" to that of the tangent line. This concept touches upon a central idea of the differential calculus as we shall soon study.

For a quadratic, a parabola, the slope changes from one point to another. The slope may be negative, zero, or positive depending on the point's location. The graph of the parabola $y = x^2$ (following figure) illustrates this change. At the point $(-2, 4)$, the slope of the tangent is -4, and at $(1, 1)$, the slope of the tangent is 2.

The slopes of the tangents at various points of the parabola $y = x^2$ are twice the value of x there. When $x < 0$, the slopes of the tangents are negative. When $x > 0$, their slopes are positive. There is a horizontal tangent (slope = 0) at the origin in this case.

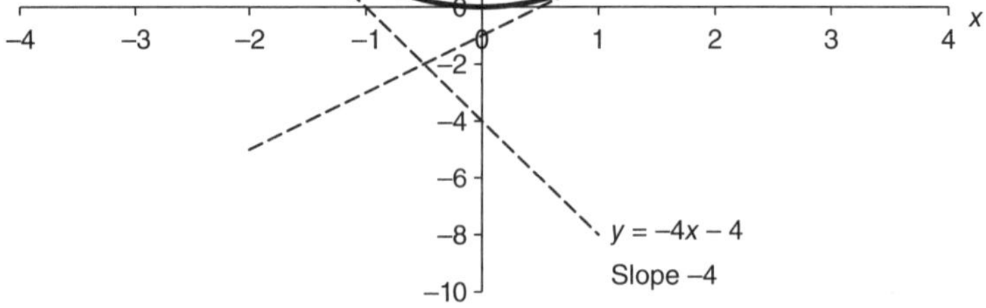

Slope of $y = x^2$

The slope of $y = x^2$ is $2x$ at any point $P = (x, y)$.

Example 2.1.1 Slopes of a Curve

What is the slope of $y = x^2$ a) at (3, 9)? b) at (−1, 1)?

Solution:

a) *The slope is twice the x coordinate, so $2x = 2(3) = 6$ at (3, 9).*

b) *The slope is $2(-1) = -2$ at (−1, 1).*

Often, the equation of the tangent line is of interest. It may be easier to express the tangent line in point-slope form before simplifying as the following example illustrates.

Example 2.1.2 Tangents to a Curve

Determine the tangent line to the curve $y = x^2$ at a) $x = 4$ b) at $x = 3/2$.

Solution:

a) *The slope is $2(4) = 8$. The point P is $(4, (4)^2) = (4, 16)$. Using the point-slope form of a line yields $y - 16 = 8(x - 4)$ or $y = 8x - 16$.*

b) *The slope is $2(3/2) = 3$. The point P is $(3/2, (3/2)^2) = (3/2, 9/4)$. Using the point-slope form of line yields $y - 9/4 = 3(x - 3/2)$ or $y = 3x - 9/4$.*

EXERCISES 2.1

1. Determine the slope of the tangent to $y = 5$.
2. Determine the slope of the tangent to $y = 12$.

In Exercises 3–6, determine the slope of the tangent to x^2 at the indicated point.

3. (5, 25).
4. (−5, 25).
5. (1/2, 1/4).
6. (−1/3, 1/9).

In Exercises 7–10, determine the equation of the tangent to $y = x^2$ at the indicated value of x.

7. $x = 5/2$.
8. $x = 9/2$.
9. $x = -7/2$.
10. $x = 5/6$.

In Exercises 11–14, locate the point on $y = x^2$ with the indicated slope.

11. 7/4
12. −3/5
13. 4/5
14. 5

15. Determine the point on $y = x^2$ where the tangent is parallel to the line $y = (2/3)x + 5$.
16. Determine the point on $y = x^2$ where the tangent is perpendicular to the line $2x - 3y = 6$.

In Exercises 17–20, determine the slope of the tangent to x^3 at the indicated point. The slope of $y = x^3$ is $3x^2$.

17. (2, 8).
18. (−1, −1).
19. (1/2, 1/8).
20. (−3, −27).

21. Determine the equation of the tangent line to $y = x^3$ at the point where $x = 4$.
22. Determine the equation of the tangent line to $y = x^3$ at the point where $x = -2$.
23. Determine the equation of the tangent line to $y = x^3$ at the point where $x = 3/4$.
24. Determine the points of $y = x^3$ where the slope is 12.
25. Determine the points of $y = x^3$ where the slope is 48.
26. Where are the tangents to $y = x^3$ perpendicular to the line $y = -x + 5$?
27. Where are the tangents to $y = x^3$ parallel to the line $4x - 3y = 6$?

2.2 LIMITS

Calculus is the mathematics of change. Sometimes, change is finite (discrete) as, for example, counting a stack of dollar bills. Other change is continuous as, for example, the trajectory and speed of a thrown ball.

For the continuous case, in which change is instantaneous, smaller and smaller intervals improve the accuracy of approximating measurements. The ball's speed can be approximated by using the time to traverse a small distance on its trajectory. The accuracy of the approximation improves as the distance traversed decreases. The successive measurements over shorter and shorter distances represent a passage to a limit.

↑ Four fundamental concepts underlie the calculus: a function, the limit, the derivative, and the integral. Functions were considered in the previous chapter, and limits and derivatives arise in this chapter; the integral is considered in Chapter 6.

The **limit** concept is fundamental to the calculus. There are three limits that we consider in this text. Two are one-sided limits. One is a limit as x approaches $x = a$ from the negative side (left) (when $x < a$). A second, when x approaches $x = a$ from the positive side (right) (when $x > a$). Third, when both one-sided limits approach the same number we say the limit as x approaches a is that value. Actually, $f(x)$ may not exist at $x = a$, and only the limit from the left and/or the right as x approaches $x = a$ may exist.

↓ To visualize these limit types consider a horizontal x-axis with an origin at $x = 0$ and a function, $f(x)$, that takes real values at every point, and $f(0)$ in particular.

Imagine a tiny bug on the left of the origin and crawling to the right toward it, never quite reaching $x = 0$. This is the left sided limit for $f(x)$ and denoted by $\lim_{x \to 0^-} f(x) = f(0)$.

Now, imagine the tiny bug on the right side of the origin and moving toward it, again, getting closer, and most important, never actually reaching $x = 0$. This right sided limit is denoted by $\lim_{x \to 0^+} f(x) = f(0)$.

Finally, when the same limit is approached, the origin here, from both left and right sides, we simply write $\lim_{x \to 0} f(x) = f(0)$ without directional qualification. Unless otherwise indicated, limits in this text are two sided.

↑ Zeno's Paradox, the legendary race of the tortoise and the hare puzzled thinkers for some 2000 years until the notion of limits was developed.

One-sided limits are not widely used in basic calculus courses but as an introduction to what is generally referred to as the limit. About notation: mathematicians use the symbolism $\lim_{x \to a} f(x) = f(a)$ to signify that the limiting value of a function $f(x)$ as x approaches a, $(x \to a)$ equals the value of the function there, $f(a)$. A similar representation applies to one-sided limits

Example 2.2.1 One-Sided Limit

Using the following graph, determine the limits:

a) $\lim_{x \to 2^-} f(x)$ *b)* $\lim_{x \to 2^+} f(x)$ *c)* $\lim_{x \to 4^-} f(x)$ *d)* $\lim_{x \to 4^+} f(x)$

⟶

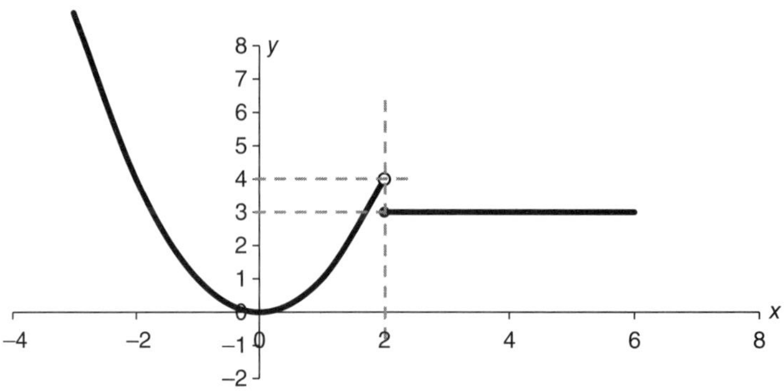

Solution:

a) *In this case, as x approaches 2 from the left (negative side), (to the open endpoint) the limit is 4.*
b) *In this case, as x approaches 2 from the right (positive side), along the horizontal, (to the closed endpoint) the limit is 3.*
c) *In this case, as x approaches 4 from the left (negative side), along the horizontal line, (to almost 4), the limit is 3.*
d) *In this case, as x approaches 4 from the right (positive side), along the horizontal line, (to almost 4), the limit is 3.*

Example 2.2.2 Graphical Limit

Continuing Example 2.2.1, determine the limits

a) $\lim_{x \to 2} f(x)$ b) $\lim_{x \to 4} f(x)$ c) $\lim_{x \to 0} f(x)$ d) $\lim_{x \to -1} f(x)$

Solution:

a) *In this case, the limit does not exist as* $\lim_{x \to 2^-} f(x) \neq \lim_{x \to 2^+} f(x)$.
b) *In this case, as both one-sided limits were equal to 3 the limit is 3,* $\lim_{x \to 4} f(x) = 3$.
c) *On either side of* $x = 0$ *the function approaches 0, so this limit is 0,* $\lim_{x \to 0} f(x) = 0$.
d) *On either side of* -1 *the function approaches 1, so this limit is 1,* $\lim_{x \to -1} f(x) = 1$.

Henceforth, all limits are assumed to be two sided unless otherwise noted. Notice that there is no limit at points on the piecewise graph when there is a break (as $x = 2$ in the previous figure). Otherwise, if there is no break in a graph near the point of interest, the limit exists there. As polynomials are continuous (unbroken) curves, a limit exists at every point. The limit is simply found by substituting $x = a$ into the polynomial $y = f(x)$.

In general, a simple way to evaluate limits of $y = f(x)$ as x approaches a is to evaluate $f(a)$. If a unique value, $f(a)$, exists, it is the limit. If it increases indefinitely, the limit does not exist.

Occasionally, an **indeterminate** form, such as 0/0 or ∞/∞, is encountered. These forms require further study. Sometimes, the cause is a vanishing factor common to the numerator and denominator. Factoring and reducing fractions sometimes remove indeterminacies of the form 0/0.

Example 2.2.3 Limits

Determine limits where they exist:

a) $\lim_{x\to 3} x^4 - 2x^3 + 3x + 5$ b) $\lim_{x\to 4} \frac{x^2 - 5x + 4}{x - 4}$ c) $\lim_{x\to 3} \frac{x^3 + 2x + 1}{x - 3}$

d) $\lim_{x\to 2} \frac{x^3 - 8}{2x + 3}$ e) $\lim_{x\to\infty} \frac{4}{3x + 1}$ f) $\lim_{x\to\infty} \frac{4x^2}{x^2 - 1}$

Solution:

a) *Substituting* $x = 3$ *yields* $(3)^4 - 2(3)^3 + 3(3) + 5 = 41$ *as the limit.*

b) *Substituting* $x = 4$ *yields an indeterminate form 0/0. After factoring the numerator as* $(x - 1)(x - 4)$ *and cancellation, the limit becomes* $\lim_{x\to 4}(x - 1) = 4 - 1 = 3$.

c) *Substituting* $x = 3$ *yields* $\frac{(3)^3 + 2(3) + 1}{3 - 3}$. *As division by zero is undefined, this limit does not exist.*

d) *Substituting* $x = 2$ *yields* $\frac{(2)^3 - 8}{2(2) + 3} = \frac{0}{7} = 0$.

e) *As* $x \to \infty$, *the fraction continuously decreases so, the limit is 0.*

f) *As* $x \to \infty$, *there is an indeterminate form. However, dividing numerator and denominator by* x^2 *and letting* $x \to \infty$, *the limit is 4.*

Sometimes, you are given the equation of a piecewise function rather than its graph (as in Example 2.2.1). Within the specified interval, you can substitute for x to determine whether the limit exists. Endpoints of intervals may pose problems for the existence of limits. This is illustrated in the following example.

Example 2.2.4 Limits of Piecewise Functions

Determine the limits if they exist:

a) $\lim_{x\to 1} f(x)$ *for* $f(x) = \begin{cases} 2x & x < 1 \\ x^3 - 1 & x \geq 1 \end{cases}$

b) $\lim_{x\to 3} f(x)$ *for* $f(x) = \begin{cases} 4 & x = 3 \\ \frac{x^2 - 2x - 3}{x - 3} & x \neq 3 \end{cases}$

⟶

Solution:

a) *In this case, both one-sided limits must be evaluated. If they are equal, the limit exists. To determine the limit from the negative side, evaluate the function when x is slightly less than 1 using the 2x portion of the piecewise function. Therefore,* $\lim_{x \to 1^-} f(x) = 2(1) = 2$.

Now, evaluate the function from the right side, when x is slightly more than 1 (the $x^3 - 1$ *portion of the function is used). Therefore,* $\lim_{x \to 1^+} f(x) = (1)^3 - 1 = 0$. *As the two one-sided limits differ, the limit does not exist. However, the one-sided limits exist.*

b) *In this case, when x is either slightly more or slightly less than 3, the second part of the piecewise function is used* ($x \neq 3$). *Therefore,*

$$\lim_{x \to 3} \frac{\cancel{(x-3)}(x+1)}{\cancel{(x-3)}} = \lim_{x \to 3} (x+1) = 4.$$

EXERCISES 2.2

In Exercises 1–3, use the graph to determine indicated limits (if they exist).

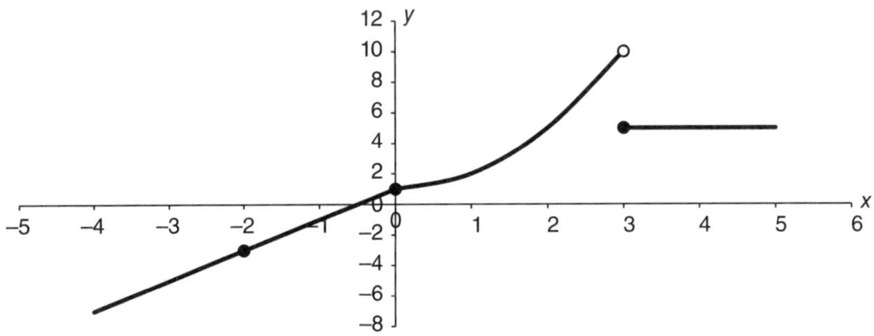

1. a) $\lim_{x \to 3^-} f(x)$ b) $\lim_{x \to 3^+} f(x)$ c) $\lim_{x \to 3} f(x)$

2. a) $\lim_{x \to 0^-} f(x)$ b) $\lim_{x \to 0^+} f(x)$ c) $\lim_{x \to 0} f(x)$

3. a) $\lim_{x \to -2^-} f(x)$ b) $\lim_{x \to -2^+} f(x)$ c) $\lim_{x \to -2} f(x)$

In Exercises 4–6, use the graph to determine indicated limits (if they exist).

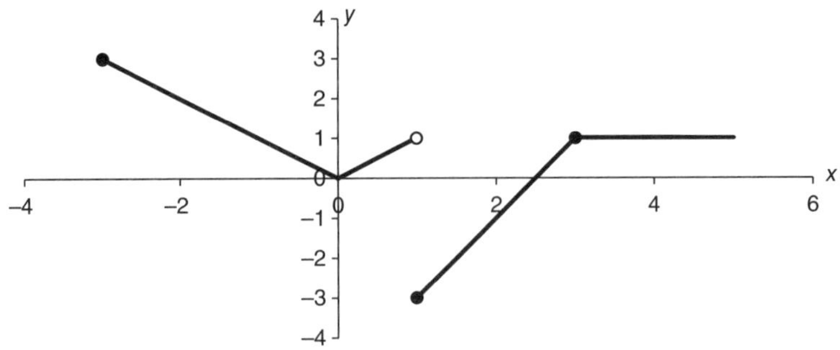

4. a) $\lim_{x\to 3^-} f(x)$ b) $\lim_{x\to 3^+} f(x)$ c) $\lim_{x\to 3} f(x)$

5. a) $\lim_{x\to 0^-} f(x)$ b) $\lim_{x\to 0^+} f(x)$ c) $\lim_{x\to 0} f(x)$

6. a) $\lim_{x\to -2^-} f(x)$ b) $\lim_{x\to -2^+} f(x)$ c) $\lim_{x\to -2} f(x)$

In Exercises 7–28, determine the indicated limits if they exist.

7. $\lim_{x\to -2} 4$
8. $\lim_{x\to 0} 12$
9. $\lim_{x\to 5} (3x + 2)$
10. $\lim_{x\to 3} (2x + 7)$
11. $\lim_{x\to 4} (9x + 5)$
12. $\lim_{x\to 6} (3x - 2)$
13. $\lim_{x\to 1} (5x^2 + 9x + 3)$
14. $\lim_{x\to 1} (8x^4 + 2x^2 + 1)$
15. $\lim_{x\to 0} (2x^7 + 3x^4 + 9x + 3)$
16. $\lim_{x\to 2} (4x^3 - 6x^2 - 7)$
17. $\lim_{x\to -1} (2x^3 + 5x^2 - 4)$
18. $\lim_{x\to 1} 3 + \frac{5}{x}$
19. $\lim_{x\to 2} \frac{3x}{x - 2}$
20. $\lim_{x\to 4} \frac{x - 4}{x + 1}$
21. $\lim_{x\to 3} \frac{x^2 - 9}{x - 3}$
22. $\lim_{x\to 5} \frac{x^2 - 25}{x - 5}$
23. $\lim_{x\to 2} \frac{x^3 - 8}{x - 2}$
24. $\lim_{x\to 4} \frac{x^2 - 3x - 4}{2x - 8}$
25. $\lim_{x\to \infty} \frac{-4}{x^3}$
26. $\lim_{x\to \infty} (x^2 + 3x + 2)$
27. $\lim_{x\to \infty} (2x^3 + 5x + 1)$
28. $\lim_{x\to \infty} \frac{3x^2 - 6x + 1}{x^2 + 8x + 3}$

In Exercises 29–35, evaluate these limits for f(x), if they exist.

$$f(x) = \begin{cases} \frac{1}{x} & x < 0 \\ 2x + 5 & 0 \leq x \leq 3 \\ x^2 - 3x - 4 & 3 < x < 5 \\ x & x \geq 5 \end{cases}$$

29. $\lim_{x\to -1} f(x)$
30. $\lim_{x\to 0} f(x)$
31. $\lim_{x\to 1} f(x)$
32. $\lim_{x\to 2} f(x)$
33. $\lim_{x\to 4} f(x)$
34. $\lim_{x\to 5} f(x)$
35. $\lim_{x\to 10} f(x)$

2.3 DERIVATIVES

The *change* in a continuous function, $y = f(x)$, between neighboring points x and $x + \Delta x$ is, of course, $f(x + \Delta x) - f(x)$, sometimes called a **differential**, (a Δ is a commonly used in calculus to denote a "small change"). Here, Δx means a "small change in x". The *rate of change* or *relative change* of such a function is the change in $f(x)$ divided by the change in x. In symbols,

$$\frac{f(x+\Delta x)-f(x)}{(x+\Delta x)-x} = \frac{f(x+\Delta x)-f(x)}{\Delta x}$$

↓ It is well to reflect on this basic concept as it is fundamental to the calculus. We stress, again, that calculus is the *mathematics of change*.

Let $f(x_0)$ be the value of $f(x)$ at $x = x_0$ and at a nearby point, $x_0 + \Delta x$, the function value is $f(x_0 + \Delta x)$. Many writers prefer to replace Δx by "h" for simplicity. Thus,

$$\frac{f(x_0+h)-f(x_0)}{(x_0+h)-x_0} = \frac{f(x_0+h)-f(x_0)}{h} \text{ is equivalent to}$$

$$\frac{f(x_0+\Delta x)-f(x_0)}{(x_0+\Delta x)-x_0} = \frac{f(x_0+\Delta x)-f(x_0)}{\Delta x}$$

The slope of a **secant line** joining the two points $x = x_0 + h$ and $x = x_0$ on the curve $y = f(x)$, at respectively $f(x_0 + h)$ and $f(x_0)$, is shown as follows

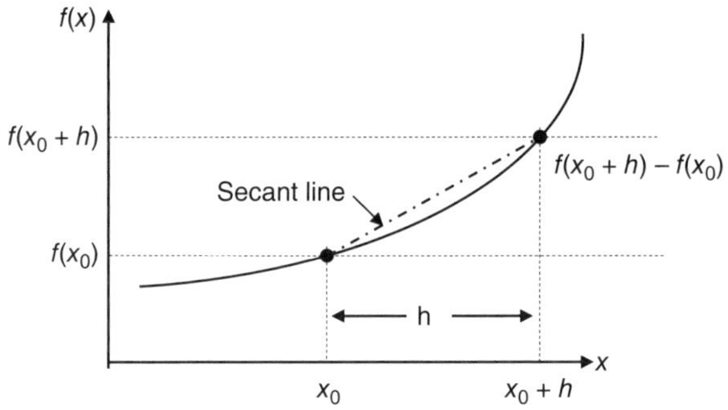

For a very small value of h, the rate of change expression shown previously approximates the slope of the curve near x_0. The approximation improves as h decreases.

↓ Imagine that as h decreases, so $x_0 + h$ moves to the left in the aforementioned figure, that $f(x_0 + h)$ also moves to the left along the curve. In the limit, the secant line approximates the tangent line.

Imagine, now, that the value of h tends to zero, that is, $h \to 0$. The limit of the change in $f(x)$ as $h \to 0$ is written as

$$\lim_{h\to 0}\frac{f(x_0+h)-f(x_0)}{h} = f'(x_0)$$

This limit denoted by $f'(x_0)$ (and read "f prime at x_0") defines the *derivative at* x_0. It is the *instantaneous rate of change* of $f(x)$ at $x = x_0$. Note that the derivative is the slope of a tangent at x_0. Therefore, henceforth, we can associate a derivative with the slope of a curve. This is important theoretically and considerably simplifies obtaining slopes of curves. Not all functions possess derivatives. Those that do are said to be **differentiable**.

The Derivative

If $f(x)$ is a differentiable function of x, the derivative of $f(x)$ at $x = x_0$ is defined by

$$f'(x) = \lim_{h \to 0} \frac{f(x_0 + h) - f(x_0)}{h}$$

The derivative is the slope of $f(x)$ at $x = x_0$.

If $f(x)$ has constant value (figure), its derivative is zero, as $f(x + h)$ and $f(x)$ have the same value. Their difference being zero, there is no change in $f(x)$.

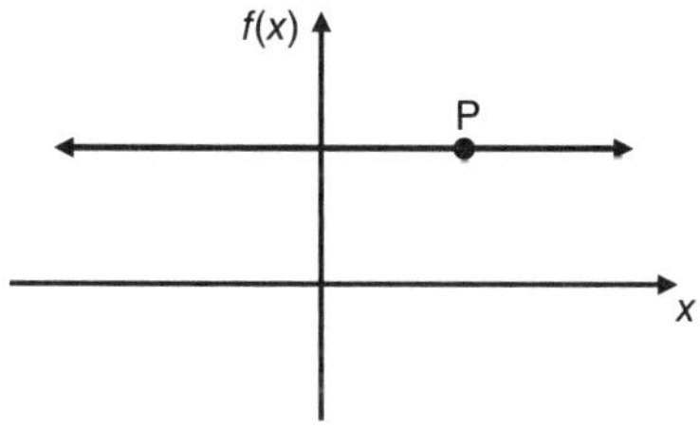

The tangent at P is the horizontal line itself with zero slope.

Example 2.3.1 Derivative of a Constant

Use the definition to show that the derivative of a constant is zero.

Solution:
Let $f(x) = b$, b being constant. The quotient

$$\frac{f(x+h) - f(x)}{h} = \frac{b - b}{h} = 0$$

In this case, no limit is necessary, as $f(x)$ does not change.

Derivative of a Constant

If $f(x) = b$,
then $f'(x) = 0$ for every x
where b is a constant

For linear functions, $f(x) = mx + b$, the tangent at P is, once again, the line itself (following figure). Its slope is m. Therefore, the derivative of any linear function is its constant slope.

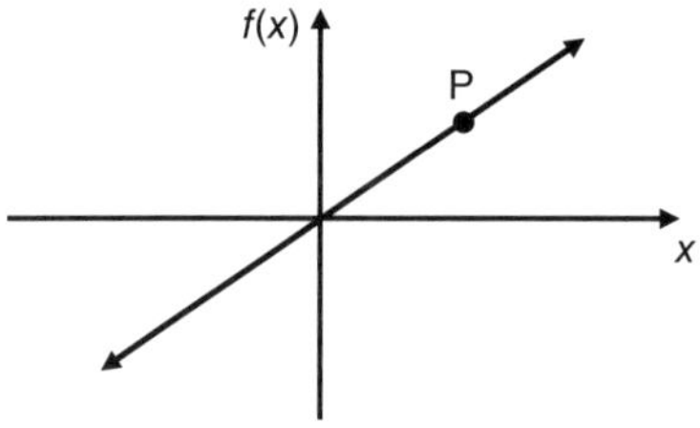

Example 2.3.2 Derivative of a Linear Function

Use the definition to prove that the derivative of a linear function is its slope.

Solution:
Let $f(x) = mx + b$; a linear function with slope m.
Firstly, using the defining quotient

$$\frac{f(x_0 + h) - f(x_0)}{h}$$

substitute values to yield:

$$\frac{[m(x_0 + h) + b] - [mx_0 + b]}{h} = \frac{m\cancel{h}}{\cancel{h}} = m.$$

Note that the slope, m, is constant independent of the value of h.

Derivative of a Linear Function

If $f(x) = mx + b$,
then $f'(x) = m$
where m is the fixed slope.

There are two ways to determine derivatives using the secant line concept. They are:

Derivatives from the Secant Line

To calculate $f'(x)$ using a secant line:

1. **Formulate the difference quotient $\frac{f(x+h) - f(x)}{h}$.**
2. **Take the limit as h approaches 0, that is, $h \to 0$.**
3. **This defines the derivative $f'(x)$.**

The difference quotient $[f(x)-f(a)]/(x-a)$ can also determine a derivative. As the distance between x and a decreases, that is, as $x \to a$, the secant line approaches the tangent and becomes the derivative.

Derivatives from a Secant Line (a Second Look)

To calculate $f'(x)$ using a secant line:

1. **Formulate the difference quotient $\dfrac{f(x)-f(a)}{x-a}$.**
2. **Take the limit as x approaches a, that is, $x \to a$.**
3. **This defines the derivative of $f(x)$ at $x = a$ where a is any value of x for which the limit exists.**
4. **As this difference quotient yields $f'(a)$, replace a by x to obtain $f'(x)$.**

Linear equations have a constant slope; nonlinear functions slopes vary with x, (such as at points P, P', and P'' in the following figure).

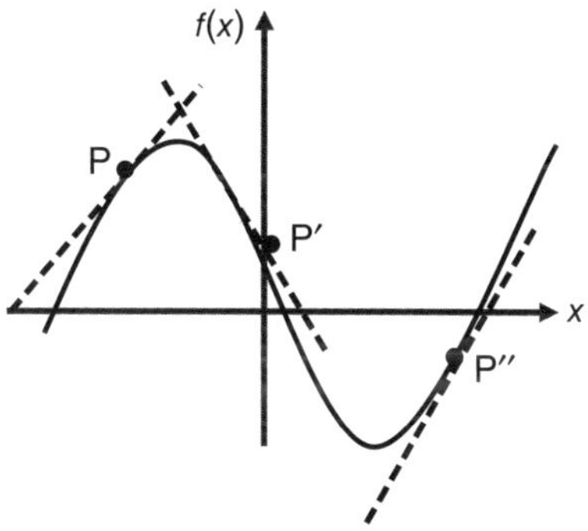

Example 2.3.3 ***Derivative of x^2***

Using the secant line definitions, find the derivative of x^2.

Solution:
One secant line approach yields,

$$\frac{f(x+h)-f(x)}{h} = \frac{(x+h)^2-x^2}{h} = \frac{(x^2+2xh+h^2)-x^2}{h} = \frac{2xh+h^2}{h}$$
$$= \frac{(h)(2x+h)}{h} = 2x+h$$

As h approaches zero, this becomes 2x. Thus, the derivative of x^2 is 2x.
The second equivalent alternate secant line approach yields,

$$\frac{f(x)-f(a)}{x-a} = \frac{x^2-a^2}{x-a} = \frac{(x-a)(x+a)}{x-a} = x+a$$

⟶

Letting x approach a yields $f'(a) = a + a = 2a$ and $f'(x) = 2x$, which agrees with the first method.

↑ The derivative is so important and so widely used in many contexts that you should be alert to relatively minor changes in symbols and notation.

It was mentioned earlier that derivative notations vary among writers. Here, for the derivatives of $y = f(x) = x^2$, these are equivalent:

$$y' = f'(x) = d/dx(x^2) = dy/dx = D_x f(x) = 2x$$

We may use these interchangeably.

There are some handy formulas for derivatives that follow from evaluating a difference quotient and taking a limit. We state them without proof.

Other Derivatives

For $f(x) = x^r, f'(x) = rx^{r-1}$, r real	**Power Rule**
For $y = k \cdot f(x), y' = k \cdot f'(x)$, k constant	**Constant Multiple Rule**
For $y = f(x) \pm g(x), y' = f'(x) \pm g'(x)$	**Summation Rule**

↑ You can easily prove any of the formulas in the shaded box by forming a difference quotient and taking a limit. For example, for

$$f(x) = x^r, \frac{f(x+h) - f(x)}{h} = \frac{(x+h)^r - x^r}{h} = \frac{x^r + rhx^{r-1} + \dfrac{r(r-1)x^{r-2}}{2} + \cdots + h^r - x^r}{h}.$$

You may recognize the numerator as the binomial expansion. Note that x^r and $-x^r$ cancel each other. Dividing by h and letting $h \to 0$ yields rx^{r-1}.

Example 2.3.4 Derivatives of Functions

Obtain derivatives of these functions:

a) $f(x) = 90$

b) $f(x) = 7x + 6$

c) $f(x) = x^{17} + 3x^{10} - 7x^5 + 9x + 30$

d) $f(x) = 5x^{7/4} + 3x + 11$

e) $f(x) = 3/x^8 + \sqrt[3]{x^2}$

Solution:

a) *The derivative of any constant is zero, so $f'(x) = 0$.*

b) *The derivative of a linear function is its slope, so $f'(x) = 7$.*

c) *Adding derivatives term wise, $f'(x) = 17x^{16} + 30x^9 - 35x^4 + 9$.*

d) *Exponents needn't be integers, $f'(x) = (5)(7/4)x^{3/4} + 3 = (35/4)x^{3/4} + 3$.*

e) *Rewrite the function as $f(x) = 3x^{-8} + x^{2/3}$ to obtain $f'(x) = -24x^{-9} + (2/3)x^{-1/3}$.*

Trade-Offs and the Basic Lot Size Model

Decision-making situations are usually characterized by two "opposing" types of "costs." One type of cost per unit time increases with increasing time (as, e.g., holding and storing quantities of inventory), while another type decreases with increasing time (as, e.g., the spreading of overhead and fixed costs over a longer time). The required decision is to balance these "opposing" costs.

The following sketch illustrates these costs for an inventory cycle in which holding costs are assessed per item per unit time held and the set up cost is fixed for the cycle.

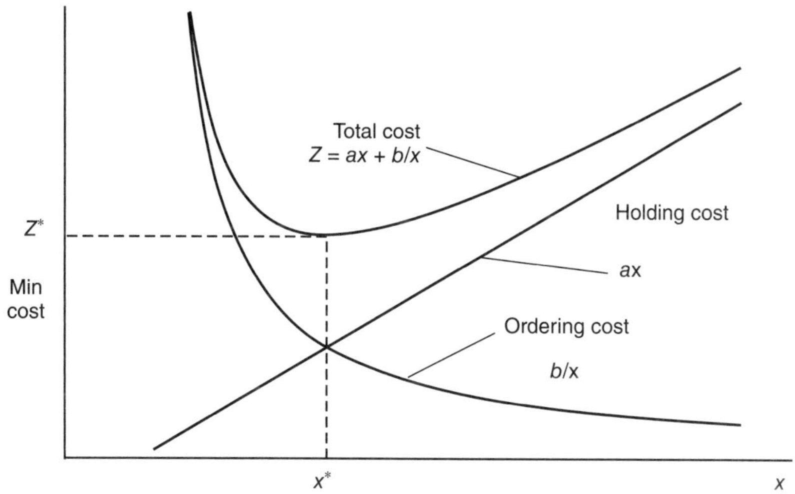

The solution to this model occurs when the holding cost is equal to the ordering cost.

◆ Also known as the Economic Order Quantity Model it has a remarkable variety of practical uses as shown in later chapters.

Example 2.3.5 The Basic Lot Size Model

Given a basic lot size model $Z = f_{inc}(x) + f_{dec}(x) = ax + b/x$, *determine* Z'. *Here,* $f_{inc}(x)$ *represents the increasing cost (holding cost), and* $f_{dec}(x)$ *represents the decreasing cost (ordering cost).*

Solution:

$$Z' = a - \frac{b}{x^2}$$

Note: we will revisit this example in Chapter 3 where the derivative will be used to determine an optimal solution.

EXERCISES 2.3

In Exercises 1–8, determine $f'(x)$ using both difference quotient approaches.

1. If $f(x) = 4x + 11$
2. If $f(x) = 9x - 3$
3. If $f(x) = x^2 + 5x + 1$
4. If $f(x) = x^2 + 7x + 5$
5. If $f(x) = x^2 - 6x + 1$
6. If $f(x) = 2x^2 - 3x + 2$
7. If $f(x) = x^3 + 5$
8. If $f(x) = x^3 + 2x + 3$

In Exercises 9–20, determine derivatives of f(x).

9. $f(x) = 9$
10. $f(x) = \sqrt{3}$
11. $f(x) = 5\pi$
12. $f(x) = 24.2$
13. $f(x) = 7x + 11$
14. $f(x) = 8x + 3$
15. $f(x) = 14x + 1$
16. $f(x) = 2x^2 + 7x - 4$
17. $f(x) = 2x^2 + 7x + 4$
18. $f(x) = -5x^2 + 4x + 13$
19. $f(x) = 10x^3 - 9x^2 + 3x + 40$
20. $f(x) = 5x^3 - 3x^2 + 8x + 3$

In Exercises 21–26, determine the indicated derivatives.

21. $\frac{d}{dx}(7x^5 - 4x^4 + 3x^2 + 40)$
22. $\frac{d}{dx}(2x^8 - 6x^7 + 3x^5 + 9x)$
23. $\frac{d}{dx}\left(9x + \frac{4}{x}\right)$
24. $\frac{d}{dx}\left(3x^6 + \frac{2}{x^3}\right)$
25. $\frac{d}{dx}\left(\sqrt[6]{x^5} + \frac{2}{x^4} + 8\right)$
26. $\frac{d}{dx}\left(\sqrt[3]{x^2} - \frac{5}{x^3} + \frac{9}{x}\right)$

27. Determine the slope of $y = x^3 + 2x^2 + 3x + 11$ at $x = 1$.
28. Determine the slope of $y = x^5 + 9x^4 + 8x + 5$ at $x = 0$.
29. If $f(x) = x^2 + 4x + 2$ determine $f(2)$ and $f'(2)$.
30. If $f(x) = x^2 + 3x + 5$ determine $f(1)$ and $f'(1)$.
31. If $f(x) = x^3 + 5x^2 + 2$ determine $f(-1)$ and $f'(-1)$.
32. If $f(x) = x^3 + 9x + 7$ determine $f(3)$ and $f'(3)$.
33. Determine the equation of the tangent line to $y = 3x^2 + 5x + 2$ at $x = 2$.
34. Determine the equation of the tangent line to $y = 5x^3 + 7x^2$ at $x = 1$.
35. Determine the equation of the tangent line to $y = 7x^5 + 8x + 25$ at $x = 0$.
36. Determine the equation of the tangent line to $y = \frac{2}{x^3} + 1$ at $x = 1$.

2.4 DIFFERENTIABILITY AND CONTINUITY

The derivative of $f(x)$ at $x = a$ has been defined as its limiting rate of change as $x \to a$. As graphs of polynomials are "smooth and unbroken" curves (i.e., they are continuous functions), their derivatives exist for all x. Polynomial functions are always **differentiable**.

↑ While graphs of polynomials, being continuous functions, are indeed "smooth and unbroken," mathematicians prefer more precise language for continuity. A function $y = f(x)$ is said to be continuous at $x = a$ if $f(a)$ exists and can be approximated to any preassigned accuracy ε (no matter how small) if, another positive number, δ can be found so that $|f(x)–f(a)| < \varepsilon$ for all x for which $|x–a| < \delta$.

This contrasts with functions whose limiting rate of change may not exist at some (or all) values of x. For example, $f(x) = 1/x$ is not differentiable at $x = 0$, because $f(x) \to \infty$ as $x \to 0$. Geometrically, $f(x)$ is not differentiable at $x = a$ when a tangent doesn't exist there. Geometrically, $f(x)$ is not differentiable at cusps or "sharp points" such as $f(x) = |x|$ at $x = 0$, or if there is a vertical tangent.

↑ To see this, consider $\dfrac{f(x+h)-f(x)}{h}$ for $f(x) = \dfrac{1}{x}$. The difference quotient is $\dfrac{\dfrac{1}{x+h}-\dfrac{1}{x}}{h} = \dfrac{\dfrac{x-(x+h)}{x(x+h)}}{h} = \dfrac{-h}{hx(x+h)} = -\dfrac{1}{x(x+h)}$ as $h \to 0$ the factor $(x+h)$ tends to x yielding a derivative of $\dfrac{-1}{x^2}$. When x is zero the derivative fails to exist so the function is not differentiable there.

For piecewise functions, differentiability is questionable at endpoints of an interval even if the function is defined there, as connecting segments of the function usually result in a **cusp** (sharp point). If segments do not join, then a limit will not exist, and, again, there can be no derivative.

Example 2.4.1 Differentiability

The function $f(x) = \begin{cases} |x| & -3 \le x < 1 \\ 2x-5 & 1 \le x \le 3 \\ 1 & x > 3 \end{cases}$ *is graphed as follows.*

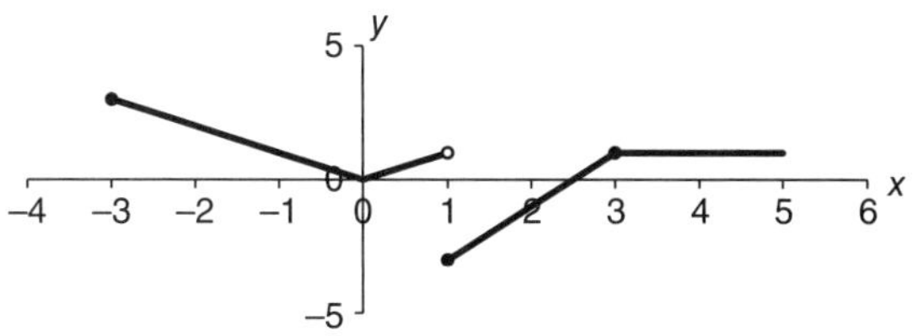

Determine whether f(x) is differentiable at these points.

⟶

a) $x = -2$ *b)* $x = 0$ *c)* $x = 1$ *d)* $x = 2$ *e)* $x = 3$ *f)* $x = 4$

Solution:

a) At $x = -2$*, the limit exists;* $f(x)$ *is differentiable there.*

b) At $x = 0$*, there is a cusp;* $f(x)$ *is not differentiable there.*

c) At $x = 1$*, there is no limit (two sided);* $f(x)$ *is not differentiable there.*

d) At $x = 2$*, the limit exists;* $f(x)$ *is differentiable there.*

e) At $x = 3$*, there is a cusp;* $f(x)$ *is not differentiable there.*

f) At $x = 4$*, the limit exists;* $f(x)$ *is differentiable there.*

Example 2.4.2 Differentiability (revisited)

Determine whether each of the following is differentiable at $x = 2$.

a) $f(x) = x^3 + 5x + 1$

b) $f(x) = \dfrac{5x}{x-2}$

c) $f(x) = \begin{cases} 3x + 5 & x < 2 \\ x^2 + 2x + 3 & x \geq 2 \end{cases}$

Solution:

a) The function, being a polynomial, is differentiable everywhere.

b) The function being undefined at $x = 2$ *is not differentiable there.*

c) While both segments of the graph join at (2, 11), they form a cusp. Hence, there is no tangent at $x = 2$*, and* $f(x)$ *is not differentiable there.*

Differentiability is linked to **continuity**. One may have continuity without differentiability but not the converse. There cannot be differentiability without continuity. It is easy to determine continuity from the graph of a function. If there is a "break" in the graph, the function is not continuous there. Basically, continuity of $f(x)$ implies that one can trace the graph of $f(x)$ without, in principle, lifting one's pencil.

Example 2.4.3 Continuity

Determine whether the graph given as follows is continuous at

a) $x = -1$ *b)* $x = 0$ *c)* $x = 1$ *d)* $x = 2$ *e)* $x = 3$

⟶

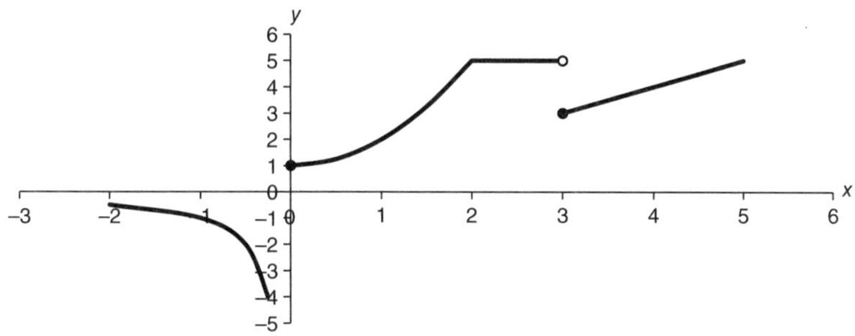

Solution:

a) The function is continuous at $x = -1$.

b) There is a break at $x = 0$, so $f(x)$ is not continuous here.

c) The function is continuous at $x = 1$.

d) The function is continuous at $x = 2$.

e) There is a break at $x = 3$, so $f(x)$ is not continuous here.

↑ To illustrate the "nuts and bolts" of establishing continuity of a function, consider $f(x) = x^2$; a relatively simple function known to be continuous at an arbitrary point, $x = 2$, $f(2) = 2^2 = 4$. Consider whether $|f(x)–f(2)|$ can be less than, say, $\varepsilon = 0.01$. Is there a value, δ, such that $|f(x)–f(2)| < \varepsilon$, where ε is arbitrary? For a $\delta < 0.002$, say, we can choose $x = 2.001$. Now, $|(2.001)^2 – 2^2| = 0.004001 < 0.01$. Similarly, for every $x < 2.001$, we will have $|x^2 - 2^2| < 0.01$. Therefore, $f(x) = x^2$ is continuous at $x = 2$. This tedious calculation is rarely necessary to establish continuity. We include it here solely to illustrate an abstract theoretical point.

To determine whether a function $f(x)$ is continuous at some point $(a, f(a))$, the following three conditions must be satisfied:

1. $f(a)$ must exist.
2. $\lim_{x \to a} f(x)$ must exist.
3. $\lim_{x \to a} f(x)$ must equal $f(a)$.

If all three conditions hold, the function is continuous at $x = a$.

Example 2.4.4 More on Continuity

Determine whether each of the following functions is continuous at $x = 3$.

a) $f(x) = x^4 + 2x + 3$

b) $f(x) = \begin{cases} 2x + 1 & x < 3 \\ x^2 - 2 & x \geq 3 \end{cases}$

c) $f(x) = \begin{cases} \dfrac{x^2 - x - 6}{x - 3} & x \neq 3 \\ 3x + 1 & x = 3 \end{cases}$

⟶

Solution:

a) *The function, being a polynomial, is everywhere continuous.*

b) *Firstly, determine whether $f(3)$ exists by substituting 3 into the second part of the function (as its interval includes $x = 3$). Substituting, $f(3) = (3)^2 - 2 = 7$. Next, determine whether the limit as x approaches 3 exists. The limit exists because both one-sided limits have a value of 7. Finally, $f(3)$ equals the limit as x approaches 3. This function is continuous at $x = 3$.*

c) *$f(3) = 3(3) + 1 = 10$. Next, the limit is an indeterminate form. However, it is factorable to yield*

$$\lim_{x\to 3}\frac{\cancel{(x-3)}(x+2)}{\cancel{(x-3)}} = \lim_{x\to 3}(x+2) = 5. \text{ Here, } f(3) \neq \lim_{x\to 3} f(x).$$

Here, $f(3) \neq \lim_{x\to 3} f(x)$. This function is not continuous at $x = 3$.

EXERCISES 2.4

Determine whether the function whose graph is shown as follows is differentiable at the following values of x.

1. $x = -1$
2. $x = 0$
3. $x = 2$
4. $x = 3$
5. $x = 1/2$
6. $x = 4$

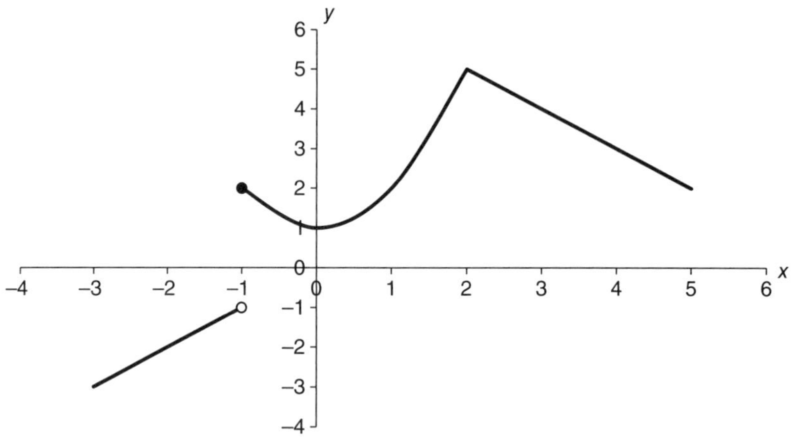

Determine whether the function whose graph is shown previously is continuous at the following values of x.

7. $x = -1$
8. $x = 0$
9. $x = 2$
10. $x = 3$
11. $x = 1/2$
12. $x = 4$

In Exercises 13–17, determine whether at x = 0, the following functions are a) differentiable b) continuous.

13. $f(x) = x^3 + 3x^2 + 5x + 1$

14. $f(x) = \sqrt{x-1}$

15. $f(x) = \begin{cases} x^2 & x > 0 \\ 2x & x \leq 0 \end{cases}$

16. $f(x) = \begin{cases} 2 & x = 0 \\ x^2 & x \neq 0 \end{cases}$

17. $f(x) = \begin{cases} x - 2 & -3 \leq x < 0 \\ x^3 - 2 & 0 \leq x < 2 \\ 6 & x \geq 2 \end{cases}$

In Exercises 18–21, determine whether at x = 1, the following functions are a) differentiable b) continuous.

18. $f(x) = 1/x$

19. $f(x) = 2x^5 + 9x + 5$

20. $f(x) = \begin{cases} 1 & x < 1 \\ x^2 & x \geq 1 \end{cases}$

21. $f(x) = \begin{cases} 2x - 3 & -2 \leq x < 1 \\ x^3 - 2 & x \geq 1 \end{cases}$

In Exercises 22–25, determine whether at x = 2, the following functions are a) differentiable b) continuous.

22. $f(x) = \sqrt[3]{x-1}$

23. $f(x) = 9x + 1$

24. $f(x) = \begin{cases} 3x - 5 & -2 \leq x < 2 \\ x^2 - 3 & x \geq 2 \end{cases}$

25. $f(x) = \begin{cases} x + 2 & -3 \leq x < 1 \\ x^2 - 1 & 1 \leq x < 2 \\ 3 & x \geq 2 \end{cases}$

In Exercises 26–30, the functions are defined for all values of x except one. Determine the limit of f(x) at this value (if possible). Use this information to define f(x) at the exceptional point in a way that has f(x) continuous for all x.

26. $f(x) = \dfrac{x^2 - 7x + 12}{x - 4} \quad x \neq 4$

27. $f(x) = \dfrac{x^2 - 36}{x + 6} \quad x \neq -6$

28. $f(x) = \dfrac{x^3 - 7x + 12}{x - 3} \quad x \neq 3$

29. $f(x) = \dfrac{x^3 - 25x}{x - 5} \quad x \neq 5$

30. $f(x) = \dfrac{(x+4)^2 - 16}{x} \quad x \neq 0$

2.5 BASIC RULES OF DIFFERENTIATION

Recall the power rule,

$$d/dx[x^r] = r[x^{r-1}]$$

Similarly, the power rule for a function's derivative, $[g(x)]^r$, is

General Power Rule

The derivative of $f(x) = [g(x)]^r$ is

$$r[g(x)]^{r-1}g'(x)$$

where f(x) and g(x) are differentiable functions of x.

↓ This is actually the power rule again. Not only is the original exponent a coefficient, but also the argument, whether x or $g(x)$, is also differentiated. This accounts for the $g'(x) = dg(x)/dx$. When differentiating x^r, the rx^{r-1} is multiplied by dx/dx, which, being unity, is omitted.

Example 2.5.1 Derivatives of [g(x)] ʳ

Find the derivatives of

a) $(x+1)^2$ b) $(3x+1)^2$ c) $(x^2+1)^2$

Solution:

a) *For the derivative, expanded as*
$x^2 + 2x + 1$. *Next,* $d/dx(x^2 + 2x + 1) = 2x + 2 = 1[2(x + 1)]$.

b) *Expand as* $9x^2 + 6x + 1$. *Next,*
$d/dx\ (9x^2 + 6x + 1) = 18x + 6 = 3[2(3x + 1)]$.

c) *Expanded as* $x^4 + 2x^2 + 1$. *Next,*
$d/dx\ (x^4 + 2x^2 + 1) = 4x^3 + 4x = 2x[2(x^2 + 1)]$.

The aforementioned example illustrates that for the derivative of a function $[g(x)]^r$, the original exponent becomes a coefficient and the new exponent has been decreased by unity and multiplied by the derivative of $g(x)$ as indicated in the shaded box previously.

Using the general power rule, one does not need to expand an algebraic expression in order to determine the derivative as in the previous example. The following examples are illustrative.

Example 2.5.2 General Power Rule

Find the derivative of $f(x) = (7x^9 + 3x^5 + 9x + 5)^{20}$.

Solution:

Using the general power rule,

$$f'(x) = 20(7x^9 + 3x^5 + 9x + 5)^{19}(63x^8 + 15x^4 + 9)$$

Example 2.5.3 General Power Rule (revisited)

Find the derivatives of

a) $f(x) = (2x^3 + 3x + 5)^{11/5}$

b) $f(x) = \dfrac{2}{\sqrt[3]{x^2 + 8x + 7}}$

Solution:

a) Using the general power rule,

$$f'(x) = (11/5)(2x^3 + 3x + 5)^{6/5}(6x^2 + 3)$$

b) Firstly, rewrite the function as $2(x^2 + 8x + 7)^{-1/3}$*. Next, using the general power rule yields*

$$f'(x) = (-2/3)(x^2 + 8x + 7)^{-4/3}(2x + 8)$$

Recall that tangent lines to quadratics and cubics and other simple functions were obtained earlier. A point and the slope there were all that was necessary to determine its equation. The same applies to more complex functions. The following example illustrates this.

Example 2.5.4 Equation of a Tangent Line

Find the tangent line to the curve $(3x - 1)^{4/3}$ *at* $x = 3$*.*

Solution:

Firstly, determine the point of tangency. Substituting, $f(3) = (3(3) - 1)^{4/3} = (8)^{4/3} = 16$*, so the point is* $(3, 16)$*. Next, seek the slope at* $x = 3$*. It is the derivative at* $x = 3$*. Using the general power rule,* $f'(x) = (4/3)(3x - 1)^{1/3}(3) = 4(3x - 1)^{1/3}$*. Therefore,* $f'(3) = 4(3(3) - 1)^{1/3} = 8$*. The equation of the line through (3, 16) with slope 8 is* $y - 16 = 8(x - 3)$*.*

EXERCISES 2.5

In Exercises 1–20, determine the indicated derivatives.

1. $f(x) = (4x^2 + 1)^4$
2. $f(x) = (9x + 3)^5$
3. $y = (5x^2 + 3)^7$
4. $y = 5(4x + 1)^2$
5. $f(x) = 7(3x^2 + 1)^4$
6. $f(x) = (7x^2 + 5x - 2)^2$

7. $f(x) = (9x^{10} + 6x^5 - x)^5$
8. $y = (8x^3 + 5x^2 - 3x + 1)^6$
9. $y = (12x^7 + 3x^4 - 2x + 5)^{10}$
10. $f(x) = \sqrt[3]{(x^4 + 3x + 1)^2}$
11. $f(x) = \dfrac{4}{(9x^3 + \sqrt{x} + 3)^5}$
12. $y = (x^3 + x^2 - x + 4)^{15}$
13. $f(x) = (7x^{8/5} + 5x + 1)^6$
14. $f(x) = (8x^{9/2} + 5x^{3/2} - 3x^{1/3} + 1)^8$
15. $f(x) = (\sqrt{x} + 1)^5$
16. $y = (\sqrt{5x} + 3)^{28}$
17. $y = \left(4x^5 + \sqrt[3]{x^2} + 1\right)^6$
18. $y = \left(7x^3 + \sqrt[5]{x^4}\right)^{15}$
19. $y = 7x^3 + 8x^2 + \dfrac{5}{(4x - 3)^7}$
20. $f(x) = 5x^4 - \dfrac{3}{x^2} + \dfrac{5}{(3x + 4)^7}$
21. Determine the slope of the tangent to $y = (5x + 1)^2$ at $x = 1$.
22. Determine the slope of the tangent to $y = (3x^2 - 6x + 2)^4$ at $x = 2$.
23. Determine the equation of the tangent to $y = (4x + 1)^{3/2}$ at $x = 2$.
24. Determine the equation of the tangent to $y = \sqrt{2x - 1}$ at $x = 5$.
25. Determine the equation of the tangent to $y = \dfrac{5}{(x^3 - 2x + 2)^3}$ at $x = 1$.

2.6 CONTINUED DIFFERENTIATION

Thus far, functions encountered have an independent variable, x, and a dependent variable, y, where $y = f(x)$. Often, letters other than x and y are used to denote variables. For example, in business applications, p and q are used for price and quantity, respectively. Often, t is used for time.

Recall, when $y = x^3 + 2x + 1$, the derivative can be denoted as dy/dx instead of using the "prime" notation. Suppose instead of y, another letter, say u, is used, so $u = x^3 + 2x + 1$. We can denote the derivative as u' or du/dx. Likewise, if another letter instead of x, say t, is used, then $y = t^3 + 2t + 1$, and the derivative is denoted dy/dt. In short, various letters can arise in various ways as long as their meaning is clear.

Rules you have learned for derivatives are valid regardless of the letters chosen to represent variables. The notation for derivatives is not standard. Several notations used in this text and elsewhere are repeated in the following table.

Prime Notation	$\frac{d}{dx}$ Notation
$f'(x)$	$\frac{d}{dx}(f(x))$
y'	$\frac{dy}{dx}$

Example 2.6.1 Derivatives for Other Variables

Determine these derivatives:

a) $\frac{d}{dt}(7t^4 + 3t^3 + 2)^{10}$ b) v' for $v = (7q - 4)^8$ c) $\frac{dr}{dp}$ for $r = 2p^5 - 7p^2 + 3p$

Solution:

a) *In this case, using the power rule,*
$\frac{d}{dt}(7t^4 + 3t^3 + 2)^{10} = 10(7t^4 + 3t^3 + 2)^9(28t^3 + 9t^2)$.

b) *In this case, using the power rule,* $v' = 8(7q - 4)^7(7) = 56(7q - 4)^7$.

c) *In this case,* $\frac{dr}{dp} = 10p^4 - 14p + 3$.

Sometimes, several variables appear in an expression, and only the derivative with respect to one of them is of interest. In this situation, differentiate with respect to the variable of interest, treating other variables as constants for the time being.

For instance, $\frac{d}{dt}(a^5t^3) = a^5[3t^2] = 3a^5t^2$ regarding "a" as a constant for the derivative with respect to t. On the other hand, $\frac{d}{da}(a^5t^3) = [5a^4]t^3 = 5a^4t^3$ when t is considered a constant.

Example 2.6.2 Derivatives of a Selected Variable

Determine the indicated derivatives:

a) $\frac{d}{dz}(a^3z^4 - b^2z^3 + 2cz^2 + 9z + 10)$ b) $\frac{d}{db}(a^3z^4 - b^2z^3 + 2cz^2 + 9z + 10)$

c) $\frac{d}{dt}(5a^6t^4 + 3t^2 + 2a + 1)^6$ d) $\frac{d}{da}(5a^6t^4 + 3t^2 + 2a + 1)^6$

Solution:

a) *In this case, a, b, and c are considered constants, so the indicated derivative is*
$a^3[4z^3] - b^2[3z^2] + 2c[2z] + 9 = 4a^3z^3 - 3b^2z^2 + 4cz + 9$.

b) *In this case, a, c, and z are considered constants, so the only terms to have a derivative contain b. The derivative is* $[-2b]z^3$.

c) *In this case, a is considered a constant. The derivative is*
$6(5a^6t^4 + 3t^2 + 2a + 1)^5[20a^6t^3 + 6t]$.

d) *In this case, t is considered a constant. The derivative is*
$6(5a^6t^4 + 3t^2 + 2a + 1)^5[30a^5t^4 + 2]$.

Second Derivative

Differentiating a function, $f(x)$, yields $f'(x)$ its slope at x. This is called the **first derivative**. If the first derivative is differentiable, it can also be differentiated. The derivative of the first derivative is the **second derivative**. It can be denoted by a double prime as $f''(x)$. Some other ways to denote the second derivative are $\frac{d^2f(x)}{dx^2}$ or $\frac{d^2y}{dx^2} = \frac{d}{dx}\left(\frac{dy}{dx}\right)$.

Important Note

Notation varies, and it is important to avoid ambiguity. The 2's in the second derivative are not exponents, despite appearances. They signify only that a second derivative, a derivative of the first derivative is intended.

Furthermore, it is common to reference derivatives using the "prime" symbol. Hence, for $y = f(x)$, $\frac{dy}{dx} = \frac{df(x)}{dx} = f'(x)$. Similarly, $\frac{d^2y}{dx^2} = \frac{d^2f(x)}{dx^2} = f''(x)$, and so on.

Example 2.6.3 Second Derivative

Determine the second derivatives of the following functions:

a) $f(x) = x^7 - 4x^4 + 3x^2 + 9x + 3$

b) $f(x) = (2x + 7)^9$

c) $f(x) = 2/x^3$

Solution:

a) *The first derivative, $f'(x)$, is $7x^6 - 16x^3 + 6x + 9$. The second derivative $f''(x)$ is $42x^5 - 48x^2 + 6$.*

b) *The first derivative, $f'(x)$, is $9(2x + 7)^8(2) = 18(2x + 7)^8$. The second derivative, $f''(x)$, is $18(8)(2x + 7)^7(2) = 288(2x + 7)^7$.*

c) *Firstly, rewrite as $f(x) = 2x^{-3}$. The first derivative $f'(x)$ is $-6x^{-4}$, and the second is $24x^{-5}$ or $\frac{24}{x^5}$.*

You may have correctly anticipated that besides first and second derivatives, there might be further differentiation of derivatives as third, fourth, or higher derivatives.

Example 2.6.4 Higher Order Derivative

Find all nonzero higher order derivatives of the following function:

$$f(x) = 2x^4 + 3x^2 + 2x + 1$$

Solution:

The first derivative, $f'(x)$ is $8x^3 + 6x + 2$. The second derivative $f''(x)$ is $24x^2 + 6$. The third derivative $f'''(x)$ is $48x$, and the fourth derivative $f^{iv}(x)$ is 48. The fifth and higher derivatives vanish.

Note that $f(x)$, is a polynomial of degree 4 and, as such, has four nonzero derivatives. In general, an nth degree polynomial will have n nonzero derivatives ($n = 1, 2, \ldots$).

The derivative of $y = f(x)$ evaluated at a specific value of x, say at $x = a$, yields $f'(a)$, the slope of the curve of $y = f(x)$ at the point $(a, f(a))$. Sometimes, the notation $\frac{dy}{dx}|_{x=a}$ is used to denote $f'(a)$. Likewise, $\frac{d^2y}{dx^2}|_{x=a}$ is used to denote $f''(a)$.

Example 2.6.5 Evaluating Derivatives at $x = a$

Evaluate these derivatives of $y = 5x^4 + 3x^3 + 2x + 1$.

a) $\frac{dy}{dx}|_{x=0}$ b) $\frac{dy}{dx}|_{x=1}$ c) $\frac{d^2y}{dx^2}|_{x=2}$ d) $\frac{d^2y}{dx^2}|_{x=-1}$

Solution:

a) $\frac{dy}{dx} = 20x^3 + 9x^2 + 2$, *so at* $x = 0$, *the derivative is 2.*
b) *The first derivative is evaluated at* $x = 1$ *to yield* $20(1)^3 + 9(1)^2 + 2 = 31$.
c) $\frac{d^2y}{dx^2} = 60x^2 + 18x$, *so at* $x = 2$, *the second derivative is 276.*
d) *The second derivative is evaluated at* $x = -1$ *to yield* $60(-1)^2 + 18(-1) = 42$.

EXERCISES 2.6

In Exercises 1–10, determine the first derivatives.

1. $f(z) = z^4 + 3z^2 + 2$
2. $f(t) = 3t^5 + 5t^3 + 2t$
3. $s = 3r^4 + r^3 + 2r^2$
4. $q = p^4 + 3p^3 + 2p - 7$
5. $f(p) = 3p^8 + 5p^6 + 2p^3 + 4p$
6. $h(y) = 2y^4 + 3y^{-2} + 5$
7. $f(t) = (5t^2 + 3t + 1)^{10}$
8. $f(z) = \sqrt[3]{5z^2 + 3z + 4}$
9. $S(p) = (3p^{10} + \sqrt{p} + 5)^{12}$
10. $T(v) = \frac{2}{(3v^2 + 4)^{10}}$
11. Determine $\frac{d}{dt}(3t^{7/5} - 5t - 4)$.
12. Determine $\frac{d}{dr}\left(\frac{4}{3}\pi r^3\right)$.
13. Determine $\frac{d}{dp}(5p^4 - 3p^{2/3})$.
14. Determine $\frac{d}{dz}(4z^3 - 3z + 2)^5$.
15. Determine $\frac{d}{dt}(2a^7t^5 - 9bt^3 + t^2 + 3)$ and $\frac{d}{db}(2a^7t^5 - 9bt^3 + t^2 + 3)$.

16. Determine $\frac{d}{dx}(4a^3x^3 - 9ax^2 - 2a + 7)$ and $\frac{d}{da}(4a^3x^3 - 9ax^2 - 2a + 7)$.

In Exercises 17–22, determine first and second derivatives.

17. $f(x) = 9x^3 + 4x + 2$

18. $f(x) = 5x^5 + 8x^3 + 2x + 50$

19. $y = 2x^3 + 3x + \sqrt[4]{x^3}$

20. $y = \frac{1}{x^2} + 50$

21. $v = 2t^3 + 18t - 4$

22. $s = (1 + 4r)^7$

23. Evaluate $\frac{d}{dx}(x^3 - 7x + 3)|_{x=1}$.

24. Evaluate $\frac{d}{dt}(2t^5 - 3t^3 + 9)|_{t=2}$.

25. Evaluate $\frac{d^2}{dt^2}(t^6 + 2t + 1)|_{t=-1}$.

26. Evaluate $\frac{d^2}{dx^2}(x^5 - 2x^3 + 9x^2 + 7x + 1)|_{x=0}$.

27. Evaluate $y''(1)$, where $y = (2x + 7)^{5/2}$.

28. Evaluate $y''(0)$, where $y = \frac{5}{(3t + 1)^4}$.

29. Determine the fourth derivative of $y = x^5 + 3x^3 + 9x + 2$.

30. Determine the third derivative of $y = \sqrt[3]{x^2}$.

2.7 INTRODUCTION TO FINITE DIFFERENCES

Say "calculus" and a listener is likely to assume you intend differential calculus, the famous relative of ***finite calculus***. With good reason! Differential calculus has an extraordinary role in the development of mathematical and physical science and engineering for more than three centuries.

Differential calculus and finite calculus are the mathematics of change and of rates of change. They differ in that in differential calculus changes are infinitesimal and rates of change are instantaneous, while they are discrete and finite in the finite calculus.

In some respects, the finite calculus better models human activity than the differential calculus. Although situations such as the continuous compounding of interest by banks theoretically changes one's bank balance from one instant to the next, the reality is that interest is reported at finite intervals. Similarly, in psychological studies, a subject periodically signals notice of a change in brightness in an illuminated lamp, although brightness may actually be changing continuously. Mileage between cities and a person's change in weight are examples given in whole numbers, although they obviously change continuously, and so on.

This section introduces elements of finite calculus. Later, the many similarities with differential calculus will be apparent. In any event, finite calculus is easier to understand and in some ways more practical than its famous relative.

Finite Difference Operations

Consider a function $f(x)$ whose independent variable, x, takes only integer values $x = 0, 1, 2, 3, \ldots$ The **first finite difference** is defined by

$$\Delta f(x) = f(x+1) - f(x) \qquad x = 0, 1, 2, \ldots$$

It is important to note that the usage of the symbol delta (Δ) here is not necessarily the same as in other areas of mathematics and, in particular, its earlier use as "a little bit of." In this case, it simply means the change in the function by a unit change in its variable.

Example 2.7.1 First Finite Difference

If $f(x) = 5x$, *find* $\Delta f(x)$.

Solution:

$$\Delta f(x) = 5(x+1) - 5x = 5$$

Example 2.7.2 Another First Finite Difference

If $f(x) = 3x^2$, *find* $\Delta f(x)$.

Solution:

$$\begin{aligned}\Delta f(x) &= 3(x+1)^2 - 3x^2 \\ &= 3(x^2 + 2x + 1) - 3x^2 \\ &= 6x + 3\end{aligned}$$

Similarly, one defines a **second difference**, analogous to a second derivative as

$$\begin{aligned}\Delta^2 f(x) &= \Delta(\Delta f(x)) = \Delta(f(x+1) - f(x)) \\ &= [f(x+2) - f(x+1)] - [f(x+1) - f(x)] \\ &= f(x+2) - 2f(x+1) + f(x)\end{aligned}$$

Example 2.7.3 Second Finite Difference

If $f(x) = 3x^2$ and $\Delta f(x) = 6x + 3$, find $\Delta^2 f(x)$.

Solution:

$$\Delta^2 f(x) = \Delta(\Delta f(x)) = [6(x + 1) + 3] - (6x + 3)$$
$$= 6$$

Alternately, using the second difference formula directly,

$$\begin{aligned}\Delta^2 f(x) &= f(x + 2) - 2f(x + 1) + f(x)\\ &= 3(x + 2)^2 - 6(x + 1)^2 + 3x^2\\ &= 3x^2 + 12x + 12 - 6x^2 - 12x - 6 + 3x\\ &= 6\end{aligned}$$

Example 2.7.4 Another Second Finite Difference

If $f(x) = x^3$, find the first and second finite difference.

Solution:
The first finite difference is

$$\begin{aligned}\Delta f(x) &= f(x + 1) - f(x)\\ &= (x + 1)^3 - x^3\\ &= (x^3 + 3x^2 + 3x + 1) - x^3\\ &= 3x^2 + 3x + 1\end{aligned}$$

The second finite difference is

$$\begin{aligned}\Delta^2 f(x) &= \Delta(\Delta f(x))\\ &= [3(x + 1)^2 + 3(x + 1) + 1) - (3x^2 + 3x + 1)]\\ &= 3x^2 + 6x + 3 + 3x + 3 + 1 - 3x^2 - 3x - 1\\ &= 6x + 6\end{aligned}$$

Generally, the $r+1^{\text{st}}$ difference of $f(x)$ is defined as

$$\Delta^{r+1}f(x) = \Delta(\Delta^r f(x)) = \Delta^r[f(x+1) - f(x)] = \Delta^r f(x+1) - \Delta^r f(x) \qquad r \geq 0$$

where $\Delta^r f(x)$ is read as "delta to the r of $f(x)$"or the "rth difference of $f(x)$." Note that $\Delta^1 = \Delta$.

We return to finite differences in Chapter 5. This section is an introduction. One might already note the similarity with the conventional derivative as introduced earlier in this chapter and also with the difference quotients from Section 1.6.

EXERCISES 2.7

In Exercises 1–8, find the first finite differences.

1. $f(x) = 6x + 4$
2. $f(x) = 9x + 1$
3. $f(x) = 4x^2$
4. $f(x) = 3x^2 + 1$
5. $f(x) = 5x^2 + 2x + 3$
6. $f(x) = 2x^3 + 7$
7. $f(x) = x^3 + 3x + 1$
8. $f(x) = x^3 + 2x^2 + 5x + 3$

In Exercise 9–14, find the second finite differences.

9. $f(x) = 9x - 1$
10. $f(x) = 8x + 3$
11. $f(x) = 2x^2 + 5$
12. $f(x) = 3x^2 + 5x + 6$
13. $f(x) = 5x^3 + 2$
14. $f(x) = x^3 + 7x + 11$

HISTORICAL NOTES

Calculus is central to mathematical analyses. Utilizing both algebra and geometry, it rests upon two major ideas. The first, differential calculus, studies rates of change, usually represented as the slope of a tangent line. The second, integral calculus, calculates areas under curves and is the mirror image of differential calculus because they are inverse operations of each other.

Apart from concepts of differentiation and integration, as the foundation for that branch of mathematics known as analysis, it is central to physics, engineering, and most branches of science.

Gottfried Leibniz (1646–1716) — German philosopher, physicist, and mathematician. He was self-taught in mathematics and is credited with independent development of the calculus. His work was published after Newton's. He is credited with the calculus notation which is still in use.

Leibniz made many contributions to the theory of differential equations. He discovered the method of separation of variables as well as a procedure for solving first-order linear equations.

Sir Isaac Newton (1642–1727) — Considered by some as the foremost scientist of all time, Newton, an Englishman, is credited with the invention of the calculus (although some credit Leibniz with an independent discovery). His intellectual prowess was not limited to mathematics. Newton made important discoveries in physics, optics, and astronomy, and a number of physical laws bear his name. He defined the laws of motion and universal gravitation, and his "Opticks" and "Principia" embody much of his scientific work.

Newton was educated at Trinity College, Cambridge. During his tenure at the College (1661–1696), he produced most of his mathematical works. In 1696, Newton was appointed Master of the Royal Mint and moved to London. He resided there until his death. Newton is buried in Westminster Abbey, the first scientist accorded the honor.

CHAPTER 2 SUPPLEMENTARY EXERCISES

1. Find the slope of the tangent to x^2 at (1/4, 1/16).
2. Find the equation of the tangent line to $y = x^3$ at the point where $x = 3/2$.
3. Find the point on $y = x^2$ where the slope is 3/4.
4. Find the points on $y = x^3$ where the slope is 75.
5. Determine $\lim_{x\to 2} 3x + \frac{4}{x}$.
6. Determine $\lim_{x\to 0} \frac{5x}{x+2}$.
7. Determine $\lim_{x\to 4} \frac{x^2 + 2x - 24}{x - 4}$.
8. If $f(x) = \begin{cases} 5x - 1 & x < 1 \\ 2x + 2 & 1 \le x \le 4 \\ x^2 + 3x + 2 & x > 4 \end{cases}$

   Determine the following limits if they exist:

   a) $\lim_{x\to 1} f(x)$ b) $\lim_{x\to 3} f(x)$ c) $\lim_{x\to 4} f(x)$

9. If $f(x) = 3x^2 + 5x + 1$, determine the derivative using both secant line methods.
10. If $f(x) = x^3 + 2x^2 + 8$, determine the derivative using both secant line methods.
11. If $f(x) = 5x^3 + 2x^2 + 3$, determine $f(1)$ and $f'(1)$.
12. Determine the slope of the curve $y = 2x^{7/5} + 4x + 5$ at $x = 1$.
13. Determine the equation of the tangent line to $y = 5x^7 - 3x^2$ at $x = 1$.
14. Determine whether $f(x) = 5x^3 + 9x + 2$ is

    a) continuous b) differentiable at $x = 1$.

15. Determine whether $f(x) = \begin{cases} 2x - 3 & x = 1 \\ \dfrac{2x^2 - 5x + 3}{x - 1} & x \neq 1 \end{cases}$ is continuous at $x = 1$.

16. The following function is defined for all but a single value of x. Determine the limit of $f(x)$ at this value (if possible). Use this information to define $f(x)$ at the exceptional point in a way that has $f(x)$ continuous for all x.

$$f(x) = \frac{x^3 - 8}{x - 2} \qquad x \neq 2$$

17. Determine $\dfrac{d}{dx}(9x^3 + 4x^2 + 3x + 1)^{25}$.

18. Determine the derivative of $y = (9x + 5)^{2/3}$.

19. Determine the derivative of $f(x) = \left(2\sqrt[3]{x^2} + 3x + 1\right)^4$.

20. If $f(x) = (5x^2 + 6x + 2)^3$, find $f(0)$ and $f'(0)$.

21. Determine the equation of the tangent to $y = (4x + 1)^{5/2}$ at $x = 6$.

22. Determine $\dfrac{d}{dz}(4z^7 - 2z + 1)^{100}$.

23. Determine $\dfrac{d}{dp}(5a^3p^4 + 3ap^2 + 2bp + c)$ and $\dfrac{d}{da}(5a^3p^4 + 3ap^2 + 2bp + c)$.

24. Determine the first and second derivatives of $y = 2x^5 - 9x^2 + 30$.

25. Evaluate $\dfrac{d^2}{dx^2}(4x^9 - 3x^7 + 7x + 9)|_{x=1}$.

26. Evaluate $\dfrac{d^2}{dt^2}(3t - 5)^{10}|_{t=2}$.

3 *Using The Derivative*

Fundamentals of Calculus, First Edition. Carla C. Morris and Robert M. Stark.

Companion Website: http://www.wiley.com/go/morris/calculus

In the previous chapter, you learned rudimentary principles of the calculus – the meaning and definition of a derivative. Already you can use your new knowledge in important ways. In this chapter, you will learn applications to display functions graphically (curve sketching) and to formulate applications of maxima and minima.

3.1 DESCRIBING GRAPHS

Graphs are akin to "pictures" of functions! More precisely, a function, $y = f(x)$, assigns a value to y for every allowed value of x. Graphs of functions are characterized by a number of properties, and it is worth an effort to understand them.

Increasing and Decreasing Functions

A function that increases (rises) as x increases (left to right) is an **increasing function**. Similarly, a function that decreases (falls) as x increases (left to right) is a **decreasing function**.

In other words, if on an interval, $x_1 < x_2$, $f(x_1) < f(x_2)$, the function is increasing and its graph has a positive slope (rises from left to right). For example, in the graph on the left, for $y = x^3 + 1$, using $x_1 = 0$ and $x_2 = 1$, $f(0) = 1$ is, indeed, less than $f(1) = 2$ signaling an increasing function. Actually, $y = x^3 + 1$ is an increasing function for all x.

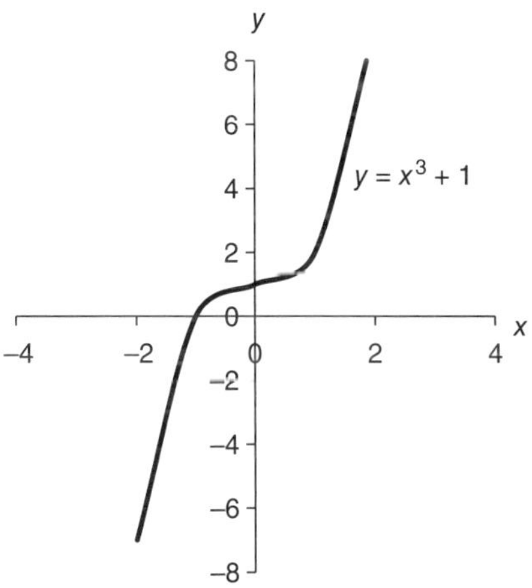

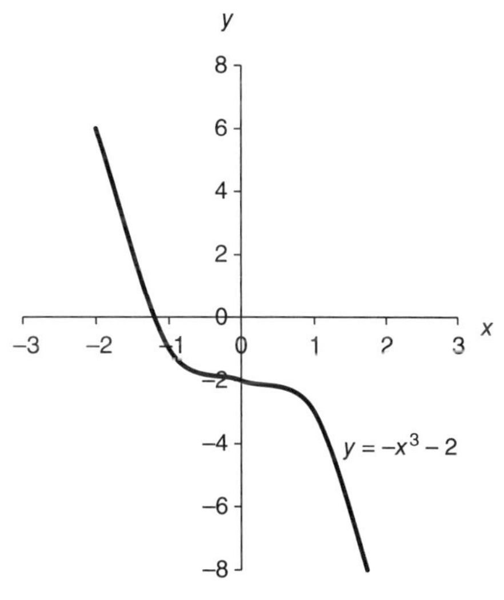

Similarly, the graph of $y = -x^3 - 2$, top right, is a decreasing function for all values of x.

Functions may increase in some intervals and decrease in others. The graph of $y = x^3 - 6x^2 + 9x$ has the function increasing for $x < 1$, for $x > 3$, and decreasing for $1 < x < 3$.

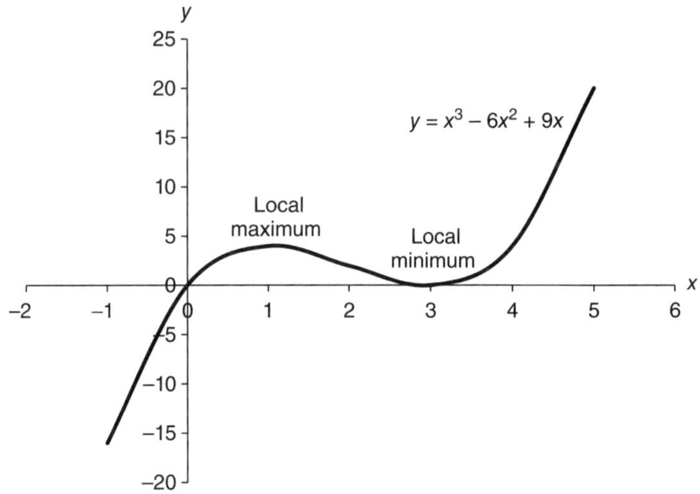

Maximum and Minimum Values

The previous figure also illustrates extreme values of $y = x^3 - 6x^2 + 9x$. Its absolute, largest and smallest, values of the function are infinite. However, of greater interest are the **local** (or relative) **maximum** (at (1, 4)) and the **local** (or relative) **minimum** (at (3, 0)). Such points of the curve possess a horizontal tangent. Note that at a local maximum the function changes from increasing (rising) to decreasing (falling), and, similarly, at a local minimum, the function changes from decreasing (falling) to increasing (rising). Local maxima and minima are also called **local** (or relative) **extrema**.

The **absolute maximum** of a function, on the other hand, is the largest value the function assumes over its domain. If it becomes infinite, the maximum is **unbounded**. Likewise, the **absolute minimum** of a function is its smallest value over its domain. Absolute maxima and minima may also have horizontal tangents and, in a sense, are local as well as absolute. Absolute extrema have functional values that are greater or lesser than any other extrema a function may have.

Usually, an absolute maximum (minimum) occurs at a local extremum. Sometimes, it occurs at an endpoint of the domain or when restrictions are placed upon the function, for example, that x must be positive valued as in many applications or when a function is defined only in the first quadrant. When the maximum (minimum) occurs at an endpoint, it is called an **endpoint extremum** and it lacks a horizontal tangent.

The following example is illustrative of these concepts.

Example 3.1.1 Identifying Extrema

Identify increasing and decreasing intervals of the function. Also, identify extrema (local, endpoint, and absolute).

⟶

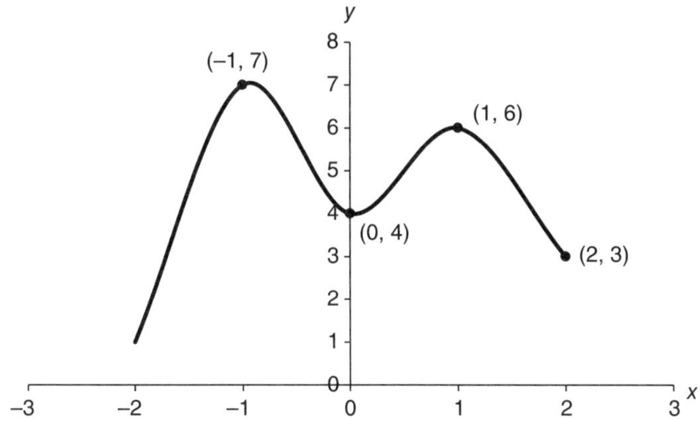

Solution:

The function increases on the intervals $(-\infty, -1)$ and (0, 1). It decreases on $(-1, 0)$ and (1, 2).

There are two local maxima: at $(-1, 7)$ and (1, 6). A local minimum exists at (0, 4).

There is an endpoint minimum at (2, 3) (signified by the closed endpoint.)

Here, the local maximum at $(-1, 7)$ is also the absolute maximum of the function.

The absolute minimum of the function is unbounded $(-\infty)$ as x decreases through negative values.

Concavity and Inflection

A function whose graph is "concave up" can "hold water." A graph that is "concave down" "spills water." Respective sketches appear as follows.

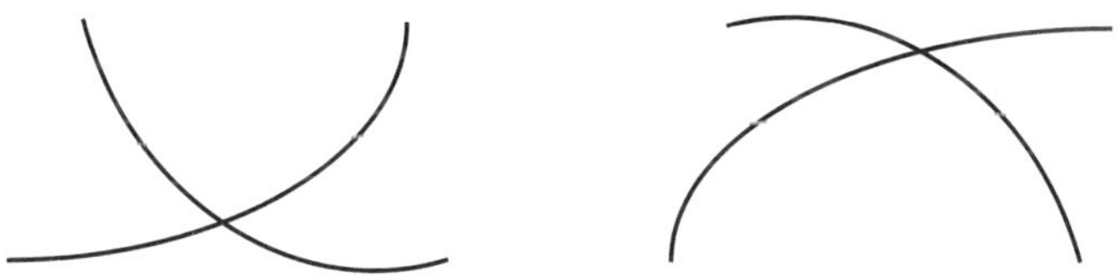

Concave up
"holds water"

Concave down
"spills water"

Where concavity changes from "up to down" or vice versa, an **inflection point** is said to occur as shown in the following example.

Example 3.1.2 More on Identifying Extrema

Indicate where the following function increases, decreases, and has extrema. Also indicate where the function is concave up, concave down, and identify any inflection points.

⟶

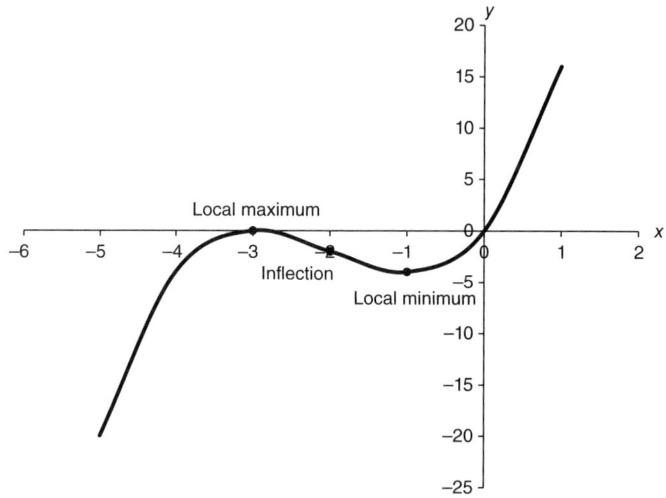

Solution:
The function increases on the intervals $(-\infty, -3) \cup (-1, \infty)$*, decreases on the interval* $(-3, -1)$*, and has a local maximum at* $(-3, 0)$ *and a local minimum at* $(-1, -4)$*. The function is concave up on the interval* $(-2, \infty)$ *and concave down on the interval* $(-\infty, -2)$*. The point* $(-2, -2)$ *is an inflection point.*

↓ Recall, when a function increases its graph rises; when it decreases its graph falls. If the graph changes from rising to falling, then a local maximum occurs at the change point. Likewise, if the function decreases and then increases, a local minimum occurs at the change point.

Key questions to consider when graphing functions are:

1. In what intervals does the function increase and/or decrease?
2. Are there local extrema?
3. In what intervals is the function concave up or concave down?
4. Are there inflection points?
5. Are there intercepts, endpoint extrema, absolute extrema, undefined points, or asymptotes?

EXERCISES 3.1

Which of the functions graphed:

1. increase for all x?
2. decrease for all x?
3. have slopes that always increases as x increases (concave up for all x)?
4. have slopes that always decreases as x increases (concave down for all x)?
5. have an inflection point?

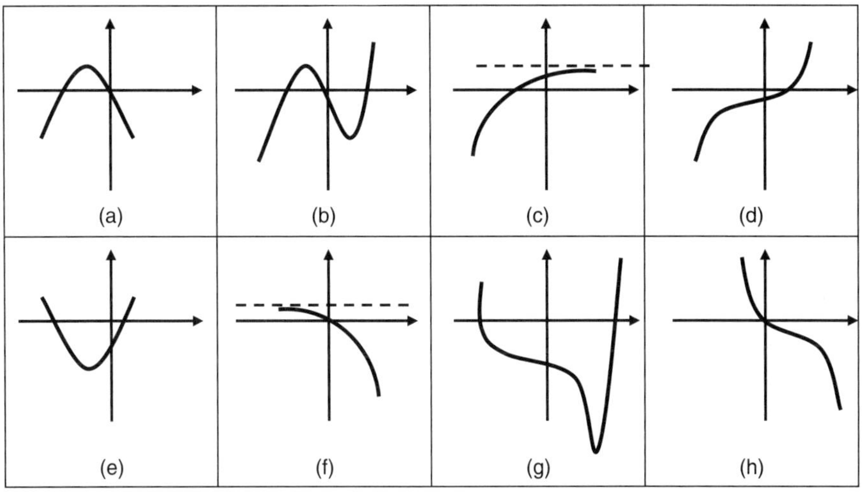

Describe the following graphs using the five criteria for graphing functions listed previously.

6.

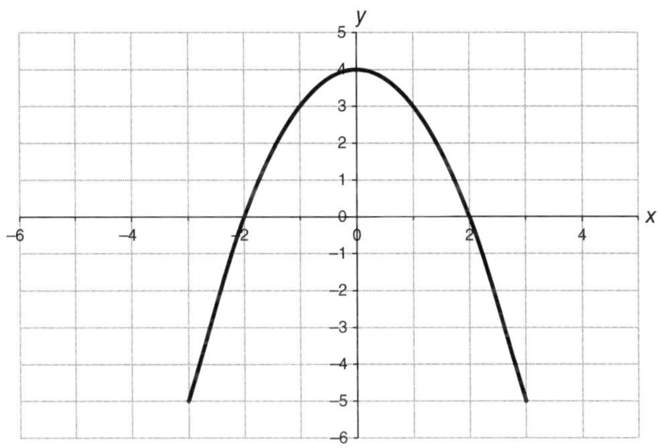

7.

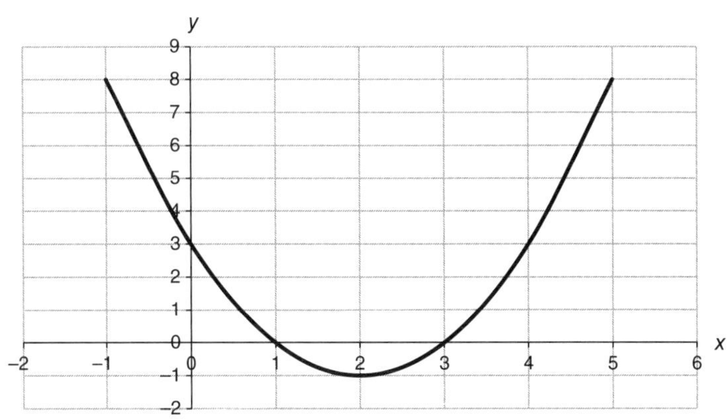

8.

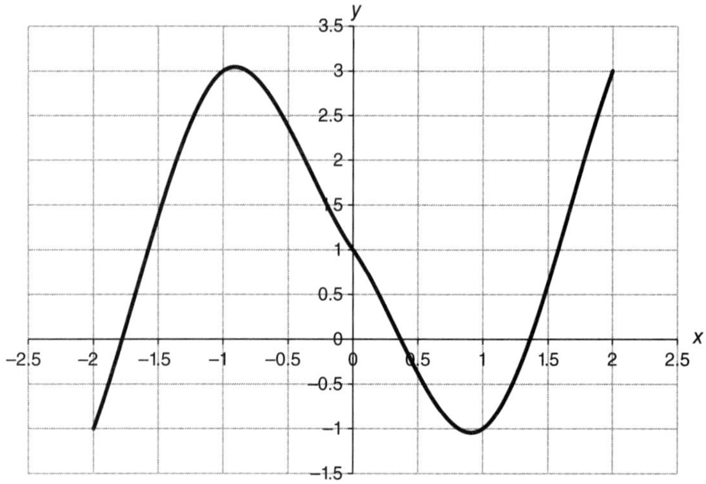
y
x
3.5
3
2.5
2
1.5
1
0.5
0
−0.5
−1
−1.5
−2.5
−2
−1.5
−1
−0.5
0
0.5
1
1.5
2
2.5

9.

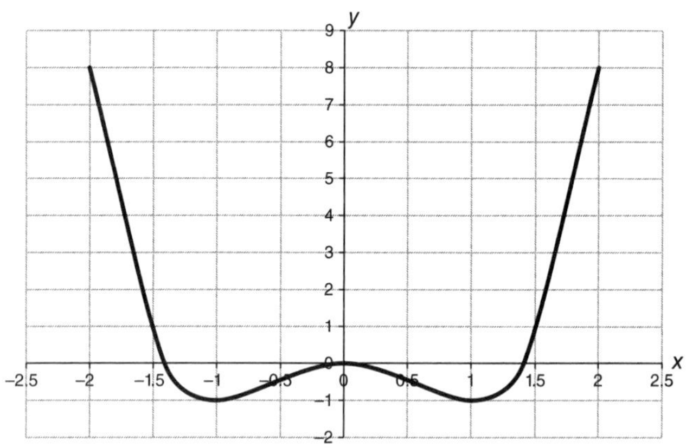
y
x
9
8
7
6
5
4
3
2
1
0
−1
−2
−2.5
−2
−1.5
−1
−0.5
0
0.5
1
1.5
2
2.5

10.

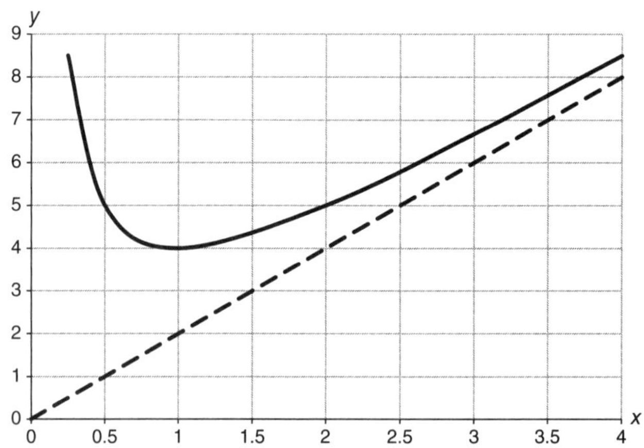
y
x
9
8
7
6
5
4
3
2
1
0
0
0.5
1
1.5
2
2.5
3
3.5
4

For Exercises 11–15, use the following graph

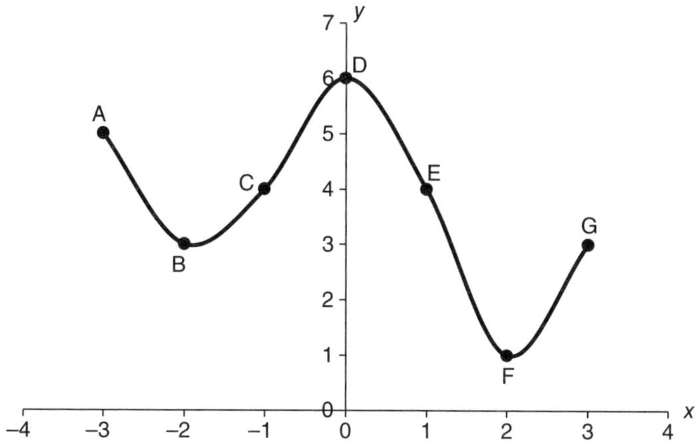

Determine at which of the labeled points the function

11. has a local or endpoint extremum.
12. is increasing.
13. has a point of inflection.
14. is concave down.
15. has an absolute extremum.

3.2 FIRST AND SECOND DERIVATIVES

Identifying the properties of graphs of functions in the previous section can be cumbersome. Derivatives, if they exist, do much of the "work" in this section!

The differential calculus is powerful for identifying the characteristics of functions. It signals where functions increase or decrease, points of **local** (or **relative**) maxima, minima, and inflection points.

At a maximum, a function (and its graph) changes from increasing to decreasing as x increases. A minimum occurs when function values change from decreasing to increasing as x increases. Both are characterized by a horizontal tangent to the curve at points of change.

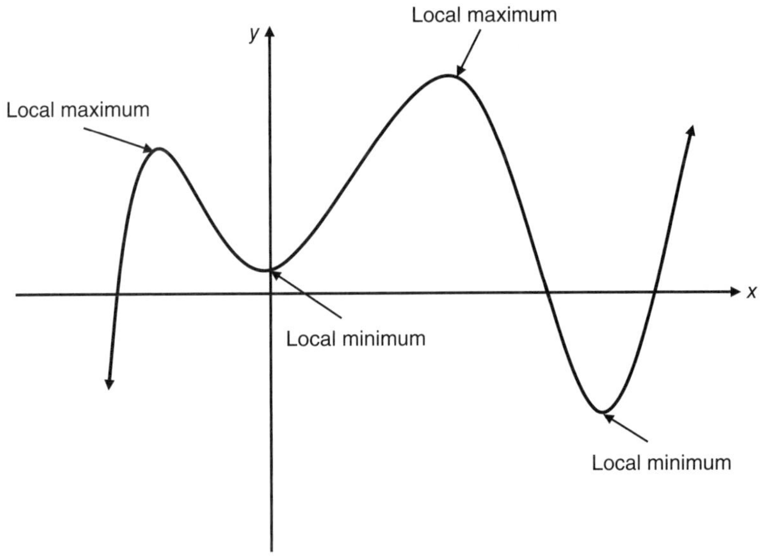

First Derivative

Where function values are increasing, the slope and derivative are positive valued. Function values decrease where its slope and derivative are negative. When the derivative is zero, it signals a horizontal tangent. The point at which the derivative vanishes is called a **stationary point** or a **critical point**.

↓ Remember that the derivative is the slope, so, naturally, it increases and decreases with the function.

Vanishing derivatives are used to locate extreme points as we will see. However, a vanishing derivative is not necessarily an extreme point. Functions may have several maxima and minima; the largest (smallest) among them being the **absolute (or global) maximum (minimum)** and the others **local maxima (minima)**.

In the graph shown previously, the global maximum is infinite (upward arrow). Likewise, the global minimum is negative infinity (downward arrow).

↓ Don't be baffled by the extensive jargon in use – "extrema," "critical point," "local," "relative," "global," "stationary," and so on. From a calculus perspective, they are usually synonyms and a signal of a horizontal tangent. Exception: they can also refer to endpoints, cusps, and such that lack derivatives and horizontal tangents and are considered in advanced texts.

At each value of x where $f(x)$ is differentiable, the instantaneous slope of its graph is its derivative there. It is also the slope of the tangent line to the curve at that point.

Example 3.2.1 Slope of a Tangent Line to a Curve

Verify that the slope of $y = x^2 + 1$ *at* $x = 2$ *is that of the tangent line* $y = 4x - 3$.

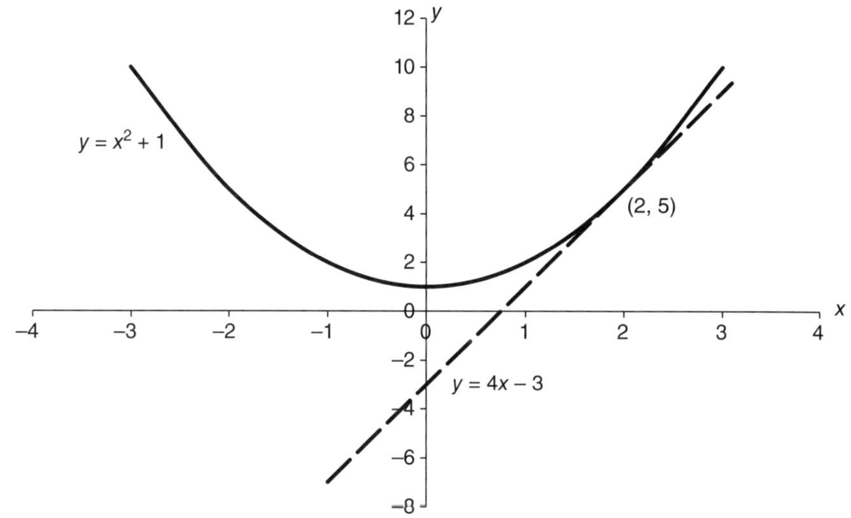

Solution:
The derivative of $y = x^2 + 1$ *is* $2x$. *Therefore, at* $x = 2$, *the slope of the graph is 4 as is the slope of the tangent line* $y = 4x - 3$.

The importance, power, and convenience of obtaining slopes of curves, and other properties of functions, $y = f(x)$, simply by differentiation are impressive (when derivatives exist).

In plain language, when a derivative is positive, the function increases as x increases in the neighborhood of x. Similarly, a negative derivative means that the function decreases as x increases. Points at which the derivative vanishes are stationary (critical) points, that is, x_0 is a stationary (critical) point if $f'(x_0) = 0$.

The First Derivative

Values of x for which $f'(x) > 0$ correspond to $f(x)$ increasing.
Values of x for which $f'(x) < 0$ correspond to $f(x)$ decreasing.
Values of x for which $f'(x) = 0$ correspond to $f(x)$ being stationary.

A first derivative that vanishes identifies a critical point as it signals a horizontal tangent (slope is zero). A horizontal tangent can signal a maximum or minimum. Sometimes, it signals an inflection point.

These following diagrams are illustrative!

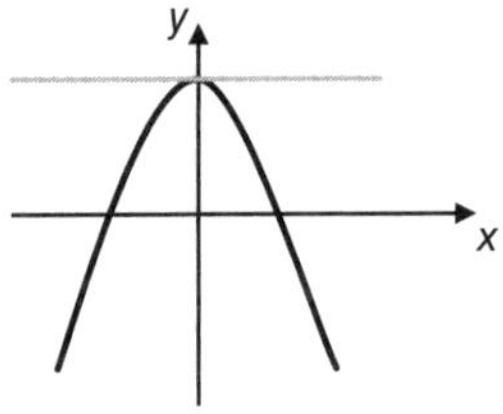

Maximum at the horizontal tangency

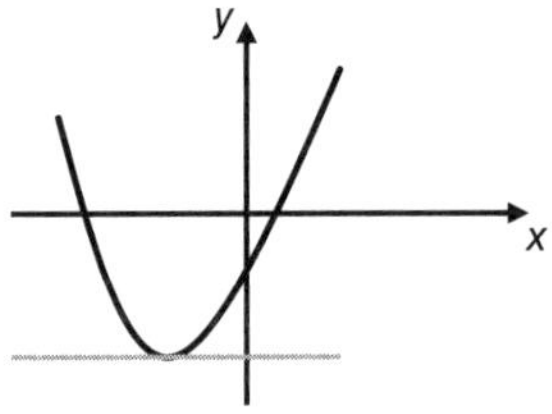

Minimum at the horizontal tangency

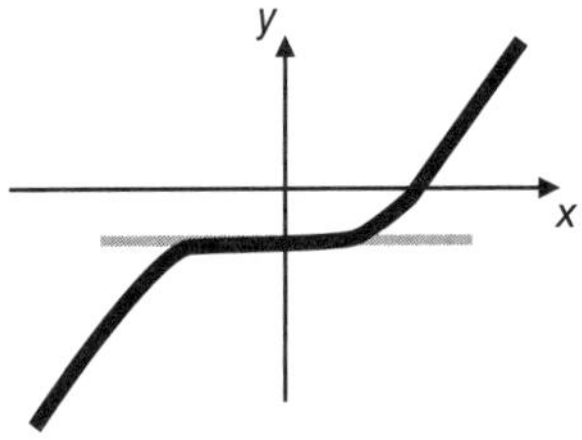

Inflection point at the horizontal tangency

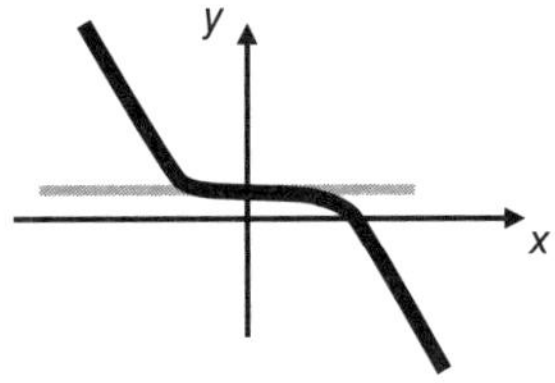

Inflection point at the horizontal tangency

Note that while the slope of $f(x)$ vanishes and its sign changes on passing through a maximum or minimum, the sign of the slope does not change on passing through an inflection point.

It is a necessity for the slope to change sign on passing through a maximum or minimum. Therefore, one cannot always depend solely on a vanishing derivative to signal a maximum or minimum. An example illustrates:

Example 3.2.2 Local Extrema

Determine all local extrema of $f(x) = 2x^3 - 3x^2 + 5$

Solution:
Firstly, determine the critical points by setting the first derivative to zero.

$$f'(x) = 6x^2 - 6x = 6x(x - 1) = 0$$

The critical points, where extrema are possible, are at $x = 0$ *and* $x = 1$. *The table organizes the signs of the factors and their product.*

	$x < 0$	$0 < x < 1$	$x > 1$
$6x$	−	+	+
$x - 1$	−	−	+
$f'(x) = 6x(x - 1)$	+	−	+

(boundaries at 0 and 1)

The sign change of $f'(x)$ (positive to negative) about $x = 0$ indicates a maximum. The change in sign of $f'(x)$ (from negative to positive) about $x = 1$ indicates a minimum.

Therefore, $f(0) = 5$ is a local maximum and $f(1) = 4$ a local minimum.

Note that while the derivative elegantly signals a local extremum, it is silent as to whether it may also be an absolute maximum or minimum. For a maximum or minimum of a function, $f(x)$, to occur at a critical point $x = a$, it is a *necessary condition* that its derivative vanishes there, that is, that $f'(a) = 0$. However, it is not a *sufficient condition* for extrema as illustrated in the following example.

Example 3.2.3 More on Local Extrema

Show that $f(x) = x^3$ lacks extrema.

Solution:

For possible extrema, set $f'(x) = 0$. Here $f'(x) = 3x^2$. Therefore, the only critical point is at $x = 0$.

Next, investigate the sign of the derivative about $x = 0$. The derivative $3x^2$ is positive (increasing function) both when $x < 0$ and when $x > 0$. Therefore, without a sign change at $x = 0$, there is neither a maximum nor a minimum. (Soon we will learn that it is an inflection point.)

Second Derivative

The derivative of a function, say $y = f(x)$, yields another function we have denoted by $f'(x)$. Now, $f'(x)$ as a function can have a derivative – called a second derivative of $f(x)$, $f''(x)$. The second derivative also aids in determining the behavior of functions. The second derivative, the "derivative of the first derivative," is the **rate of change** of the first derivative. Graphically, it is the rate of change of the slope of the tangent line. Unfortunately, it is not easy to depict in a sketch.

The slopes of the tangents, being first derivatives, their rate of change is the second derivative (the rate of change of the first derivative). Consider the local maximum, depicted in the concave down curve that follows, "it spills water." Observe the change in slopes of tangents as they pass through the maximum. Clearly, their slopes continuously decrease as x increases. Here, in a maximum, the rate of change of the slope of the tangent is negative.

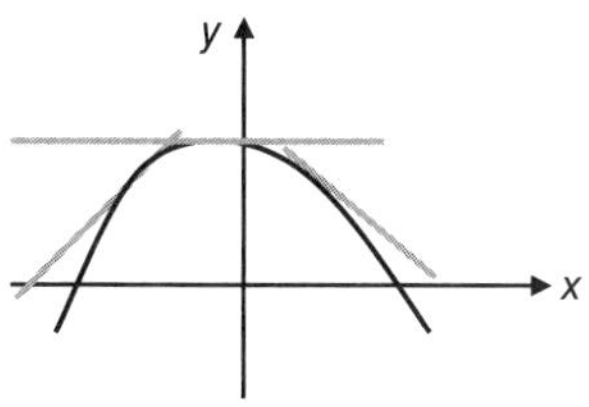

Maximum
Second derivative is negative

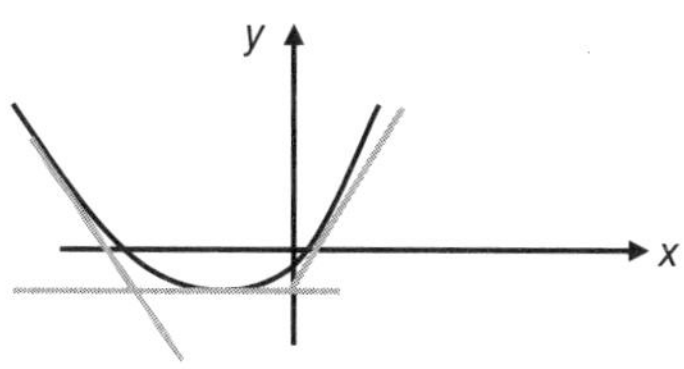

Minimum
Second derivative is positive

Similarly, for a local minimum (concave up curve on the top right), the changes in slopes of tangents increase as x increases. The second derivative is positive for a minimum, "it holds water."

↓ Note the decreasing slope of the tangent (left figure) from positive values to zero and then to negative values. Similarly, note the increasing slope of the tangent, from negative values to zero to positive values (right figure). It is correct to associate the second derivative with the change in slope of the tangent line.

Concavity

Values of x for which $f''(x) > 0$ correspond to $f(x)$ concave up ("holds water").
Values of x for which $f''(x) < 0$ correspond to $f(x)$ concave down ("spills water").

Inflection Points

Sometimes, a second derivative vanishes at a point. This means its sign has changed from positive to negative (or vice versa) as x increases through the vanishing point. For the second derivative to change in sign from positive to negative (or vice versa), it must pass through zero. It is a signal for an **inflection point** as in the following example.

Example 3.2.4 Points of Inflection

Show that $y = x^3 + 1$ (sketch in Section 3.1) has a horizontal tangent at its inflection point at (0, 1).

Solution:
The first and second derivatives are $f'(x) = 3x^2$ and $f''(x) = 6x$, respectively. Both derivatives vanish at $x = 0$. However, as the first derivative does not change sign (positive for all x here), there can neither be a maximum nor a minimum at $x = 0$.

Clearly, for $x < 0$, the second derivative is negative and for $x > 0$, it is positive. The sign change in the second derivative on passing through $x = 0$ signals a change in concavity. It corresponds to an inflection point at (0, 1).

While all inflection points have tangents, they are not always horizontal. Consider $y = x^3 - 4x$. Here, $y'(x) = 3x^2 - 4$ and $y''(x) = 6x$. The second derivative vanishes at $x = 0$, and as $y''(x)$ changes sign as x changes from negative to positive values, there is an inflection point at $x = 0$. Now, at $x = 0$, y is also zero, so the curve passes through the origin. Its slope at the origin $y'(0) = -4$; which is also the slope of the tangent there as shown in the figure.

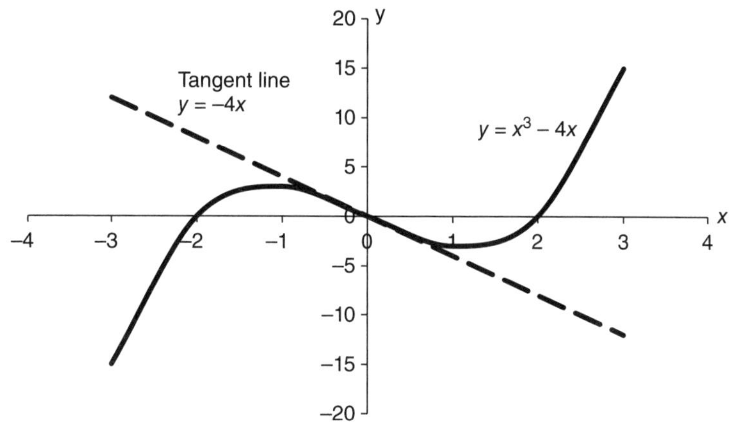

The maxima and minima in the preceding examples were easy to identify from their graphs, if not from the functions themselves. However, in practical situations, graphs may be neither available nor convenient. The question arises as to whether one can identify a horizontal tangent with certainty as a maximum, a minimum, or an inflection point. Remarkably, the answer is yes! And a (sufficient) test is simple!

The Second Derivative and Local Extrema

Let $f(x)$ be differentiable on an open interval containing $x = a$ and $f'(a) = 0$.

If $f''(a) > 0, f(x)$ has a local minimum at $x = a$
If $f''(a) < 0, f(x)$ has a local maximum at $x = a$.

(If $f''(a) = 0$, the test fails and the changing slope of the first derivative about $x = a$ determines the character of the function there).

Example 3.2.5 The Second Derivative and Local Extrema

Locate and identify maxima and minima of $y = x^3 - 6x^2 + 9x$ using its derivatives. (A graph appears early in the previous section.)

Solution:
In this case, $y' = 3x^2 - 12x + 9$. Setting y' to zero (and factoring) yields $y' = 3(x - 1)(x - 3) = 0$. Hence, the critical points (corresponding to horizontal tangents) occur at $x = 1$ and $x = 3$.

The second derivative is $y'' = 6x - 12$. At $x = 1, y'' < 0$. This negative second derivative indicates a maximum as the curve is "concave down" about $x = 1$.

⟶

At $x = 3$, the second derivative is positive. This indicates a minimum as the curve is "concave up" about $x = 3$.

Therefore, there is a local maximum at (1, 4) and a local minimum at (3, 0).

Example 3.2.6 Quadratics and Local Extrema

Locate and identify extrema of the quadratic $y = ax^2 + bx + c$. $(a \neq 0)$

Solution:

In this case, $y' = 2ax + b$. Setting y' to zero yields $x = \dfrac{-b}{2a}$ as a possible extremum.

The second derivative is $y'' = 2a$. Therefore, when $a < 0$, the function is concave down at $x = \dfrac{-b}{2a}$ and the function is a maximum there. When $a > 0$, the function is concave up at $x = \dfrac{-b}{2a}$ and the function is a minimum there. You may recall this as the axis of symmetry and parabola opening according to the sign of a.

EXERCISES 3.2

In Exercises 1–4, identify graphs that depict

1. a positive first derivative for all x.
2. a negative first derivative for all x.
3. a positive second derivative for all x.
4. a negative second derivative for all x.

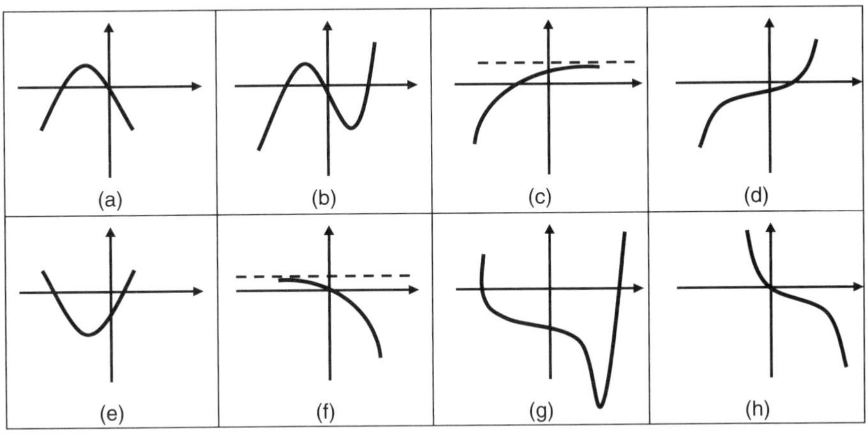

In Exercises 5–8, sketch the described function.

5. The graph is concave down for all x, (2, 5) is on the graph, $f(0) = 1$ and $f'(2) = 0$.
6. $f''(x) < 0$ for all x, $f(0) = -1$, $f'(1) = 0$, and (1, 0) on the graph.

7. $f(1) = 4/3, f(3) = 0, f'(x) > 0$ on $(-\infty, 1) \cup (3, \infty)$, and $f'(x) < 0$ on $(1, 3)$.
8. $f'(x) > 0$ on $(-\infty, 0) \cup (2, \infty), f'(x) < 0$ on $(0, 2)$. $(2, -2)$ and $(0, 2)$ are on the graph.
9. Using the following sketch, enter +, −, or 0 in the table as appropriate. Assume a local minimum at B and inflection point at C.

	f	f'	f''
A			
B			
C			
D			

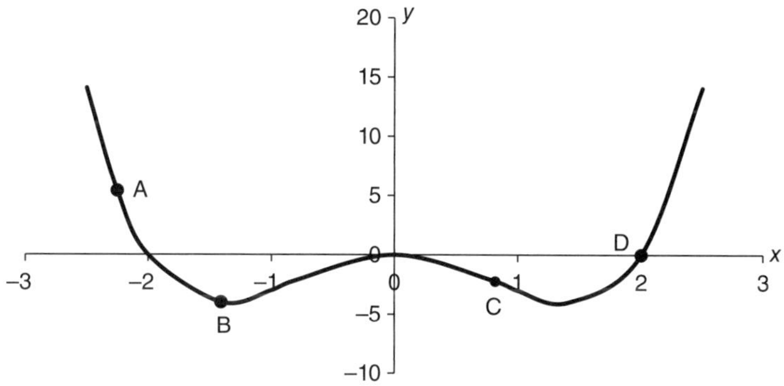

In Exercises 10–15, use the following graph

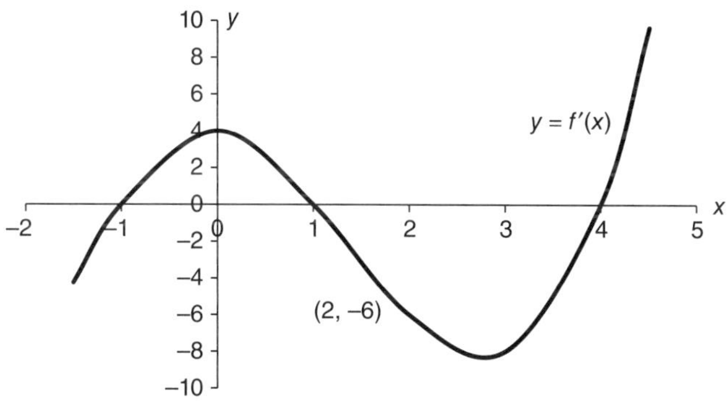

10. Why must $f(x)$ be decreasing at $x = 2$?
11. Why must $f(x)$ be increasing at $x = 5$?
12. Why must $f(x)$ have an inflection point at $x = 0$?
13. Where does $f(x)$ have a local maximum?
14. Determine the two values of x where $f(x)$ has local minima.
15. If $f(2) = -2/3$ determine the equation of the tangent to the curve at $x = 2$.

3.3 CURVE SKETCHING

Calculus furnishes useful ways to sketch graphs of functions.

> ***Curve Sketching Guide***
>
> 1. **Obtain the first and second derivatives of $y = f(x)$.**
> 2. **Set $f'(x) = 0$ to determine critical points.**
> 3. **Identify intervals where the function increases, decreases, and has local extrema.**
> 4. **Set $f''(x) = 0$ to determine its critical points. Identify intervals of concavity. Be alert for inflection points.**
> 5. **Identify x- and y-intercepts of $y = f(x)$ and additional points needed for a sketch.**
> 6. **Polynomials of degree n have at most $n - 1$ "turns," are differentiable (smooth and unbroken), and tend to infinity at extremes.**

Guard against mistakes! Check that your sketch has requisite characteristics. Sketching functions is a useful skill that improves with practice. The examples help.

Example 3.3.1 Curve Sketching – A Quadratic

Sketch $y = x^2 - 2x - 15$ guided by derivatives.

Solution:

1. *The first derivative is $y' = 2x - 2$ and the second $y'' = 2$.*
2. *Setting y' to zero yields $x = 1$ as a critical point.*
3. *The derivative (slope of the tangent) is negative when $x < 1$ and positive when $x > 1$. There is a local minimum at $(1, -16)$.*
4. *The second derivative, $y'' = 2$, indicates that the function is concave up for all x, as it is always positive, and because the second derivative never vanishes, there is no prospect of an inflection point.*
5. *The two x-intercepts obtained by factoring are at $x = -3$ and $x = 5$. The y-intercept is $y = -15$. Some additional points are plotted to aid sketching (as the derivatives yielded less information here).*
6. *As y is a polynomial of second degree, there can only be one "turn."*

See sketch.

⟶

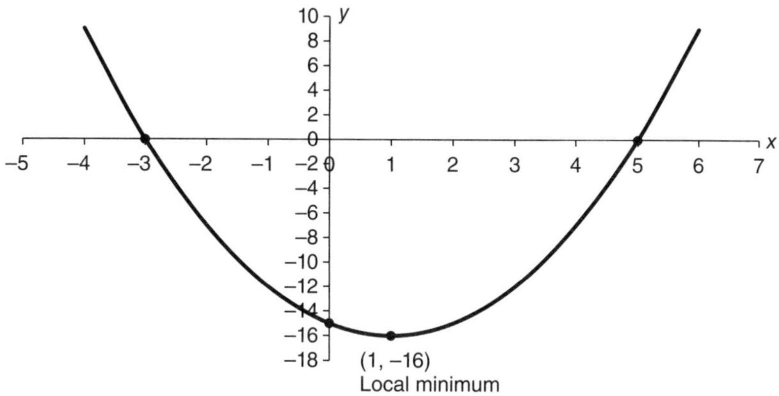

Setting the derivative to zero also identifies the axis of symmetry for any quadratic, $f(x) = ax^2 + bx + c$. So, $f'(x) = 2ax + b = 0$. Solving for x yields an extremum at $x = -b/2a$, which locates the vertical axis of symmetry.

First and second derivatives are even more useful in curve sketching for higher degree polynomials as the following examples illustrate.

Example 3.3.2 *Curve Sketching – A Cubic*

Use derivatives to aid in a sketch $y = x^3 + 6x^2 + 9x$.

(the Curve Sketching Guide helps (last shaded box))

Solution:

In this case, $y' = 3x^2 + 12x + 9 = 3(x + 1)(x + 3) = 0$

The critical points are at $x = -1$ *and* $x = -3$.

A critical point table organizes sign changes for the first derivative.

$x + 1$	−	−	+
$x + 3$	−	+	+
$y' = 3(x + 1)(x + 3)$	+	−	+
	−3	−1	

The sign change (positive to negative) about $x = -3$ *suggests a maximum. The change in sign (negative to positive) about* $x = -1$ *suggests a minimum.*

The second derivative, $y'' = 6x + 12$, *is zero when* $x = -2$.

It is negative for $x < -2$ *and positive for* $x > 2$, *indicating an inflection point at* $x = -2$.

The origin is the y-intercept.

Collecting, a sketch of $y = x^3 + 6x^2 + 9x$ *emerges (following figure).*

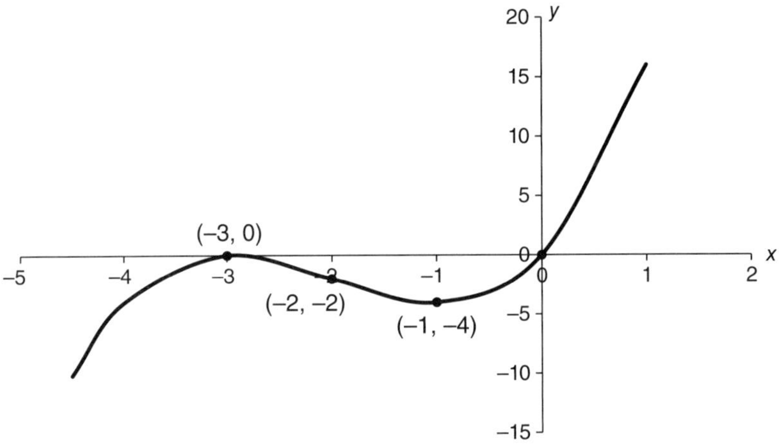

A fourth-degree polynomial is investigated in the following example.

Example 3.3.3 Curve Sketching – A Quartic

Sketch $y = x^4 - 4x^2$ using derivatives.

Solution:

In this case, $y' = 4x^3 - 8x = 4x(x^2 - 2) = 0$

The critical points are $x = 0$ and $x = \pm\sqrt{2}$.

A critical point table organizes the signs of the factors and their product.

$4x$	−	−	+	+
$x^2 - 2$	+	−	−	+
$y' = 4x(x^2 - 2)$	−	+	−	+
	$-\sqrt{2}$	0	$\sqrt{2}$	

The sign change (positive to negative) about $x = 0$ suggests a maximum. The change in sign (negative to positive) about $x = \pm\sqrt{2}$ suggests two symmetric local minima.

The local maximum is at (0, 0), and the local minima are at $(\pm\sqrt{2}, -4)$.

The second derivative is $y'' = 12x^2 - 8$. The function is concave up on $(-\infty, \frac{-\sqrt{6}}{3}) \cup (\frac{\sqrt{6}}{3}, \infty)$. It is concave down on $(\frac{-\sqrt{6}}{3}, \frac{\sqrt{6}}{3})$ and has inflection points at $(\frac{\pm\sqrt{6}}{3}, \frac{-20}{9})$. The x-intercepts are at 0, 2, and −2. (Note: the second derivative can also be used to determine the nature of the extema. For example, at $x = 0$, the second derivative is −8 indicating a maximum.)

Collecting this information a sketch emerges.

⟶

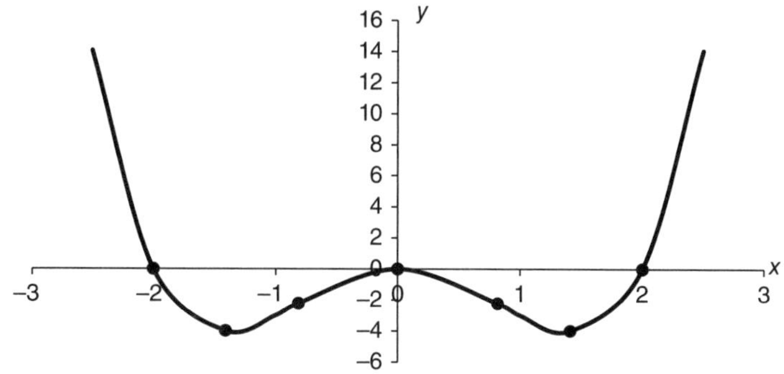

EXERCISES 3.3

1. Identify the local extremum of $y = x^2 - 4x - 5$.
2. Identify the local extremum of $y = -x^2 + 6x - 5$.
3. Show that $y = x^3 - 3x^2 + 3x - 1$ has no local extrema.
4. Show that $y = x^3 + 6x^2 + 12x + 8$ has no local extrema.

In Exercises 5–18, sketch the functions

5. $f(x) = x^2 + 8x + 7$
6. $f(x) = 2x^2 + 7x - 15$
7. $f(x) = x^3 - 4x$
8. $f(x) = 2x^3 + 9x^2 + 12x - 10$
9. $f(x) = 4x^3 - 6x^2 + 3x + 4$
10. $f(x) = -x^3/3 + 3x^2/2 - 2x + 4$
11. $f(x) = -x^3 + (3/2)x^2 + 3$
12. $f(x) = x^3 - 3x^2 - 5x$
13. $f(x) = -(1/3)x^3 - 2x^2 + x + 6$
14. $f(x) = x^4 - x^2$
15. $f(x) = x^4 - 2x^2 - 5$
16. $f(x) = -x^3 + 6x^2 + 1$
17. $f(x) = x + 1/x$
18. $f(x) = 2x + 2/x$

3.4 APPLICATIONS OF MAXIMA AND MINIMA

An American Heritage dictionary defines an optimum as the "best or most favorable condition for a particular situation." Mathematical optimization or, simply, optimization, has come to describe the mathematics of determining maxima and minima of functions. Such situations arise in a multitude of practical instances – natural and otherwise.

Optimization

Derivatives aid curve sketching and help to locate *optima*. Local (or relative) maxima and minima imply optima in a region about the point in question. Absolute maxima and minima refer to the largest and smallest, respectively, in the entire domain – not necessarily at horizontal tangents.

A series of examples follow.

Example 3.4.1 Maximum Value of a Product

Find the maximum value of the product of two numbers whose sum is 10.

Solution:

Let x and $10 - x$ represent the numbers. Their product is the quadratic:

$$f(x) = x(10 - x) = 10x - x^2$$

Now, setting the derivative to zero gives $f'(x) = 10 - 2x = 0$ and, therefore, $x = 5$. Notice that when $x < 5$, the derivative is positive, and when $x > 5$, the derivative is negative. Therefore, there is a maximum about $x = 5$. If $x = 5$, then $10 - x$ is also 5. The desired numbers are both equal to 5!

The maximum value of their product is $25(= 10(5) - (5)^2)$.

(You may remember a "rule of thumb" from a pre-calculus course. The maximum of the product of two positive numbers, whose sum is fixed, occurs when they are equal).

Quadratics have many applications as in the previous example and the following one.

Example 3.4.2 Maximum Garden Area

What is the maximum possible area of a rectangular garden using 40 feet of fencing?

Solution:

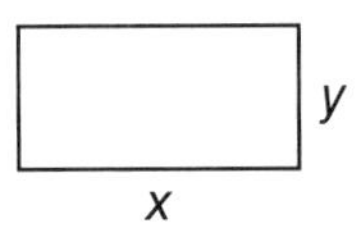

Let x and y represent the sides.

The perimeter, P, is $2x + 2y(= x + y + x + y)$, and the area is $A = xy$.

We seek to maximize xy subject to the perimeter constraint $P = 2x + 2y = 40$.

An immediate obstacle is the appearance of two variables. However, use the constraint to express one of them in terms of the other. To express area, A, in terms of a single variable, say x, replace y by $20 - x$.

The area, expressed as a function of a single variable, becomes

$$A(x) = (20 - x)(x) = 20x - x^2$$

→

↓ *We have replaced A by A(x) simply to signal it is now a function of a single variable, x.*

The derivative $A'(x) = 20 - 2x$ *is set to zero to yield a critical point at* $x = 10$. *As the derivative changes sign (positive when* $x < 10$ *and negative when* $x > 10$*), there is a maximum at* $x = 10$. *Alternatively, the second derivative being negative indicates a maximum at the critical point* $x = 10$ *from the first derivative.*

When $x = 10$, *y is also 10. Therefore, the rectangle of maximum area is a square.*

The maximum area is 100 square feet for the 40-foot perimeter garden.

(Note a similarity to the previous example.)

Example 3.4.3 Pipe Cutting

A pipe of length L is to be cut into four sections to form a parallelogram that encloses a maximum area. How shall cuts be made?

Solution:

Let x and y represent the sides of the parallelogram and A, the enclosed area. Using the fact that A is proportional to xy, write

$$A = kxy$$

k being a constant.

↓ *Interestingly, the precise formula for the area is not necessary – only its variable dependence.*

Also, $L = 2x + 2y$. *Substitution for, say, y yields*

$$A = kx\left(\frac{L - 2x}{2}\right) = \frac{1}{2}kxL - kx^2$$

Differentiating,

$$\frac{dA}{dx} = \frac{1}{2}kL - 2kx = 0$$

so $kL = 4kx$ *and* $x = \dfrac{L}{4}$.

The derivative changes sign (positive to negative) about $x = L/4$ *indicating a maximum. Therefore, y is also L/4 indicating the maximum area occurs when the pipe is cut to form a square. (Alternatively, the second derivative being* $-2k$ *also indicates a maximum at the critical value of the first derivative,* $x = L/4$.*)*

↓ *Note, again, that the value of k is only necessary to compute the area, not its dimensions.*

Example 3.4.4 Parcel Post

Postal restrictions limit parcel size to a length plus girth that cannot exceed 108 inches. What is the largest volume of a package with square cross-sectional area that can be mailed?

⟶

Solution:

Consider a package of length L and square cross-sectional area with side X. We seek to maximize a volume, X^2L, subject to length plus girth constraint $L + 4X = 108$.

In terms of a single variable, the volume is $V(X) = X^2(108 - 4X) = 108X^2 - 4X^3$.

Its derivative is $V'(x) = 216X - 12X^2$.

Setting the derivative to zero suggests $X = 18$ as a possible extremum. The derivative being positive for $0 < X < 18$ and negative for $X > 18$ indicates a maximum when $X = 18$ and $L = 108 - 4(18) = 36$. (Alternatively, the second derivative, $216 - 24x$ being negative at $x = 18$, indicates a maximum there.) The largest package volume that can be sent under these conditions is $18'' \times 18'' \times 36''$ for a maximum volume of $V = 11{,}664$ cubic inches.

Example 3.4.5 Parcel Post Revisited

A Rock group will design a poster to be printed and mailed in large quantities for display. Before ordering cylindrical mailing tubes in which posters can be sent, a check is made of postal regulations. These require that parcels not exceed 108″ in length plus girth. What size mailing tubes should be ordered to maximize the number of posters that can be shipped in each tube?

Solution:

Let r be the tube radius and h its height. Then, the volume $V = \pi r^2 h$ (the volume of a cylinder is its base area times its height). The girth is the circumference plus the cylinder height,

$$2\pi r + h = 108$$

The problem is to determine r and h so that the volume is a maximum. That is, Maximize: $V = \pi r^2 h$ Subject to: $2\pi r + h = 108$.

Again, there appears to be two variables, r and h. However, they are related by the girth constraint. One variable can be substituted in terms of the other. Here, it is easier to substitute for h in V. That is,

Maximize: $V = \pi r^2(108 - 2\pi r) = 108\pi r^2 - 2\pi^2 r^3$.

Differentiating and setting to zero, $\frac{dV}{dr} = 216\pi r - 6\pi^2 r^2 = 0$.

Factoring, $6\pi r(36 - \pi r) = 0$ so $r = 0$ or $r = 36/\pi$. So $r = 36/\pi$ is the solution; $r = 0$ being extraneous. As the derivative changes sign (positive to negative) about $r = 36/\pi''$, it is a maximum. Next, $h = 108 - 2\pi(36/\pi) = 36''$.

Thus, a cylindrical mailing tube of height 36″ (1 yard) and circumference 72″ (2 yards) has the largest volume that satisfies postal regulations.

◆ The 108″ postal restriction dates to the early 1900s. Actually, the optimal length of 36″ applies regardless of the cross-sectional shape. The exercises suggest that you prove this.

Example 3.4.6 A Cautionary Example

What point on the circle $x^2 + y^2 = 4$, is closest to (1, 0)? Obviously, the answer is the point (2, 0) – but don't get ahead of the story!

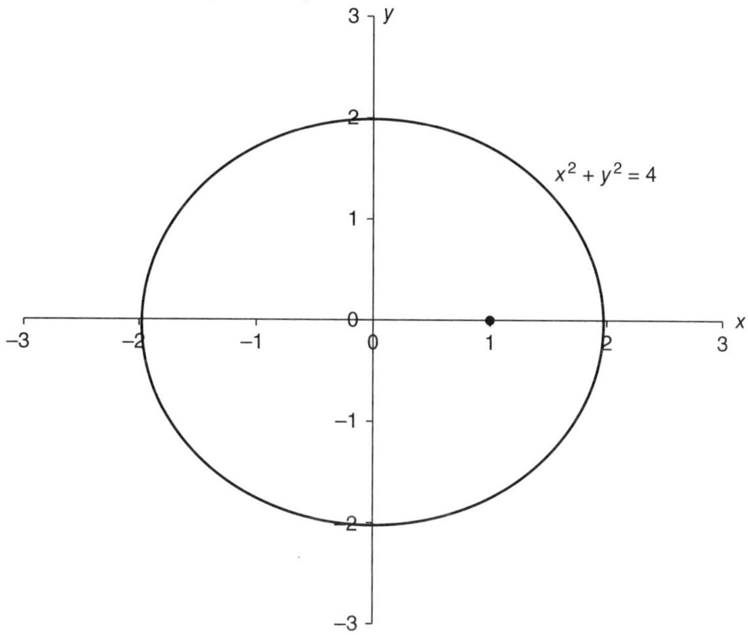

Solution:

Let (x, y) be coordinates of the desired point. The square of the distance, D, from (1, 0) is $D^2 = (x - 1)^2 + y^2$. (Using the squared distance avoids inconvenient radicals).

Again, there are two variables of which one can be eliminated by using the equation of the circle. Eliminating y appears easiest. So we seek x to minimize $D^2 = (x-1)^2 + (4 - x^2)$.

Differentiating the expression yields $2(x - 1) - 2x$, which simplifies to -2. Oh! Oh! The derivative equal to -2 cannot be set to zero.

Let us try again, this time substituting for x. From the equation of the circle, $x = \pm\sqrt{4 - y^2}$, so
$D^2 = (\pm\sqrt{4 - y^2} - 1)^2 + y^2 = (4 - y^2 + 1 - 2\sqrt{4 - y^2}) + y^2 = 3 - 2\sqrt{4 - y^2}$.

The derivative is $-2(\frac{1}{2})(4 - y^2)^{-1/2}(-2y) = \dfrac{2y}{\sqrt{4 - y^2}}$.

Setting this to zero yields $y = 0$ and $x = \pm 2$. Clearly, (2, 0) is closer to (1, 0).

This time, the usual procedure of setting the derivative to zero worked!

The results were obvious from an inspection of the figure. Less obvious is the cause of the failure of the substitution for x.

The purpose of this example is twofold: firstly, it alerts you to the possibility that the choice of variable for elimination may not be arbitrary and, secondly, that mathematicians have found a way to avoid such "disasters" by using Lagrange Multipliers. They are studied in Chapter 8.

Example 3.4.7 Bus Shelter Design

A "see-through" bus shelter is required to enclose a volume $V = 486$ cubic feet. The roof and parallel sides are of equal area with a square back panel.

What dimensions minimize the required area, A, of "see-through" plastic?

Solution:

Let h represent the shelter height and b its depth. Then, the total material area is $A = 3hb + h^2$.

Note that the total area A involves just two variables, as the back panel is square. Using the volume $V = 486\ feet^3 = bh^2$, it is easiest to solve for b and substitute for it in A. That is, $b = V/h^2$, so

$$A(h) = 3h\left(\frac{486}{h^2}\right) + h^2 = \frac{1458}{h} + h^2$$

Setting the derivative of A(h) with respect to h to zero yields

$$\frac{dA}{dh} = -\frac{1458}{h^2} + 2h = 0 \ \ or \ \ 2h^3 = 1458 \ \ and \ \ h = 9 \ \ feet.$$

When $0 < h < 9$, the first derivative is negative, and when $h > 9$, the first derivative is positive indicating a minimum at $h = 9$.

Solving for b yields a shelter that is 6 feet deep with a minimum material usage of 243 square feet.

Example 3.4.8 Sizing Canned Foods

About 12 ounces of fruit juice requires a volume of 113 cubic inches for its packaging. How should a cylindrical can be sized to minimize the amount of material required?

Solution:

Let h and r represent the height and radius, respectively, of a can to be fabricated from sheet metal. Then, its surface area, S, can be written as

$$S = 2\pi rh + 2\pi r^2$$

where $2\pi rh$ is the rectangular area of the unrolled cylinder and $2\pi r^2$ the area of the can's ends.

The volume $V = 113\ in^3$ can be written as

$$V = \pi r^2 h = 113$$

It appears easiest to substitute for h so

$$Minimize \ \ S = 2\pi r\left(\frac{113}{\pi r^2}\right) + 2\pi r^2 = \frac{226}{r} + 2\pi r^2$$

⟶

Differentiating and equating to zero,

$$\frac{dS}{dr} = -\frac{226}{r^2} + 4\pi r = 0$$

$$4\pi r = \frac{226}{r^2}$$

$$r = \left(\frac{113}{2\pi}\right)^{1/3} \approx 2.62 \text{ inches.}$$

And, therefore,

$$h = \left(\frac{113\,(4)}{\pi}\right)^{1/3} \approx 5.24 \text{ inches.}$$

As h is approximately 2r (the diameter), the minimum amount of material is used when the height to diameter ratio is about 1 (a so called "square can").

♦ The large variety of sizes and shapes of canned foods on supermarket shelves belies our "square can" result in the previous example. Condensed soup cans have a height to diameter ratio of about 3/2, while tuna cans usually have a height to diameter ratio of 1/2. Clearly, other factors are at work! Among these factors are the waste in cutting circular ends from square metal sheets, fabrication costs in securing the ends and seams, and marketing and product considerations. (You can explore these interesting matters further in an article in College Mathematics Journal by P.L. Roe, May 1993 among others).

Example 3.4.9 The Basic Lot Size Model

In Chapter 2, we encountered a basic "lot size" model $Z = f_{inc}(x) + f_{dec}(x) = ax + b/x \quad a, b > 0$. *Determine* x^* *and* Z^*, *the optimal value of x and Z, respectively, where the total cost Z is a minimum.*

Solution:

The sketch of the lot size model earlier and in Example 2.3.5 indicates that one should expect the optimum resource level that minimizes total cost to occur in the relatively shallow portion of this curve. The simplest and most frequent example of the lot size model takes the form $Z = ax + b/x$.

The first derivative must be zero at a possible optimum. The derivative encountered in Example 2.3.5 was $Z' = a - \frac{b}{x^2}$. *Setting the derivative to zero yields* $x^* = \sqrt{\frac{b}{a}}$ *as a possible minimum, and as the first derivative is negative below this value and positive above, it is a local minimum. Alternatively, the second derivative is* $Z'' = \frac{2b}{x^3}$ *and being positive indicates a minimum at* x^*. *The minimum total cost is* $Z^* = 2\sqrt{ab}$.

Note: the value for $x*$ *balances the opposing costs of the increasing and the decreasing components of the total cost function. It occurs when* $f_{inc}(x) = f_{dec}(x)$.

Example 3.4.10 The Lot Size Model $2x^2 + 32/x$

Consider another lot size model $Z = f_{inc}(x) + f_{dec}(x) = 2x^2 + 32/x$. Determine x^* and Z^* the value of x and Z, respectively, where the total cost, Z, is a minimum.

Solution:
In this case, $Z' = 4x - \frac{32}{x^2}$. Setting the derivative to zero for critical points yields $4x = \frac{32}{x^2}$, so $x^* = 2$ and $Z^* = 24$. It is easy to verify that this corresponds to a minimum.

EXERCISES 3.4

1. For what value of x does $f(x) = x^2 - 6x + 5$ have a minimum?
2. What is the minimum of $f(x) = x^2 - 2x - 8$ and where does it occur?
3. What is the maximum value of $f(x) = -x^4 + 3x^2$?
4. What is the minimum value of $f(x) = 8x^3 - 6x^2 + 7$ for $x > 0$?
5. What is the maximum of $f(x) = 16x - x^2$ and where does it occur?
6. Where is $f(x) = \sqrt{x}$ closest to $(1, 0)$?
7. The sum of two numbers is 8. What is the maximum value of their product?
8. The sum of two numbers is 10. What numbers maximize their product?
9. The sum of two numbers is 20. What is the minimum value of the sum of their squares?
10. Find two positive numbers x and y whose product is 64 and whose sum is a minimum.
11. What number exceeds its square by the largest amount?
12. What number exceeds twice its square by the largest amount?
13. What is the largest rectangle whose perimeter is 60 feet?
14. Verify that the largest rectangle of fixed perimeter is a square.
15. What is the largest area of a right triangle in which the sum of the lengths of the shorter sides is 10 cm?
16. A river bounds a field on one side. How can 400 feet of fencing enclose a maximum area?
17. Find the minimum of $x^2 + y^2$ subject to the constraint $x + y = 8$.
18. A rectangular garden of 300 square feet is fenced on three sides by material that costs \$6 per linear foot. The fourth side uses a material at \$10 per linear foot. Find garden dimensions that minimize cost.

19. What is the maximum area of a rectangular garden that can be fenced for \$120 if fencing on three sides of the garden cost \$5 per linear foot and on the fourth side at \$7 per linear foot?

20. A Norman Window is a rectangular window capped by a semicircular arch. Find the value of r so the window perimeter is 20 feet and has maximum area. Let h represent the height, and r the radius of the semicircle. Therefore, the width will be $2r$.

21. Suppose that three sides of a rectangular pen to be built with material that costs \$10 per linear foot and a less visible fourth side uses a material that costs \$5 per linear foot. If \$2400 is available to build the pen, what are the dimensions for largest area?

22. An enclosure of 288 square yard area is to be built. Three sides of the enclosure cost \$16 per running yard, while the more visible fourth side costs \$20 per running yard. Find the dimensions of the enclosure that will minimize the cost of building and state what this minimum is.

23. A closed rectangular box with a square base is to be formed using 60 square feet of material. What dimensions maximize volume?

24. Show that the optimal length for a mailing container of maximum volume with the length plus girth restriction of 108″ is 36″ when the cross-section is

    (a) an equilateral triangle.
    (b) an arbitrary cross-section geometry.

    Hint: Assume that the cross-sectional area A, is proportional to a dimension x, so $A = kx^2$, and the perimeter, p, is proportional to x, say, $p = bx$ where k and b are proportionality constants.

25. Suppose that in Example 3.4.8 the circular ends are cut from a large sheet that has been divided into squares of side $2r$. Obtain the optimal height to diameter ratio of the mailing tube.

26. To minimize the waste in cutting circular ends from square pieces, a suggestion is made to divide the large sheet into hexagons (as a honeycomb). Show that the optimal height to diameter ratio becomes $\dfrac{2\sqrt{3}}{\pi} \approx 1.1$.

    Hint: The area of a regular hexagon is $\dfrac{\sqrt{3}}{2}W^2$ where W is its width and can diameter.

27. A vertical cylindrical tank of given volume is to be constructed. The required material for the top and bottom costs twice as much per square foot as that for the sides. Find the best height to diameter ratio.

28. A strip of length L is to be cut into 12 smaller strips to form a parallelepiped with square base and of maximum volume. How shall the strip be divided?

3.5 MARGINAL ANALYSIS

Marginal analysis has a distinctive role in business and in economics. It is the study of the effect of unit monetary or product changes: "changes at the margin." The inverse relationship of supply and demand with price is a basic example. Usually, demand decreases as price increases. Calculus aids in answering useful questions for a functional relation between price, p, and demand, x, (assumed to be a continuous and differentiable).

Economists define several functions of demand, x, such as unit price of an item as a function of demand, $p(x)$; total revenue, $R(x)$; and total cost, $C(x)$; among others. The derivatives of these functions, being rates of change, are their *marginals*. For example, the **marginal cost (MC)** is the derivative of the cost function, $C'(x)$. Economists interpret MC as the *cost of an additional unit of demand*. This is distinct from *average cost*, which is the total cost divided by the number of items, $C(x)/x$.

Similarly, as revenue is the unit price multiplied by demand, a total revenue function is $R(x) = xp(x)$. **Marginal revenue (MR)** is the derivative, $R'(x)$. It is the *change in revenue accompanying a unit change in demand.*

It is important to understand why economists use the derivative as the MC. Imagine a small change, Δx, in demand, x, from x to $x + \Delta x$. Then, the total cost changes from $C(x)$ to $C(x + \Delta x)$. The change in cost per unit change, Δx, in demand is

$$\frac{C(x + \Delta x) - C(x)}{\Delta x}$$

You may recognize (from Chapter 2) this as the difference quotient whose limit as $\Delta x \to 0$ is the derivative of $C(x)$. Understandably, calculus has a vital role in such studies.

Example 3.5.1 Maximizing Revenue

The price of an item as a function of demand, x, is $p(x) = 5 - x/4$ *dollars. What production level maximizes revenue? What is the maximum revenue?*

Solution:

Total revenue, demand multiplied by unit price, is $R(x) = (x)(5 - x/4) = 5x - x^2/4$.

Maximizing revenue implies vanishing marginal revenue, $R'(x) = 0$.

Therefore, setting $R'(x) = 0 = 5 - x/2$ *yields* $x = 10$.

The derivative is positive for $x < 10$ *and negative for* $x > 10$ *so at* $x = 10$, $R(10) = 25$.

A demand of 10 units maximizes revenue at \$25.

Example 3.5.2 Marginal Cost

Obtain and interpret the MC for the cost function $C(x) = x^3 - 12x^2 + 36x + 10$ *and a demand of x units.*

Solution:

The derivative of the cost function is the MC, so

$$MC(x) = C'(x) = 3x^2 - 24x + 36$$

It is the change in the total cost for a unit change in demand at level x.

Example 3.5.3 Marginal Revenue and Marginal Cost

For $R(x) = \frac{-x^2}{2} + 150x$ *and* $C(x) = \frac{x^3}{30} - 2x^2 + 110x + 1600$ *determine the marginal revenue and marginal cost functions.*

Solution:

The derivative of the revenue function is the marginal revenue, so

$$MR(x) = -x + 150$$

The derivative of the cost function is the marginal cost, so

$$MC(x) = C'(x) = \frac{x^2}{10} - 4x + 110$$

It is the change in the total cost for a unit change in demand when demand is at level x.

Note: we will revisit this again in Chapter 6. There, using marginal functions and initial conditions, help obtain revenue and cost functions.

Example 3.5.4 Minimizing Marginal Cost

For $C(x) = 2x^3 - 12x^2 + 40x + 60$*, determine minimum marginal cost.*

Solution:

Firstly, obtain MC(x), the derivative of C(x) as $MC(x) = C'(x) = 6x^2 - 24x + 40$. *For a minimum, set the derivative of MC(x) to zero. That is,*

$$MC'(x) = C''(x) = 12x - 24 = 0$$

When $x < 2$*, the derivative of MC is negative, and when* $x > 2$*, its derivative is positive. This indicates a minimum at* $x = 2$*. The minimum marginal cost is*
$MC(2) = 6(2)^2 - 24(2) + 40 = \$16$.
(A common mistake uses C(2) instead of MC(2)!)

Example 3.5.5 Maximizing Profit

Demand, x, for an item at price, p, D(p), is

$$D(p) = x = 50 - (p/3)$$

What production level and price maximizes profit?
The cost function is

$$C(x) = (2/3)x^3 + 2x^2 - 150x + 1200$$

⟶

Solution:
To start, investigate restrictions on values of x and p, as there could be endpoint extrema. Increasing price decreases demand, and conversely.

Therefore, seek intercepts in the first quadrant.

From the demand function, D(p), conclude that demand cannot exceed 50 units (when $p = 0$) and price cannot exceed \$150 (or demand becomes negative). That is,

$$0 \leq x \leq 50 \ \text{and} \ 0 \leq p \leq 150$$

Next, to obtain R(x), solve for price as a function of demand, x. Therefore,

$$p(x) = 150 - 3x \ \text{and} \ R(x) = x \quad p(x) = 150x - 3x^2$$

Subtracting cost from revenue yields the profit function Pr(x),

$$\begin{aligned} Pr(x) = R(x) - C(x) &= (150x - 3x^2) - ((2/3)x^3 + 2x^2 - 150x + 1200) \\ &= (-2/3)x^3 - 5x^2 + 300x - 1200 \end{aligned}$$

To maximize the profit, set its derivative to zero as

$$\begin{aligned} Pr'(x) &= -2x^2 - 10x + 30 = 0 \\ &= -2(x + 15)(x - 10) = 0 \end{aligned}$$

While there are two mathematical possibilities: either $x = -15$ or $x = 10$, only $x = 10$ satisfies the constraint $x \leq 50$ so $Pr(10) = \$633.33$; a possible maximum.

However, there may be endpoint extrema. Evaluating Pr(x) at $x = 0$ and $x = 50$ yields:

$$Pr(0) = -\$1200$$

$$Pr(50) = -\$82{,}033.33$$

Clearly, these are not usable. There are no endpoint extrema of concern here.

The optimal option is to produce 10 units for sale at a unit price of \$120(= 150 − 3(10)) for a maximum profit of \$633.33.

EXERCISES 3.5

1. What is the marginal cost for $C(x) = 2x^3 + 9x + 25$?
2. What is the marginal cost for $C(x) = x^5 - 5x^2 + 20$?
3. If $C(x) = x^7 - 5x^3 + 20x + 25$, what is the marginal cost function when $x = 1$? (hint: find $C'(1)$)
4. When marginal cost will be zero for $C(x) = 2x^4 - 36x^2 + 70$?
5. When marginal cost will be zero for $C(x) = x^3 - 9x^2 + 15x + 120$?
6. Determine the minimal marginal cost for $C(x) = 4x^3 - 24x^2 + 50x + 28$.

7. Determine the minimal marginal cost for $C(x) = x^3 - 12x^2 + 60x + 20$.

8. When the revenue function is $R(x) = 200\sqrt{x}$, what is the marginal revenue function, $R'(x)$?

9. When the revenue function is $R(x) = -x^2 + 30x$, what is the marginal revenue function, $R'(x)$?

10. If price is $p = 6 - (1/2)x$, determine the revenue and marginal revenue functions.

11. If $p(x) = (-1/4)x + 12$, find the revenue and marginal revenue functions.

12. Consumer demand for an item as a function of its price p is $x = D(p) = 50 - (p/3)$.
    Determine the production level and price that maximizes profit when the cost function is $C(x) = (2/3)x^3 + 2x^2 - 150x + 1000$.

13. Consumer demand for an item as a function of its price p is $x = D(p) = 40 - (p/6)$.
    Determine the production level and price that maximizes profit if the cost function is $C(x) = x^3 + 9x^2 - 360x + 2000$.

CHAPTER 3 SUPPLEMENTARY EXERCISES

1. Which of the following functions are increasing for all x?

2. Which of the following functions are decreasing for all x?

3. Which of the following functions have the property that the slope always increases as x increases (concave up all x)?

4. Which of the following functions have the property that the slope always decreases as x increases?

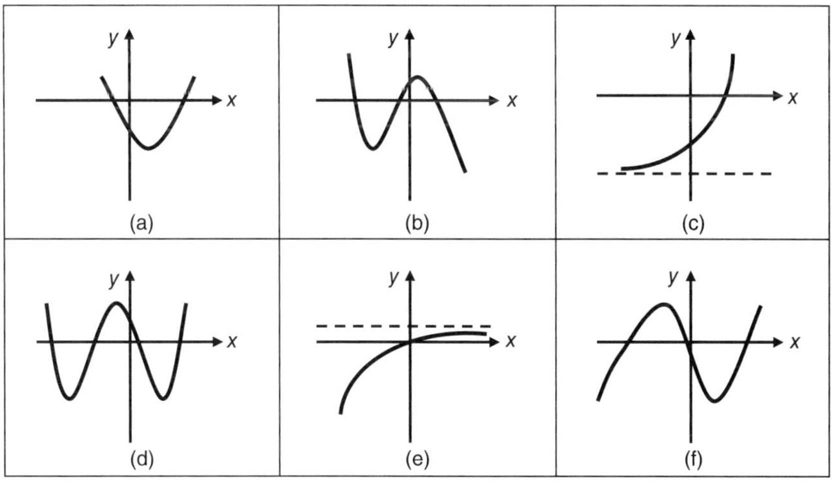

5. Show that the function $y = 1250 + 500(x + 1)^{-1}$ decreases for $x \geq 0$.

6. Find the minimum value of $f(x) = 8x^3 - 6x^2 + 10$ for $x > 0$.

7. Sketch the graph of $f(x) = 2x^2 - x - 6$ using derivatives as a guide.

8. Sketch the graph of $f(x) = x^3 - 3x^2$ using derivatives as a guide.

9. Sketch the graph of $f(x) = x^4 - 9x^2$ using derivatives as a guide.

10. Sketch the graph of $f(x) = \frac{1}{4}x^4 - \frac{1}{3}x^3 - 3x^2 + 6$ using derivatives as a guide.

11. Find any points of inflection for the curve $y = 3x^4 - 4x^3$.

12. Find any inflection points for the curve $f(x) = 2x^3 - 12x^2 + 40x + 60$.

13. There is \$480 to build a rectangular enclosure. One pair of opposite sides costs \$10 per running yard, while the other pair costs \$15 per running yard. Find the dimensions of the enclosure that maximizes its area.

14. A rectangular animal enclosure is to be built along a pond that serves as a side. The budget for materials is \$1000. Fencing costs \$10 per foot. What is the largest enclosed area possible?

15. A rectangular box with a square base is to have a volume of 64 cubic inches. Find the dimensions of the box requiring the least amount of material if

   (a) the box is closed. (b) the box is open.

16. The sum of two non-negative numbers is fixed. Find the numbers such that the sum of their squares is a minimum.

17. Statisticians, among others, measure error and variability by estimates of the variance, s^2. Sometimes, variances can be reduced by larger sample sizes (or product design). Such reductions entail added expense.

    Suppose a device's precision can be improved by reductions in the variability of two of its components. A variance is inversely proportional to the sample size. Let $V = \frac{a_1}{x_1} + \frac{a_2}{x_2}$ represent the total variance, V, of the two components, x_1 and x_2 are sample sizes and a_1 and a_2 are constants.

    However, sampling can be expensive, and there is a budget of B dollars. Let c_1 and c_2 be the unit costs of sampling, so total sampling cost is $c_1x_1 + c_2x_2 = B$.

    The problem becomes
    Minimize $= V = \frac{a_1}{x_1} + \frac{a_2}{x_2}$ subject to $c_1x_1 + c_2x_2 = B$.

18. A projectile is fired with an initial velocity $v_0 = 1000$ feet/s at a 30° elevation. Its coordinates after t seconds are $x = \frac{\sqrt{3}}{2}v_0t$ and $y = \frac{-1}{2}gt^2 + \frac{1}{2}v_0t$ (where the acceleration of gravity, $g = 32$ feet/s^2). What is the maximum height and the time for the projectile to return to earth?

19. If $R(x) = 4000x - 2x^2$, find the largest revenue possible.

20. If $p(x) = x^2 - 120x + 3600$, find the values of p and x for maximum revenue.

21. Suppose that the demand $x = D(p)$ for an item at price p is $x = D(p) = 40 - (p/6)$. Determine the production level and price that maximizes profit.

    The cost output function is $C(x) = x^3 - 21x^2 - 360x + 8350$.

4 Exponential and Logarithmic Functions

4.1 EXPONENTIAL FUNCTIONS

Functions as $f(x) = b^x$ are **exponential functions**. Their distinguishing feature is that a real variable, x, is the exponent. The base, b, is any positive real number (other than 1).

Fundamentals of Calculus, First Edition. Carla C. Morris and Robert M. Stark.

Companion Website: http://www.wiley.com/go/morris/calculus

Laws of Exponents, as in Chapter 1, appear again as follows.

Laws of Exponents

1. $b^0 = 1$ **5.** $(b^x)^y = b^{xy}$

2. $b^{-x} = \dfrac{1}{b^x}$ **6.** $(ab)^x = a^x b^x$

3. $b^x b^y = b^{x+y}$ **7.** $\left(\dfrac{a}{b}\right)^x = \dfrac{a^x}{b^x}$

4. $\dfrac{b^x}{b^y} = b^{x-y}$ **8.** $\left(\dfrac{a}{b}\right)^{-x} = \left(\dfrac{b}{a}\right)^x = \dfrac{b^x}{a^x}$

Example 4.1.1 Using the Laws of Exponents

Simplify

a) $(2^{3x}2^{5x})^{1/2}$ *b)* $27^{x/3}9^{2x}$ *c)* $\dfrac{5^{2x-1}5^{4x+5}}{5^{3x+1}}$ *d)* $\dfrac{14^{2x}}{7^{2x}}$

Solution:

a) The bases being the same, exponents add to yield $(2^{8x})^{1/2}$*. Next, multiply exponents by 1/2 to yield,* 2^{4x}*.*

b) The common base is 3, as 9 and 27 are powers of 3. Therefore, $(3^3)^{x/3}(3^2)^{2x} = 3^x 3^{4x} = 3^{5x}$*.*

c) Firstly, add numerator exponents to yield $\dfrac{5^{6x+4}}{5^{3x+1}}$*. Next, subtract the denominator exponent so* $5^{(6x+4)-(3x+1)} = 5^{3x+3}$*.*

d) As exponents are alike, the expression can be rewritten as $\left(\dfrac{14}{7}\right)^{2x} = 2^{2x}$*.*

Often, solving **exponential equations** is achieved by equating bases or equating exponents. If a common base, exponents are equated. Likewise, if exponents are equal, their bases are equated. These are:

Additional Laws of Exponents

When $b^x = b^y$**, then** $x = y$**. When** $a^x = b^x$**, then** $a = b$**.**

Example 4.1.2 Simplifying Exponential Equations

Solve for x:

a) $4^x = 128$ b) $3^{-x} = 27$ c) $7^{2x-3} = \sqrt[3]{49}$ d) $125 = x^3$

Solution:

a) *Firstly, equate bases as the variable is an exponent. The common base is 2, so* $(2^2)^x = 2^7$ *or* $2^{2x} = 2^7$. *Therefore,* $2x = 7$, *so* $x = 7/2$.
b) *Rewriting:* $3^{-x} = 3^3$ *yields* $-x = 3$ *so* $x = -3$.
c) *Firstly, rewrite* $\sqrt[3]{49} = (7^2)^{1/3} = 7^{2/3}$. *Therefore,* $7^{2x-3} = 7^{2/3}$, *so* $2x - 3 = 2/3$ *and* $x = 11/6$.
d) *In this case, the variable is the base, so exponents are equated. Rewriting as* $(5)^3 = (x)^3$ *yields* $x = 5$.

The graphs of exponential functions have several properties of interest.

a. Their domain is the real numbers, while their range is the positive real numbers.
b. There is no x-intercept. The y-intercept is unity.
c. The x-axis is a horizontal asymptote at $y = 0$.
d. When the base is between 0 and 1, the function decreases. The function increases when the base exceeds unity.

The graphs of the exponential functions $y = (1/2)^x$ and $y = 2^x$ illustrate these properties.

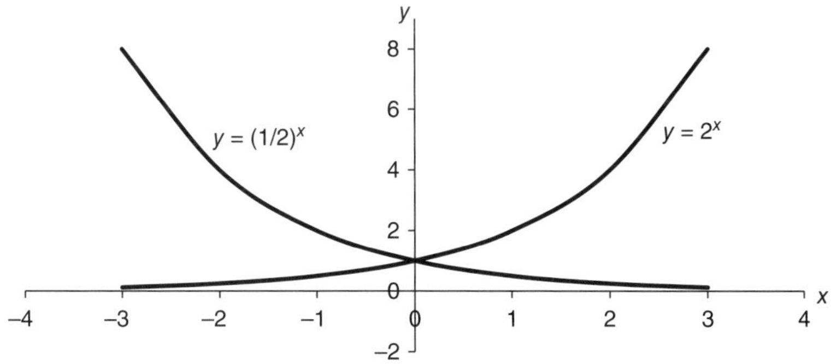

The transcendental number, **e** ≈ 2.718, is an important base for the exponential function e^x. The domain of $y = e^x$ is the real numbers, the range is the positive real numbers, the x-axis is an asymptote, and the y-intercept is unity. There is no x-intercept. Finally, $y = e^x$ is an increasing function as shown.

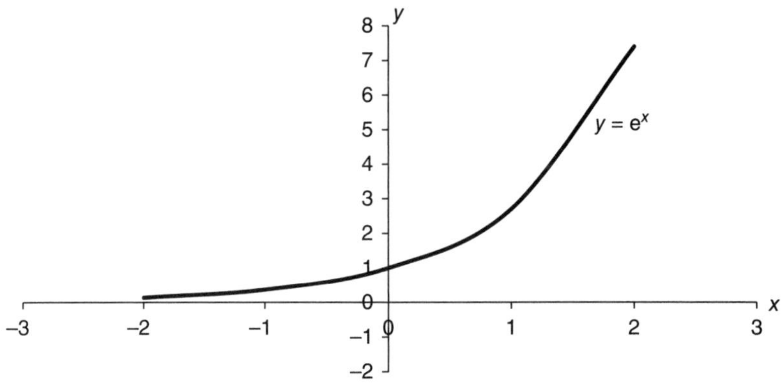

Example 4.1.3 Simplifying Exponents Involving e^x

Simplify the following:

a) $(e^3)^x$ b) $\frac{e^{5x-1}}{e^{x+2}}$ c) $\left(\frac{1}{e}\right)^{5x}$

Solution:

a) *Removing parenthesis and multiplying the exponents yields* e^{3x}.
b) *Subtracting exponents yields* $e^{(5x-1)-(x+2)} = e^{4x-3}$.
c) *Rewrite as* $(e^{-1})^{5x} = e^{-5x}$.

EXERCISES 4.1

In Exercises 1–6, rewrite the expressions with bases 2^{kx} or 3^{kx}.

1. a) $(8)^{3x}$ b) $(27)^{2x}$ c) $(16)^{5x}$

2. a) $(32)^{-2x}$ b) $(16)^{-x}$ c) $(81)^{-3x}$

3. a) $\left(\frac{1}{8}\right)^{-4x}$ b) $\left(\frac{1}{9}\right)^{6x}$ c) $\left(\frac{1}{27}\right)^{-2x}$

4. a) $\left(\frac{1}{4}\right)^{3x}$ b) $\left(\frac{1}{16}\right)^{x}$ c) $\left(\frac{1}{81}\right)^{-3x}$

5. a) $\frac{10^{5x}}{5^{5x}}$ b) $\frac{32^{2x}}{16^{2x}}$ c) $\frac{4^{3x}}{12^{3x}}$

6. a) $\frac{20^{3x}}{5^{3x}}$ b) $\frac{18^{7x}}{6^{7x}}$ c) $\frac{6^{x}}{24^{x}}$

In Exercises 7–12, simplify the expressions using the laws of exponents.

7. $\frac{7x^3x^5y^{-2}}{x^2y^4}$

8. $\left(\frac{2x^3}{y^4}\right)^{-2}$

9. $\frac{x^3}{y^{-2}} \div \frac{x}{y^5}$

10. $\left(\frac{5a^3b^4c^0}{10abc^2}\right)^{-3}$

11. $\frac{2^{5x+3}4^{x+1}}{8(2^{3x-1})}$

12. $\frac{9^{4x-1}27^{x+2}}{81^{x+3}}$

In Exercises 13–20, solve for x.

13. $7^{3x} = 7^{15}$

14. $10^{-3x} = 1{,}000{,}000$

15. $2^{7-x} = 32$

16. $3^{2x}3^{x+1} = 81$

17. $(1 + x)5^x + (3 - 2x)5^x = 0$

18. $(x^2)4^x - 9(4^x) = 0$

19. $(x^2 + 4x)7^x + (x + 6)7^x = 0$

20. $(x^3)5^x - (7x^2)5^x + (12x)5^x = 0$

In Exercises 21–25, the expressions can be factored. Find the missing factor.

21. $2^{3+h} = 2^h(\ \ )$

22. $3^{5+h} = 3^5(\ \ )$

23. $7^{x+5} - 7^{2x} = 7^{2x}(\ \ )$

24. $5^{2h} - 9 = (5^h - 3)(\ \ )$

25. $7^{3h} - 8 = (7^h - 2)(\ \ )$

4.2 LOGARITHMIC FUNCTIONS

Logarithmic functions are inverses of exponential functions and conversely!

The relationship between logarithms and exponentials is:

Logarithms and Exponentials
$\log_b x = y$ **implies** $b^y = x$ **and conversely**

Any logarithm can be expressed as an exponential and vice versa. For instance, $\log_2 x = y$ can be expressed as $2^y = x$. This format is convenient to determine the ordered pairs for the following graph.

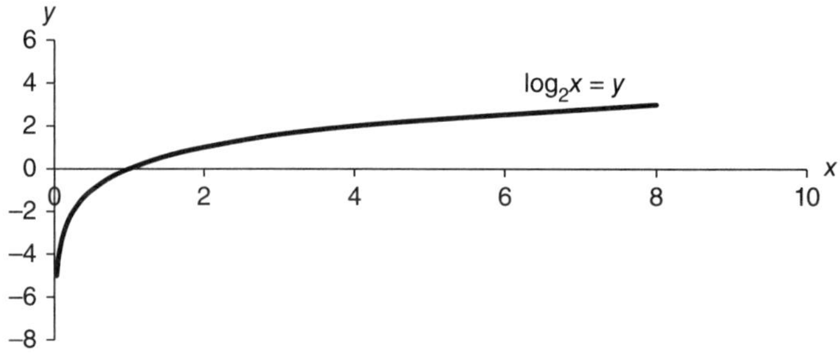

While the domain for x is $(0, \infty)$, the range for y is all real numbers. The x-intercept is at $x = 1$. There is no y-intercept.

Example 4.2.1 Exponential and Logarithmic Forms

Convert exponentials to logarithms or vice versa.

a) $3^4 = 81$ b) $2^{11} = 2048$ c) $\log_4 1024 = 5$ d) $\log_2 1/8 = -3$

Solution:

a) *The exponential is converted to the logarithm* $\log_3 81 = 4$.
b) *The exponential is converted to the logarithm* $\log_2 2048 = 11$.
c) *The logarithm is converted to the exponential* $4^5 = 1024$.
d) *The logarithm is converted to the exponential* $2^{-3} = 1/8$.

Most calculators have two logarithmic keys: $\log x$ and $\ln x$. The $\log x$ has a base 10 ($\log_{10} x$) and $\ln x$ has a base e ($\log_e x$). Base 10 logarithms are called **common logarithms,** while base e logarithms are called **natural logarithms**. Note that the base numerals 10 and e are identified by the symbolic spelling *log* and *ln*, respectively.

↑ Recall, any positive number can be an exponential base. For instance, $\log_2 8 = 3$ as $2^3 = 8$. Likewise, $\log_6 36 = 2$ as $6^2 = 36$.

Useful properties of common logarithms and natural logarithms for algebraic functions, $f(x)$ and $g(x)$, are:

General Properties of Logarithms

$\log_b 1 = 0$	$\log_b(f(x)g(x)) = \log_b f(x) + \log_b g(x)$
$\log_b b = 1$	$\log_b(f(x)/g(x)) = \log_b f(x) - \log_b g(x)$
$\log_b b^x = x$	$\log_b[f(x)^r] = r\log_b f(x)$
$b^{\log_b x} = x$	

Natural Logarithms

$\ln 1 = 0$	$\ln f(x)g(x) = \ln f(x) + \ln g(x)$
$\ln e = 1$	$\ln(f(x)/g(x)) = \ln f(x) - \ln g(x)$
$\ln e^x = x$	$\ln[f(x)^r] = r \ln f(x)$
$e^{\ln x} = x$	

A logarithm is in *reduced* or simplest form when component products, quotients, or exponents are irreducible. It takes practice to become thoroughly familiar with these properties of exponents and logarithms.

Example 4.2.2 Simplifying Logarithms

Express these logarithms in reduced form.

a) $log_3(x+1)(3x-5)$ *b)* $\log_7\left(\frac{(x+2)^3(x-4)}{(x-5)^2}\right)$ *c)* $ln(x^2+2)^3(2y-1)^4(z+3)^2$

Solution:

a) In this case, the logarithm is written as a sum of logarithms, as $log_3(x+1) + log_3(3x-5)$

b) Firstly, decompose as $log_7(x+2)^3 + log_7(x-4) - log_7(x-5)^2$. *Next, place the exponents as coefficients:* $3log_7(x+2) + log_7(x-4) - 2log_7(x-5)$. *The absence of products, quotients, or exponents completes the simplification.*

c) Firstly, expand the logarithms as $ln\ (x^2+2)^3 + ln(2y-1)^4 + ln(z+3)^2$. *Then, using exponents as coefficients,* $3ln(x^2+2) + 4ln(2y-1) + 2ln(z+3)$. *Again, a simplest form.*

♦ John Napier discovered logarithms in the early seventeenth century. They are suggested by the exponents of 10, 10^0, 10^1, 10^2, 10^3, … ; the base of the decimal system. Any positive number can be a logarithmic base. Later discoveries led to a preference for the "log naturalis" or natural logarithms that have the base *e*.

Logarithms were especially important for arithmetic computation before modern calculators and computers. They formed the basis for the slide rule; every engineering student's companion until the 1970s. Logarithms often appear in mathematical models of natural phenomena. (Historical Notes)

Example 4.2.3 Combining Logarithms

Express as a single logarithm:

a) $log_3(x + 1) + 2log_3(x + 4) - 7log_3(2x + 5)$
b) $5ln(x + 3) - 3ln(y + 2) - 2ln(z + 5)$

Solution:

a) *As bases are the same, use the coefficients as exponents to yield:*

$$log_3(x + 1) + log_3(x + 4)^2 - log_3(2x + 5)^7$$

Next, a sum (difference) of logarithms becomes a product (quotient) as $\log_3 \frac{(x + 1)(x + 4)^2}{(2x + 5)^7}$.

b) *Rewriting, coefficients become exponents as:*

$$ln(x + 3)^5 - ln(y + 2)^3 - ln(z + 5)^2.$$

The two terms with negative coefficients form the denominator as

$$ln \frac{(x + 3)^5}{(y + 2)^3(z + 5)^2}.$$

♦ The widely quoted Richter scale is one example of a widely used base 10 logarithmic measure. It was devised by Charles Richter in 1935 to compare earthquake magnitudes. As you know from news reports, on the Richter scale, an earthquake of, say, magnitude 5 is ten times a magnitude 4, and so on.

A major earthquake has magnitude 7, while a magnitude 8 or larger is called a Great Quake.

A Great Quake can destroy an entire community. The Indian Ocean Quake of December 2004, magnitude 9.0 on the Richter scale, resulted in Tsunamis that caused massive destruction.

The Great Chilean Earthquake of 1960, magnitude 9.5 on the Richter scale, is the strongest earthquake on record.

Other logarithmic scales include the decibel in acoustics, the octave in music, f-stops in photographic exposure, and entropy in thermodynamics. The pH scale chemists use to measure acidity and the stellar magnitude scale used by astronomers to measure the star brightness are also examples of logarithmic scales.

Logarithms and exponentials, being inverses, sometimes enable one to solve exponential equations as in the following examples.

Example 4.2.4 Solving an Exponential Equation (Base 10)

Solve using logarithms.

a) $10^x = 9.95$ *b)* $10^{2x+1} = 1050$

Solution:

a) As $10^0 = 1$ *and* $10^1 = 10$*, it follows that* $0 < x < 1$*. Note that x is close to 1 as 9.95 is close to 10. The logarithm of the equation (base 10, here) yields:*

$\log_{10} 10^x = \log_{10} 9.95$ *and, using a calculator,*

$$x = \log_{10} 9.95 \approx 0.9978$$

in agreement with our estimate.

b) As $10^3 = 1000$ *and* $10^4 = 10{,}000$*, an estimate of x can be obtained from* $3 < 2x + 1 < 4$*. This yields* $1 < x < 1.5$*. Taking the logarithm of the equation (base 10, here) yields:*

$$\log_{10} 10^{2x+1} = \log_{10} 1050 \quad so$$
$$2x + 1 = \log_{10} 1050$$
$$x = \frac{-1 + \log_{10} 1050}{2} \approx 1.0106$$

we could have anticipated that $x \approx 1$ *without calculation as* $x = 1$ *yields*

$$10^3 = 1000 \approx 1050.$$

Example 4.2.5 Determining an Exponent (Base e)

Solve using properties of logarithms

$$5e^{4x-1} = 100$$

Solution:

Note that $e^{4x-1} = 20$*. Using a calculator,* $e^3 \approx 20$*. Therefore,* $4x - 1 \approx 3$*, so x is close to 1. Taking the natural logarithm of both sides (base e, here) yields:*

→

$$\mathit{ln}e^{4x-1} = \mathit{ln}20$$

$$4x - 1 = \mathit{ln}20$$

$$x = \frac{1 + \mathit{ln}\,20}{4} \approx 0.9989$$

The value for x agrees with our preliminary estimate.

EXERCISES 4.2

1. Graph $y = \log_3 x$
2. Graph $y = \ln x$

In Exercises 3–12, evaluate the logarithms.

3. $\log_{10} 1{,}000{,}000$
4. $\log_3 243$
5. $\log_2 64$
6. $\log_5 125$
7. $\log_2 \frac{1}{32}$
8. $\log_3 \frac{1}{81}$
9. $\ln e^3$
10. $\ln e^7$
11. $\ln e^{7.65}$
12. $\ln e^{-3.4}$

In Exercises 13–20, evaluate the expressions.

13. $\ln(\ln e)$
14. $e^{\ln 1}$
15. $\log_9 27$
16. $\log_{25} 125$
17. $\log_4 32$
18. $\log_5 625$
19. $\log_2 128$
20. $\log_4(1/64)$

In Exercises 21–36, solve for x.

21. $\log_x 27 = 3$
22. $\log_x 64 = 3$
23. $\log_3(5x + 2) = 3$
24. $\log_2(x^2 + 7x) = 3$
25. $\ln 5x = \ln 35$
26. $e^{3x}e^{2x} = 2$
27. $\ln(\ln 4x) = 0$
28. $\ln(7 - x) = 1/3$
29. Write in reduced form: $\log_4 \frac{(x+1)^2(x-3)^6}{(3x+5)^3}$.
30. Write in reduced form: $\ln \frac{(x-1)^4}{(2x+3)^5(x-4)^2}$.
31. Write as a single logarithm: $2\ln x - 3\ln(y+1) + 4\ln(z+1)$.

32. Write as a single logarithm: $\ln 2 - \ln 3 + \ln 7$.

33. $10^{2x-1} = 105$

34. $10^{3x-1} = 100{,}100$

35. $3e^{x-1} = 4$

36. $e^{3x+1} = 22$

37. In June 2004, an earthquake of magnitude 4.1 struck Northern Illinois. Its effects were felt from Wisconsin to Missouri and from western Michigan to Iowa. Another earthquake in the Midwest, the 1895 Halloween Earthquake, is estimated at 6.8 on the Richter scale. How much stronger was the Halloween Quake than the Northern Illinois Quake?

4.3 DERIVATIVES OF EXPONENTIAL FUNCTIONS

Consider the exponential function $f(x) = e^x$, where x is real and e is that ubiquitous transcendental number. In this section, we seek its derivative.

The derivative of e^x is itself e^x. Among continuous functions, $f(x) = e^x$ has the unique property that

$$\frac{df(x)}{dx} = e^x = f(x)$$

Example 4.3.1 Derivatives of Exponentials

Find derivatives of:

a) $f(x) = e^{3x}$ *b)* $f(x) = 8e^{7x}$

Solution:

a) To use the aforementioned exponential rule, consider e^{3x} *as* $(e^x)^3$. *Applying the power rule yields* $\frac{d}{dx}((e^x)^3) = 3(e^x)^2(e^x) = 3e^{2x}e^x = 3e^{3x}$.

b) To use the exponential rule, consider e^{7x} *as* $(e^x)^7$. *Applying the power rule,* $\frac{d}{dx}((e^x)^7) = 7(e^x)^6(e^x) = 7e^{6x}e^x = 7e^{7x}$. *Therefore,* $f'(x) = 8[7e^{7x}] = 56e^{7x}$.

It is of interest to explore this result. The derivative of $y(x) = e^x$ evaluated at $x = 0$ is $e^0 = 1$. Consider the graph of $y = e^x$.

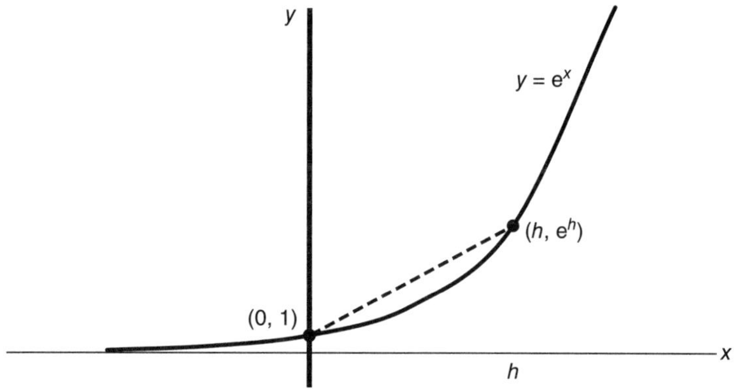

The slope of the secant line connecting (0, 1) and an arbitrary point at $x = h$, (h, e^h), has slope $\frac{e^h - 1}{h}$. The limit as $h \to 0$ is the derivative evaluated at $h = 0$. Therefore, the derivative at $h = 0$ is unity. That is, $f'(0) = \lim_{h \to 0} \frac{e^h - 1}{h} = 1$.

The rates of change of exponential functions exceed those for polynomial functions. This will become clearer in Chapter 9. As noted, e^x is the unique (nontrivial) function that remains unchanged when differentiated, that is, $f'(x) = f(x)$. More generally, if $f(x) = e^{g(x)}$, where $g(x)$ is differentiable, its derivative is

A General Exponential Rule

If $f(x) = e^{g(x)}$, then $f'(x) = e^{g(x)}g'(x)$

***g(x)*, a differentiable function.**

Example 4.3.2 More on Exponential Derivatives

Differentiate:

a) $f(x) = e^{x^3+5x^2-6}$.

b) $f(x) = 6e^{x^4+5x^3+3}$.

Solution:

a) Using the general rule, $f'(x) = e^{x^3+5x^2-6}[3x^2 + 10x]$.

b) Using the general rule, $f'(x) = (6)[e^{x^4+5x^3+3}(4x^3 + 15x^2)]$.

EXERCISES 4.3

In Exercises 1–4, rewrite the functions in the form e^{kx} using the laws of exponents.

1. $(e^3)^x$, $\left(\frac{1}{e^5}\right)^x$
2. $(e^5)^{2x}$, $\left(\frac{1}{e^4}\right)^{-2x}$
3. $e^{5x-3}e^{2x+3}$, $(e^{3x}e^{7x})^{1/2}$
4. $\dfrac{e^{4x-5}e^{3x+7}}{e^{-x+2}}$, $(e^{5x}e^{11x})^{1/4}$

In Exercises 5–18, differentiate the functions.

5. $f(x) = e^2$
6. $f(x) = 6e^3$
7. $f(x) = e^{2x}$
8. $f(x) = e^{9x}$
9. $f(x) = 5e^{6x}$
10. $f(x) = 4e^{10x}$
11. $f(x) = 5e^{25x}$
12. $f(x) = 3e^{30x}$
13. $f(x) = 6e^{8x^9}$
14. $f(x) = 8e^{7x^5}$
15. $f(x) = e^{2x^3+5x+1}$
16. $f(x) = e^{5x^4-3x^2+7}$
17. $f(x) = 6e^{8x^7+9x^4}$
18. $f(x) = 9e^{1/x}$

19. Determine the tangent to $y = e^{5x} + 3x + 1$ at $x = 0$.
20. Determine the tangent to $y = 3e^x$ at $x = 0$.

4.4 DERIVATIVES OF NATURAL LOGARITHMS

Having differentiated exponential functions in the previous section, we consider differentiation of logarithmic functions in this section.

> ***Logarithmic Derivatives***
>
> **If $f(x) = \ln x$, then $f'(x) = 1/x$**
>
> **If $f(x) = \ln[g(x)]$, then $f'(x) = g'(x)/g(x)$**
>
> ***($g(x)$ is a differentiable function of x)***

Example 4.4.1 Derivatives of ln(kx)

Verify the derivative of $f(x) = ln(kx)$ as $1/x$ where $x, k > 0$.

Solution:

Setting $g(x) = kx$, it follows from the "shaded box" rule, $\frac{d}{dx}(ln(kx)) = \frac{g'(x)}{g(x)} = \frac{k}{kx} = \frac{1}{x}$.

↓ *$ln(kx)$ can be rewritten as $ln\ k + ln\ x$. The derivative of $ln(k)$ is zero (it is a constant), while the derivative of $ln\ x$ is $1/x$.*

Example 4.4.2 Logarithmic Derivatives

Determine the derivatives of:

a) $f(x) = ln(x^7 - 4x^3 + 3x + 9)$ *b)* $f(x) = ln(7x^5 + 5x + 1)$

Solution:

a) Using the derivative rule, $f'(x) = \frac{7x^6 - 12x^2 + 3}{x^7 - 4x^3 + 3x + 9}$

b) Using the derivative rule, $f'(x) = \frac{35x^4 + 5}{7x^5 + 5x + 1}$

↓ *Notice that the derivative of the argument of the logarithm is* $\frac{d}{dx}(7x^5 + 5x + 1) = 35x^4 + 5$.

EXERCISES 4.4

In Exercises 1–16, determine the derivative.

1. $f(x) = \ln 25x$
2. $f(x) = \ln 16x$
3. $f(x) = x^3 + 5x^2 + \ln 3x$
4. $f(x) = x^2 + e^{4x} + \ln 8x$
5. $f(x) = \ln(4x + 7)$
6. $f(x) = \ln(9x - 1)$
7. $f(x) = \ln(5x^3 + 6x^2 + 9x + 2)$
8. $f(x) = \ln(7x^4 + 3x + 1)$
9. $f(x) = \ln(9x^5 + 6x^4 + 8x)$
10. $f(x) = \ln(4x^6 + 3x^3 - 5x^2 + 15)$

11. $f(x) = \ln \sqrt{5x + 1}$
12. $f(x) = \ln \sqrt[3]{x^2 - 3x + 1}$
13. $f(x) = (17x^2 + 15x + \ln 2x)^8$
14. $f(x) = (10x^9 + \ln(3x^3 - 5x))^{20}$
15. $f(x) = \ln(e^{3x} + 2)$
16. $f(x) = \ln(e^{5x} + 4x + 2)$
17. Determine the tangent to $y = \ln x$ at $x = e$.
18. Examine $x \ln x$ for extrema.
19. For the cost function, $C(x) = 2500 + 100 \ln(3x + 1)$, obtain its marginal cost function.
20. Simplify and differentiate $\ln[(7x + 5)^3(9x^2 - 4x + 1)]$.
21. Simplify and differentiate $\ln[(3x + 5)^5(4x^3 - 7x + 1)^2]$.

4.5 MODELS OF EXPONENTIAL GROWTH AND DECAY

Populations grow and decline. Intuitively, population changes are generally proportional to population size. While populations of people, animals, insects, bacteria, and such quickly come to mind, radioactive decay, consumer demand, bodily drug absorption, attenuation of light beams, electronic signals, and dollar interest on principal are examples of the principle.

♦ A mathematical "model" is a symbolic representation of a reality as in the examples cited previously.

A basic mathematical model for growth (or decay) assumes that at a time, t, for a population size, y, the rate of change of population with time, dy/dt, is proportional to the population at that instant.

In symbols,

$$\frac{dy}{dt} = \alpha y$$

where α is a coefficient of proportionality.

↑ The previous equation, containing a derivative, is called a **differential equation**. Differential equations are of basic importance in the modeling of phenomena involving change.

We seek the population as a function of time, $y(t)$. The solution to this elementary differential equation is

$$y(t) = y_0 e^{\alpha t}$$

where y_0 is the value of $y(t)$ when $t = 0$. You can check this solution by substitution into the differential equation. This relatively simple model has amazingly wide application. When α is positive, $y(t)$ increases with increasing time, t; a model for exponential increases and growth. When α is negative, $y(t)$ decreases with increasing t; a model for exponential decline and decay.

Basic exponential models typically assume that reproduction is continuous, that organisms are identical, and that environmental conditions are constant. There are many applications in the biological sciences such as the prediction of fish populations and carbon dating. The exponential model is fairly robust, meaning that it provides reasonable estimates even when assumptions are approximations.

Example 4.5.1 Population Growth

An exponentially increasing population numbered 50,000 in 1980 and 75,000 in 2000. Forecast the population, y(t), in 2020.

Solution:

Set $t = 0$ for the year 1980. Then, $t = 20$ for the year 2000. Use these to solve for α in the exponential equation.

$$75,000 = 50,000e^{20\alpha} \quad or \quad 3/2 = e^{20\alpha}$$

Using logarithms,

$$ln(3/2) = 20\alpha \quad and \quad \alpha = 0.020273$$

Therefore, $y(t) = 50,000e^{0.02073t}$
In 2020, when $t = 40$, $y(40) = 50,000e^{0.020273(40)} \approx 112,500.$

◆ Thomas Robert Malthus (1766–1834) prophesied mankind's eventual doom. He contended that human populations increase geometrically, while areas under cultivation (agriculture) can only grow arithmetically. The widely known "Malthusian Doctrine" has, so far, been a "victim" of the agricultural revolution. The coefficient, α, has been called the "Malthusian Parameter."

◆ Consider several "circles of friends." To contact a friend of a friend of a friend, there are a potentially large number of choices. A variant of this idea is the basis for the Kevin Bacon board game.

In the June 4, 1998 issue of *Nature, a* mathematical model described the "six degrees of separation" (or "small world thesis") devised by mathematicians Steven Strogatz and Duncan Watts. Their shortcuts help to explain why strangers often find they share common acquaintances and why epidemics can spread quickly.

Radioactive Decay

Many natural elements are radioactive; their intensity measured by a Geiger Counter as they decay. A popular measure of decay is an element's *half-life*. The half-life is the time required for an elements' radioactive intensity to fall to one-half of its initial value. Physicists use exponential models for radioactive decay.

Negative exponential growth, exponential decay, occurs when the proportionality factor is negative valued. The model becomes $y'(t) = -\alpha y(t)$ $(\alpha > 0)$ whose solution is

$$y(t) = y_0 e^{-\alpha t} \qquad \alpha > 0$$

↓ It is easy to verify that $y(t) = y_0 e^{-\alpha t}$ satisfies $y'(t) = -\alpha y(t)$. Simply differentiate $y(t)$ to obtain $y'(t)$.

As before, $y(t)$ is the radioactive intensity at a time, t; y_0, the initial intensity; and α, the decay constant (varies among radioactive elements). For the half-life, set $y(t) = (1/2)\, y_0$ to yield

$$(1/2)y_0 = y_0 e^{-\alpha t}$$

Using logarithms,

$$\begin{aligned} \ln(y_0 e^{-\alpha t}) &= \ln(1/2 y_0) \\ \ln\, y_0 + \ln\, e^{-\alpha t} &= \ln\, y_0 + \ln 1/2 \\ \ln\, e^{-\alpha t} &= \ln\, 1/2 \\ -\alpha t &= \ln\, 1/2 \end{aligned}$$

♦ Half-lives can vary from millionths of a second to billions of years. For instance, the half-life of Thorium-232 is 14 billion years. Tritium, a radioactive isotope of hydrogen, composed of one proton and two neutrons, has a half-life of about 12.5 years. Iodine-131 has a half-life of about 8 days. Recently discovered elements have half-lives of a fraction of a second.

Example 4.5.2 Radioactive Decay

Dairy cows that eat hay with higher levels of radioactive Iodine-131 produce milk unfit for human consumption. Suppose that a dairy farmer has purchased hay that has about 20 times the allowable limit of Iodine-131 present. The half-life for this isotope is 8.04 days. How long should the dairy farmer store the hay before use?

Solution:

Firstly, calculate the decay constant α.

⟶

Using the model $y(t) = y_0 e^{-\alpha t}, \quad \alpha > 0$

$$
\begin{aligned}
y_0 e^{-8.04\alpha} &= (0.50)y_0 \\
e^{-8.04\alpha} &= 0.50 \\
-8.04\alpha &= \ln(0.50) \\
\alpha &\approx 0.0862 \\
\therefore y(t) &= y_0 e^{-0.0862t}
\end{aligned}
$$

The dairy farmer seeks the Iodine-131 to be 1/20 of its present level so:

$$
\begin{aligned}
(1/20)y_0 &= y_0 e^{-0.0862t} \\
(1/20) &= e^{-0.0862t} \\
\ln(0.05) &= -0.0862t \\
t &\approx 35 \text{ days}
\end{aligned}
$$

$\therefore$ *the hay should be stored for 35 days before use.*

Some amazing recent scientific findings are attributed to carbon dating. Carbon-14 is a radioactive isotope of the stable Carbon-12 atom with a half-life of 5730 years. Knowing this, the antiquity of objects is estimated from their current levels of Carbon-14.

Example 4.5.3 Carbon-14 Dating

Develop a model for the remaining Carbon-14 in an object after t years. The half-life of Carbon-14 is 5730 years.

Solution:

Using the exponential decay model, $y(t) = y_0 e^{-\alpha t}$*, and the half-life for Carbon-14, estimate* α *as:*

$$
\begin{aligned}
y_0 e^{-5730\alpha} &= (0.50)y_0 \\
e^{-5730\alpha} &= 0.50 \\
-5730\alpha &= \ln(0.50) \\
\alpha &\approx 0.00012097
\end{aligned}
$$

Therefore, remaining Carbon-14 is $y(t) = y_0 e^{-0.00012097t}$.

♦ Chemical compounds are composed of varied elements, each having a distinct atomic weight. Many elements have unstable isotopes that differ in properties of atomic weight and decay rates. Archeologists use the isotope Carbon-14 to estimate the age of artifacts. Ages of artifacts older than 50,000 years have been estimated from Carbon-14 dating. Elements that have longer half-lives, such as some isotopes of uranium or potassium, have been used to date geological events millions or billions of years old.

♦ The prehistoric art of the Lascaux Caves in France is one of the famous revelations attributed to Carbon-14 dating. Discovered in the 1940s, these caves, and others in the region, have led to conjectures about early people. For this and other fascinating stories related to dating of early objects such as the Shroud of Turin and Stonehenge, begin a web search with Lascaux Caves.

Not all exponential models for growth and decay have the natural base e, although it is most common and most important. Many empirical models have bases other than e.

Consider an exponential model, base b, for a population, $y(t)$,

$$y(t) = y_0 b^{at} \qquad b > 0$$

where t is time; base, b, a positive real number; a, a constant; and $y_0 = y(0)$.

Example 4.5.4 Bacteria In a Culture

A culture of 250 bacteria doubles hourly. Without depletions, what is the population after 5 hours?

Solution:

As the population doubles hourly, it grows as 2^t, *t in hours. The model is* $y(t) = 250(2)^t$.

After 5 hours, $250(2)^5 = 8000$. *There are 8000 bacteria after 5 hours.*

Example 4.5.5 Growth of Escherichia coli Bacteria

If E. coli bacteria double their number every 20 min, how many are there in a day's growth?

Solution:

There are 72 twenty-minute periods every 24 hours. Therefore, after 24 hours, a single bacterium becomes

$$y(72) = 1(2)^{72} \approx 4.7 \times 10^{21}.$$

An astronomic number!

◆ "The mathematics of uncontrolled growth are frightening: A single cell of the bacterium *E. coli* would, under ideal circumstances, divide every twenty minutes. That is not particularly disturbing until you think about it, but the fact that bacteria multiply geometrically: one becomes two, two becomes four, four becomes eight and so on. In this way in a single day one cell of *E. coli* could produce a super colony equal in size and weight to the entire planet Earth."

Michael Crichton (1969) The Andromeda Strain, Dell, N. Y. p 247.

EXERCISES 4.5

1. A city experiences exponential growth according to $y(t) = 50{,}000e^{0.03t}$. Using the year 2010 as a base ($t = 0$), estimate the city population for 2015.
2. The population of a town in 2010 was 300 and is expected to increase to 400 by 2020. If exponential growth applies, forecast the population for 2030.
3. The size of a bacteria culture doubles hourly. If their initial number is 500, forecast the number after 7 hours.
4. A bacteria culture increases daily by 25%. Initially 1000 are present. How many bacteria are present after 2 weeks?
5. An initial population of 250 bacteria grows at a daily rate of 6%. When will the population have doubled?
6. An isotope of sodium (sodium-24) has a half-life of approximately 15 hours. If 1000 g of the isotope is present initially, how many grams will there be 5 days hence?
7. How many grams of Stontium-90 remain from 500 g after 200 years? Strontium-90 has a half-life of 28 years.
8. Plutonium is a heavy radioactive man-made metallic element. How long before 300 mg of Plutonium-239 decays to 100 mg? The half-life of Plutonium-239 is 24,000 years?
9. Argon-37 is produced from the decay of Calcium-40 whose half-life is 35 days. How long before 1/5 of the original amount remains?
10. As aortic valves close, aortic pressure at time t, $P(t)$, is modeled by $P(t) = 95e^{-0.491t}$. What is the initial aortic pressure? The pressure at $t = 0.1$ sec? When will the aortic pressure reach 80 mm Hg?
11. In Carbon-14 decay what remains of 2.5 g after 10,000 years?
12. From 1967 to 1980 the California gray whale population, $N(t)$, was described by the model $N(t) = N_0e^{0.025t}$. In fact, their growth to an estimated 21,000 whales in this period and into the early 1990s was such that in 1993 gray whales were removed from an endangered species list. Assuming the given model applies, estimate the number of California gray whales for 1967 ($t = 0$).

4.6 APPLICATIONS TO FINANCE

Interest is monies paid (or received) for the use of money. When you borrow money, your promise to pay interest is the inducement for lenders to lend. If you have a bank savings or money market account, you are the lender and the bank credits you with interest.

The borrowed amount is the **principal**. An **interest rate** is expressed in dollars per dollar per unit time. Interest rates are usually expressed in decimal format. For example, an annual interest rate of 6%, say, means that 6 cents (0.06 dollars) is paid on each dollar of principal after 1 year. To help compare interest rates, it is customary to express them on an annual basis.

In the simplest situation, the interest rate is proportional to principal, time, and interest rate.

Simple Interest

$$I = P \times r \times t$$

I = total interest	**r = interest rate (as a decimal)**
P = principal	**t = time**

Interest is usually paid periodically. Sometimes, interest is received or paid at various times in the loan's duration and not only at maturity. Therefore, when funds (interest plus principal) remain invested, at the next payment period, interest is paid not only on the principal, as before, but also on the interest earned earlier. This is **compound interest**.

Letting P_0 be an initial principal invested at rate r for a time t, in years, then the principal in successive years $P_1, P_2, \ldots, P_t$ is

$$P_1 = P_0 + rP_0 = (1 + r)P_0.$$

In words, the new principal after 1 year is the original principal P_0, plus, rP_0, the interest on that principal.

Similarly, for subsequent years,

$$P_2 = P_1 + rP_1 = (1 + r)P_1 = (1 + r)[(1 + r)P_0] = (1 + r)^2\, P_0$$
$$P_3 = P_2 + rP_2 = (1 + r)\, P_2 = (1 + r)\, [(1 + r)^2 P_0] = (1 + r)^3 P_0$$
$$\vdots$$

A pattern is clear (and proved by induction) as

$$P_t = (1 + r)^t P_0 \qquad t = 1, 2, \ldots$$

Annual Compound Interest

$$A = (1 + r)^t P_0 = P_t \qquad t = 1, 2, \ldots$$

$A = P_t$ = the total amount accumulated

P_0 = principal $\quad t$ = time

r = interest rate (expressed as a decimal)

Example 4.6.1 Compound Interest

a) What compound interest accrues on \$5000 at 6% annually for 3 years?
b) On \$10,000 at 5% annually for 48 months?

Solution:

a) After 3 years, the total is $5000(1.06)^3$ = \$5955.08. Subtract the original principal, \$5000, to yield interest of \$955.08. Notice that this exceeds the \$900 simple interest that would have been earned as compound interest always yields more than simple interest.
b) After 4 years, the total amount is $10{,}000(1.05)^4$ = \$12,155.06 and the interest is \$2155.06.

The example above assumes that interest is compounded annually. However, sometimes, interest is compounded quarterly or monthly. Let P_0 be the original principal; r, the annual interest rate (as a decimal); n, the number of periods per year that interest is compounded, the time, t, in years. Then, the amount accumulated is $P_t(= A)$ is $P_0(1 + r/n)^{nt}$.

When $n = 1$, the above-mentioned formula matches the annual compound interest formula. Note that if there are n periods per year, then the relevant interest per period is r/n and the number of compoundings in t years is nt. With these alterations the expression for P_t follows.

Compound Interest Formula

The principal $A = P_t$ (the sum of P_0 and earned interest) after t years when interest is compounded n times a year is

$$A = P_t = P_0(1 + r/n)^{nt} \qquad t = 0, 1, \ldots$$

r = interest rate (expressed as a decimal)

Example 4.6.2 Effects of Compounding Periods

a) What interest accrues on \$5000 at 6% interest for 3 years when compounded quarterly?

b) When compounded monthly?

Solution:

a) Here, $n = 4$ and

$$P_t = 5000\ (1 + 0.06/4)^{4\times 3} = 5000(1.015)^{12} = \$5978.09.$$

Therefore, the accrued interest is \$978.09.

b) Now, $n = 12$ so

$$P_t = 5000\ (1 + 0.06/12)^{12\times 3} = 5000(1.005)^{36} = \$5983.40.$$

Therefore, the accrued interest is \$983.40.

Notice that both amounts exceed the interest accrued in the previous examples when interest was compounded annually. In general, the more frequent the compounding, the greater the accrued interest.

The previous example illustrates the effect of increased compounding frequency. The amount realized increases as the frequency of compounding increases. That is, as the period between interest payments decreases. Clearly, this is due to the additional sums earned by interest on interest already accrued. That is, by compounding interest.

The "*Rule of 72*" is well known and widely used in financial circles. The rule permits one to approximately calculate either the time or the interest rate for a sum to double. At a 6% rate, for example, about $72/6 = 12$ years is required. The annual interest rate required to double a sum in 3 years is about $72/3 = 24\%$, and so on.

The previous example raises the question of instantaneous or continuous compounding. It is easy to jump to the guess that continuous compounding might result in huge sums. Alas, that is not the case! The following formula applies for continuous compounding.

Continuous Compounding Formula

The principal $A = P_t$ (P_0 plus interest earned) at time t (years) when compounding continuously is

$$P_t = P_o e^{rt}$$

$r =$ interest rate (expressed as a decimal)
$e \approx 2.71828$ (see Chapter 1)

↑ Students familiar with the curious limit $\lim_{n\to\infty}\left(1+\frac{x}{n}\right)^n = e^x$ will recognize the compound interest e^{rt} as the limit of P_t as $n \to \infty$.

Example 4.6.3 Compounding Continuously

What is the interest on \$5000 at 6% for 3 years compounded continuously?

Solution:
Using the continuous compounding formula, we have

$$P_t = 5000\ e^{0.06\times 3} = 5000\ e^{0.18} = \$5986.09.$$

As noted earlier, compounding continuously doesn't yield an infinite sum. However, it does realize a larger return than any other compounding period.

Most variables in economic theory vary continuously with time. For small time changes, the derivative is an estimate of the variables' time rate of change. The percentage change in the GDP between reporting periods is a simple notion of a growth rate.

↓ A logarithmic scale often provides a clearer view of long-term growth. The derivative of a logarithmic function with respect to time is the growth rate.

Recall that the derivative of a natural logarithmic function, $f(t)$, is $f'(t)/f(t)$. This is the **relative rate of change** of $f(t)$ per unit change of t. The **percentage rate of change** is the relative rate of change multiplied by 100%.

Percentage Rate of Change

The percentage rate of change is

$$\frac{f'(t)}{f(t)} \times 100\%$$

Example 4.6.4 Percentage Rate of Change

A country's GDP is $f(t) = 3.6 + 0.03t + 0.02t^2$. What is the percentage rate of change of GDP at $t = 2$ years?

⟶

Solution:

$$\frac{f'(t)}{f(t)} \times 100\% = \frac{0.03 + 0.04t}{3.6 + 0.03t + 0.02t^2} \times 100\%$$

At $t = 2$,

$$\frac{0.03 + 0.04(2)}{3.6 + 0.03(2) + 0.02(2)^2} \times 100\% = \frac{0.11}{3.74} \times 100\% = 2.94\%$$

Economists express demand, q, as a function of price, p, by the **demand function** $q = f(p)$ in which q decreases as p increases. The ratio of the relative rate of change of demand to the relative rate of change of price is the **elasticity of demand**.

Elasticity of Demand

The elasticity of demand at a price *p*, *E*(*p*), is

$$E(p) = \frac{-pf'(p)}{f(p)} \times 100\%$$

When $E(p) < 1$, demand is inelastic
When $E(p) > 1$, demand is elastic
When $E(p) = 1$, there is unitary elasticity

When demand is *elastic* at price p, the change in demand is opposite to the change in price (hence the negative sign in the definition). When demand is *inelastic,* the change in demand is in the same direction as the change in price.

Example 4.6.5 Elasticity of Demand

Show that the demand function $f(p) = 50\ p^{-2}$ is elastic with respect to price.

Solution:

$E(p) = \dfrac{-p(-100p^{-3})}{50p^{-2}} = \dfrac{100p^{-2}}{50p^{-2}} = 2$. *As $E(p) > 1$, demand is elastic. As $E(p)$ does not depend on the value of p, demand remains elastic regardless of the price.*

Another application of exponential and logarithmic functions deals with sales decay. Some marketing studies have demonstrated that if a product ceases to be promoted that sales for the product will decrease in proportion to current sales. The following example illustrates the concept.

Example 4.6.6 Sales Decay

A corporation finds that daily sales, S(t), are falling as $S(t) = 50,000(3^{-0.10t})$ *items, where t is the number of days since January* 1st.

a) What is a sales estimate for January 11?
b) When are sales expected to fall below 5000 items?

Solution:

a) January 11 implies that t = 10. So

$$50,000(3^{-(0.10)(10)}) = 50,000(3^{-1}) \approx 16,667 \text{ items.}$$

b)

$$\begin{aligned} 50,000(3^{-(0.10t)}) &= 5000 \\ 3^{-0.10t} &= (1/10) \\ -0.10t \ln 3 &= \ln(0.10) \\ t &\approx 21 \text{ days} \end{aligned}$$

On January 22, sales are expected to be below 5000.

◆ In one marketing study, the sales $S(t)$ at time t for an unpromoted product is $S(t) = S(0)e^{-\lambda t}$, where λ is referred to as the exponential sales decay constant. There are variations of this model.

M. Vidale and B. Wolfe, "An Operations-Research Study of Sales Response to Advertising", Operations Research, Volume 5, Issue 3, June 1957, page 371.

EXERCISES 4.6

1. If \$50,000 is compounded quarterly at 8% for 4 years, what is the sum at maturity?
2. If \$250,000 is compounded monthly at 6% for 3 years, what is the sum at maturity?
3. If \$500,000 is compounded continuously at 7% for 6 years, what is the sum at maturity?
4. Which is greater: \$75,000 at 10% compounded quarterly for 4 years or \$60,000 compounded continuously at 4% for 8 years?
5. Which is greater: \$250,000 at 3% compounded annually for 5 years or \$225,000 compounded continuously at 5% for 4 years?

6. For \$20,000 to double in 7 years with continuous compounding, what interest rate is required?
7. When will \$25,000 compounded continuously at 4% increase to \$100,000?
8. When will \$50,000 compounded continuously at 5% triple?
9. A sales decay model is $S(t) = S_0 e^{-\alpha t}$ where $\alpha = 0.2$. When will sales drop to 30% of initial value? (t is in weeks)

In Exercises 10–15, determine the percentage rate of change of the functions at the indicated point.

10. $f(t) = 2t^2$ at $t = 2$
11. $f(t) = 3t^2 + 2t + 1$ at $t = 3$
12. $f(t) = 3e^{0.2t}$ at $t = 6$
13. $f(t) = e^{0.4t}$ at $t = 5$
14. $f(t) = 2/(3t + 8)$ at $t = 4$
15. $f(t) = 1/(t + 3)$ at $t = 2$
16. $f(p) = 600 - 4p$ elastic at $p = 75$?
17. In what interval is demand $f(p) = 500e^{-0.25p}$ elastic?
18. Consider a demand function $f(p) = 450p - p^2, 0 < p < 450$. Verify that demand is elastic for $0 < p < 300$ and inelastic for $300 < p < 450$.

HISTORICAL NOTES

Leonhard Euler (1707–1783) – Perhaps the most prolific of mathematicians, he authored 866 books and papers and won the Paris Academy Prize 12 times. Born in Switzerland, he began as a theology student and changed to mathematics under the influence of Johann Bernoulli. Euler made significant contributions in differential calculus, mathematical analysis, and number theory. He introduced the symbols e, i, $f(x)$, π, and the sigma summation sign. Remarkably, most of Euler's publications appeared in the last 20 years of his life when he was totally blind.

John Napier (1550–1617) – Napier was born in Edinburgh, Scotland and little is known of his early years. He entered St. Andrews University at 13 years. Napier's study of mathematics was only a hobby. He is best known for inventing logarithms and introducing decimal notation. He also invented "Napier's Bones," which are a mechanical means to multiply, divide, and take square roots and cube roots. (Also, Section 4.2 diamond box.)

CHAPTER 4 SUPPLEMENTARY EXERCISES

1. Simplify the following expressions

a) $\dfrac{16^{3x}}{8^{3x}}$ b) $\dfrac{50^{5x-1}}{10^{5x-1}}$ c) $\dfrac{25^{2x+1}5^{3x+2}}{125^x}$

2. Use the laws of exponents to simplify $\left(\dfrac{3x^4y^{-2}z^5}{2x^2y^3z^{-2}}\right)^{-3}$

3. Solve for x:

   a) $5^{-2x} = 625$ b) $2^{3x-1} = 32$

4. Solve for x: $x^3(5^x) - 9x^2(5^x) + 8x(5^x) = 0$
5. Factor: $3^{2x} - 25$

In Exercises 6–10, determine the derivatives of the functions.

6. $f(x) = e^{7x}$
7. $f(x) = 4e^{9x}$
8. $f(x) = e^{7x^4}$
9. $f(x) = e^{4x^3}$
10. $f(x) = 3e^{8x^2+5x+3}$

11. Find the tangent to $e^{6x} + 1$ at $x = 0$.

In Exercises 12–17, evaluate the logarithms.

12. $\log_{10} 10,000$
13. $\log_3 729$
14. $\log_4 \dfrac{1}{16}$
15. $\log_9 27$
16. $\log_4 32$
17. $\ln e^{7.23}$

18. Solve: $\ln(4 + x) = 2$
19. Express as simpler logarithms:

$$\ln (2x + 3)^4(x + 1)^2(4x + 5)^7$$

20. Express as a single logarithm:

$$5 \ln(x + 2) - 3 \ln(y + 1) + \ln(z + 2)$$

21. The Great Hanshin Earthquake in Japan in January 1995 measured 6.9 on the Richter scale and caused over 100 billion dollars in damage. The devastating Great San Francisco Earthquake and Fire of 1906 measured 8.3 on the Richter scale. How much stronger was the California quake than the Great Hanshin Earthquake?

In Exercises 22–23, determine the derivatives.

22. $f(x) = \ln(5x^3 - 9x + 1)$
23. $f(x) = \ln(5x^3 + 2x^2 + 8x + 7)$
24. If $C(x) = 400 + 3 \ln(5x + 1)$ find the marginal cost.
25. The population in a small town grew from 250 in 2000 to 650 in 2005. If $y(t) = y_0 e^{\alpha t}$ applies to the town's growth, what is an estimate for the population in 2010?

26. Use the link (http://www.census.gov/cgi-bin/ipc/popclockw) for projections of the world population from the US Census Bureau. Calculate projections for the next two months and use these as $t = 0$ and $t = 1$. Assume the exponential growth model $y(t) = y_0 e^{\alpha t}$
    a) Determine the value for α.
    b) Predict the population in 3 months ($t = 2$) using the value of α you calculated in part (a).
27. A bacteria culture triples hourly. If 400 bacteria are initially present, how many will there be in 5 hours?
28. Which is worth more: \$150,000 for 5 years compounded annually at 4%, \$125,000 compounded quarterly for 4 years at 5%, or \$120,000 for 6 years compounded continuously at 3%?
29. The price of a commodity at time t is $f(t) = 0.3 + 0.03t + 0.02e^{-t}$. What is the percentage rate of change of $f(t)$ at $t = 0$? $t = 2$?
30. Determine the percentage rate of change of $f(t) = 5t^3 + 4t^2 + 3$ at $t = 5$.

5 *Techniques of Differentiation*

Fundamentals of Calculus, First Edition. Carla C. Morris and Robert M. Stark.

Companion Website: http://www.wiley.com/go/morris/calculus

So far, differentiation has been of relatively simple functions of a single variable. Often, there is a need to differentiate more complicated compound functions such as

$$(x^3 + x^2 + 3)(x^2 - 1), \quad \frac{x^2}{(2x^5 + 3x + 1)^7}, \quad x^3e^x, \quad \text{and } x(\ln x)$$

In this chapter, you learn rules that simplify differentiating compound expressions. In addition, we illustrate the similarity of the useful finite calculus introduced earlier (Chapter 2, Section 7).

5.1 PRODUCT AND QUOTIENT RULES

A very useful rule for the derivative of a product of functions $y = f(x)g(x)$ is

$$\frac{dy}{dx} = \frac{d}{dx}(f(x)g(x)) = f(x)g'(x) + g(x)f'(x)$$

In words, to differentiate the product, multiply the first function by the derivative of the second and add the product of the second and the derivative of the first.

↓ In meter, "*Derivative of the second times the first plus the derivative of the first times the second.*"

Product Rule

The derivative of the product $y = f(x)g(x)$ is

$$y' = f(x)g'(x) + g(x)f'(x)$$

f(x) and g(x) are differentiable functions of x.

Later, we derive the product rule. For now, we concentrate on examples.

Example 5.1.1 The Product Rule

Find derivatives of:

a) $f(x) = (5x^2 + 4)(2x + 3)$.
b) $f(x) = (x^2 + 7x + 5)(x + 3)^5$.

⟶

Solution:

a) The product rule yields:

$$f'(x) = (5x^2 + 4)[2] + (2x + 3)[10x]$$

b) The product and power rules yield:

$$f'(x) = (x^2 + 7x + 5)[5(x + 3)^4] + (x + 3)^5[(2x + 7)]$$

Note the use of the brackets to clarify derivatives. They help organize derivatives of complicated expressions.

Example 5.1.2 More on the Product Rule

Find derivatives of:

a) $f(x) = (3x^2 + 6x + 7)^{25} e^{9x}$.
b) $f(x) = (5x + 6)^{12} \ln(x^3 + 5x^2 + 2x + 3)$.

Solution:

a) The product rule yields:

$$f'(x) = (3x^2 + 6x + 7)^{25}[9e^{9x}] + e^{9x}[25(3x^2 + 6x + 7)^{24}(6x + 6)]$$

b) The product and power rules yield:

$$(5x + 6)^{12}\left[\frac{3x^2 + 10x + 2}{x^3 + 5x^2 + 2x + 3}\right] + \ln(x^3 + 5x^2 + 2x + 3)[12(5x + 6)^{11}(5)]$$

Quotient Rule

A rule for differentiating the quotient $\frac{f(x)}{g(x)}$ is

$$\frac{d}{dx}\left(\frac{f(x)}{g(x)}\right) = \frac{g(x)f'(x) - f(x)g'(x)}{[g(x)]^2}$$

Firstly, form the denominator by squaring the ("bottom") function. The desired numerator for the quotient rule is the denominator ("bottom") times the derivative of the numerator ("top") minus the numerator ("top") times the derivative of the denominator ("bottom").

↓ One wit placed it in musical meter: "Derivative of the top times the bottom minus derivative of the bottom times the top – all divided by the bottom squared."

Hint: note that because of the negative sign the order of differentiation matters. A common mistake reverses the numerator expression resulting in an error in sign.

Quotient Rule

The derivative of $\frac{f(x)}{g(x)}$ **is**

$$\frac{g(x)f'(x) - f(x)g'(x)}{[g(x)]^2}$$

f(x) and g(x) are differentiable functions of x.

Example 5.1.3 The Quotient Rule

Find the derivative of $f(x) = \frac{(5x + 3)}{(7x^2 + 3x + 1)}$.

Solution:
The quotient rule yields:

$$f'(x) = \frac{(7x^2 + 3x + 1)[5] - (5x + 3)[14x + 3]}{(7x^2 + 3x + 1)^2}$$

When the numerator function is a constant, it is easier to find the derivative by rewriting the expression as a reciprocal with a negative exponent.

Example 5.1.4 The Quotient Rule Revisited

Find the derivative of

$$f(x) = \frac{3}{(5x^4 + 6x + 7)^5}$$

Solution:
The quotient rule yields:

$$f'(x) = \frac{(5x^4 + 6x + 7)^5[0] - 3[5(5x^4 + 6x + 7)^4(20x^3 + 6)]}{(5x^4 + 6x + 7)^{10}}$$

⟶

which simplifies to

$$\frac{-15(20x^3 + 6)}{(5x^4 + 6x + 7)^6}$$

Note, however, that the original function could have been rewritten as $3(5x^4 + 6x + 7)^{-5}$. *Taking the derivative using the general power rule yields*

$$3(-5)(5x^4 + 6x + 7)^{-6}(20x^3 + 6) \quad or \quad \frac{-15(20x^3 + 6)}{(5x^4 + 6x + 7)^6}$$

Example 5.1.5 Differentiation of Products and Quotients of Logarithms

Find the derivatives of:

a) $f(x) = (5x^4 - 3x^2 + 2)^{20} \ln 7x$ b) $f(x) = \dfrac{\ln(5x + 2)}{(4x^3 + 3x + 5)^6}$.

Solution:

a) *Using product and logarithmic rules and differentiating:*

$$f'(x) = (5x^4 - 3x^2 + 2)^{20}[7/7x] + \ln 7x[20(5x^4 - 3x^2 + 2)^{19}(20x^3 - 6x)]$$

b) *Using quotient and logarithmic rules and differentiating:*

$$f'(x) = \frac{(4x^3 + 3x + 5)^6 \left[\dfrac{5}{5x + 2}\right] - \ln(5x + 2)[6(4x^3 + 3x + 5)^5(12x^2 + 3)]}{(4x^3 + 3x + 5)^{12}}$$

Conceivably, one might need product, quotient, and power rules to differentiate in a single expression. Separating a differentiation into stages eases the task for complicated expressions.

For example, some functions are **compound functions** of products and quotients such as

$$\frac{f(x)g(x)}{h(x)}$$

where all are differentiable functions of x.

Although the basic rules apply, the execution tends to be more complicated as shown in the following example.

Example 5.1.6 Combining the Rules

Find the derivative of

$$f(x) = \frac{(x^3 + 5)e^{10x^2}}{(2x^7 + 1)^3}$$

Solution:
The quotient rule yield for $f'(x)$

$$\frac{(2x^7 + 1)^3[(x^3 + 5)20xe^{10x^2} + e^{10x^2}(3x^2)] - (x^3 + 5)e^{10x^2}[(3)(2x^7 + 1)^2(14x^6)]}{(2x^7 + 1)^6}$$

Note the use of both the quotient rule and the product rule in the numerator.

EXERCISES 5.1

Differentiate the functions in Exercises 1–34.

1. $f(x) = (2x^2 + 3x + 8)(5x - 1)$
2. $f(x) = 3x^2(2x + 5)^4$
3. $f(x) = (x^2 + 5x + 3)^4(3x^2 - 2x + 6)^3$
4. $f(x) = (2x + 3)^9 \ln(5x^3 - 2x + 1)$
5. $f(x) = (5x) \ln(x^2 + 5)$
6. $f(x) = e^{6x} \ln(6x^3 + 2x + 7)$
7. $f(x) = x^3e^{5x} + \ln 5x + 2x^{5/7} + 8$
8. $f(x) = 4x + 2x^3e^{8x} + 5x^9 \ln 3x$
9. $f(x) = \dfrac{3}{(x^5 + 7x - 3)^8}$
10. $f(x) = \dfrac{5x - 7}{3x^2 + 2}$
11. $f(x) = \dfrac{3x + 2}{5x^2 + 7x + 3}$
12. $f(x) = \dfrac{6x - 5}{(2x^3 + 4x + 1)^7}$
13. $f(x) = \dfrac{4x^3 + 2x + 1}{5x^2 + 7x + 3}$
14. $f(x) = \dfrac{(3x^2 + 4)^3}{(x^5 + 7x - 3)^8}$
15. $f(x) = \dfrac{(5x^6 + 3x - 4)^{11}}{(7x^3 + 5x + 3)^4}$
16. $f(x) = \left(\dfrac{3x - 7}{x^4 + 5x + 3}\right)^{50}$
17. $f(x) = \left(\dfrac{30}{\left(x^7 - 5x^4 + 3\right)^3}\right)^{100}$
18. $f(x) = \left(\dfrac{(5x - 7)^3}{(5x^3 + 9x^2)^7}\right)^{50}$
19. $f(x) = \dfrac{\ln 7x}{2x - 5}$
20. $f(x) = \dfrac{\ln 3x}{(4x - 7)^{10}}$
21. $f(x) = (x^3 + 7x + \ln 8x)^{25}$
22. $f(x) = e^{7x} \ln(x^2 + 1)$
23. $f(x) = \dfrac{x^3e^{5x}}{(2x + 7)^3}$
24. $f(x) = \dfrac{e^{x^4} \ln 2x}{(5x^3 + 9x + 1)^{10}}$
25. $f(x) = \left(\dfrac{\sqrt{3x - 7}}{x^4 + 5x + 3}\right)^4$

26. $f(x) = \dfrac{\sqrt[3]{x^2 + 5x - 1}}{7x}$

27. $f(x) = (3x + 1)^4\left(\sqrt[5]{x^3} + 2x\right)^{10}$

28. $f(x) = (6x^3 + 9x^2 - 8x + 3)^{15}e^{8x}$

29. $f(x) = (6x^3 - \sqrt{x})^8 e^{x^3}$

30. $f(x) = \left(5x^2 - 2x + \sqrt[3]{x^2}\right)^4 e^{2x^5}$

31. $f(x) = \dfrac{e^{3x}}{(2x + 1)^5}$

32. $f(x) = \dfrac{e^{5x}}{(2x^7 + 5x^3 + 3)^4}$

33. $f(x) = \dfrac{e^{4x}}{e^x + 1}$

34. $f(x) = \dfrac{xe^{8x}}{(x^2 + 5x + 3)^4}$

In Exercises 35–39, find the equation of the tangent line to f(x) at the indicated point.

35. $f(x) = (5x - 4)(3x + 7)$ at $(1, 10)$.

36. $f(x) = \dfrac{3x - 1}{x + 2}$ at $(5, 2)$.

37. $f(x) = (x - 1)^4(2x + 3)$ at $x = 2$.

38. $f(x) = (3x - 8)^7(5x + 1)$ at $x = 3$.

39. $f(x) = (4x - 3)^{3/2}(x + 1)$ at $x = 3$.

40. Find all points (x, y) on the curve $y = (2x + 3)(x - 5)^2$ where the tangent is horizontal.

41. Find all points (x, y) on the curve $y = \dfrac{x^3}{(2x - 1)}$ where the tangent is horizontal.

42. Total revenue is $R(x) = \dfrac{1,500x}{(\ln(5x + 3))}$. Find the marginal revenue function.

5.2 THE CHAIN RULE AND GENERAL POWER RULE

In Section 4 of Chapter 1, we learned that a function can be a function of still another function. In symbols, $f(g(x))$ is a **composite function**.

↓ As an example, for $f(x) = x^2$ and $g(x) = x^3 + 2$, $f(g(x)) = (x^3 + 2)^2$, x is replaced by $g(x)$ in $f(x)$.

The derivative of $f(g(x))$ is

$$\frac{d}{dx}[f(g(x)] = f'(g(x))g'(x) \qquad \textbf{Chain Rule}$$

In words, the derivative of $f(g(x))$ is its derivative with respect to $g(x)$ times the derivative of $g(x)$ with respect to x as illustrated in the following example.

Example 5.2.1 The Composite Chain Rule

Let $f(x) = x^3 + 5x + 1$ and $g(x) = 2x^4 + 3x + 2$

a) Determine $f(g(x))$

b) Obtain $f'(x)$ and $g'(x)$

⟶

c) Now, obtain $\frac{d}{dx}(f(g(x)))$ *using the chain rule.*
(Compare the result using earlier rules.)

Solution:

a) $f(g(x)) = f(2x^4 + 3x + 2) = (2x^4 + 3x + 2)^3 + 5(2x^4 + 3x + 2) + 1$
Note that in f(x), x has been replaced by g(x).

b) $f'(x) = 3x^2 + 5$ *and* $g'(x) = 8x^3 + 3$

c) Firstly, determine the derivative of f(g(x)) with respect to g(x) as

$$f'(g(x)) = 3(2x^4 + 3x + 2)^2 + 5$$

Next, multiply by the derivative of g(x) to yield:
$f'(g(x))g'(x) = [3(2x^4 + 3x + 2)^2 + 5](8x^3 + 3)$
or $3(2x^4 + 3x + 2)^2(8x^3 + 3) + 5(8x^3 + 3)$.
(The result is the same as the derivative using earlier rules).

The Chain Rule for Composite Functions f(g(x))

1. **Determine derivatives $f'(x)$ and $g'(x)$.**
2. **Substitute $g(x)$ for x in $f'(x)$.**
3. **Multiply by $g'(x)$.**

Again, the derivative of a composite function, $f(g(x))$, is its derivative with respect to its argument, $g(x)$, multiplied by the derivative of the argument, $g'(x)$.

The chain rule permits the differentiation of composite functions with considerably more ease as the previous examplc illustrated.

↓ One can ask, "why learn the Chain Rule if the same result can be reached by our earlier methods?" Often, one method or the other is appreciably simpler to execute. If $f(x)$ and $g(x)$ are complicated functions of products or quotients, the chain rule will be advantageous.

Another aid to obtain the derivative of a composite function $f(g(x))$ is to replace $g(x)$ by a new variable. That is, let $u = g(x)$ so

$$y = f(g(x)) = f(u)$$

Now, use the chain rule to differentiate as

$$\frac{dy}{dx} = \frac{dy}{du} \cdot \frac{du}{dx} = f'(u)u' = f'(g(x))g'(x) \text{ since } u' = g'(x)$$

The Chain Rule – Alternative Method

If y is differentiable in u, and u is differentiable in x, then

$$\frac{dy}{dx} = \frac{dy}{du} \cdot \frac{du}{dx}$$

Example 5.2.2 More on the Chain Rule

Use the chain rule to obtain dy/dx when $y = u^4 + 2u + 3$ and $u = 2x^3 + 3x + 4$.

Solution:

$\frac{dy}{du} = 4u^3 + 2$ *and* $\frac{du}{dx} = 6x^2 + 3$. *Therefore,*

$\frac{dy}{dx} = \frac{dy}{du} \cdot \frac{du}{dx} = (4u^3 + 2)(6x^2 + 3)$, *which must be rewritten in terms of x.*

The result is $\frac{dy}{dx} = [4(2x^3 + 3x + 4)^3 + 2](6x^2 + 3)$

with u replaced by $2x^3 + 3x + 4$.

EXERCISES 5.2

In Exercises 1–4, compute f(g(x)) where f(x) and g(x) are the following:

1. $f(x) = x^2 + 2x + 4, g(x) = 3x - 1$
2. $f(x) = 2x^3 + 5x + 4,\ g(x) = 2x^2 + 9x - 3$
3. $f(x) = x^4 + 2x^3 + 5x, g(x) = 9x^2 + 5x + 7$
4. $f(x) = \frac{x^2 + 3}{2x + 5}, g(x) = 3x - 1$

In Exercises 5–8, express h(x) as a composite of f(x) and g(x) so h(x) = f(g(x)).

5. $h(x) = (5x + 1)^4$
6. $h(x) = (3x^2 + 2x + 1)^5 - 5(3x^2 + 2x + 1)^2 + 2$
7. $h(x) = (2x + 1)7 + \frac{2}{(2x + 1)^3}$
8. $h(x) = (4x^3 + 2x + 3)^{2/3}$

In Exercises 9–14, differentiate the functions using the chain rule.

9. $f(x) = (x^2 + 5x + 3)^{19}$
10. $f(x) = (2x^3 - 7x + 4)^8$
11. $f(x) = (5x^7 - 3x^4 + 3)^5$
12. $f(x) = \sqrt[3]{x^3 + 5x + 1}$
13. $f(x) = \sqrt[5]{2x^4 + 3x^2 + 3}$
14. $f(x) = \dfrac{4}{2x - 3}$

In Exercises 15–20, determine dy/dx using the alternative chain rule method.

15. $y = u^{5/2},\ u = 8x - 3$
16. $y = u^4,\ u = x^2 + 9x + 1$
17. $y = u(2u + 3)^5,\ u = 2x^3 + 7x + 1$
18. $y = 3u + u^{-1/2},\ u = 3x + 2$
19. $y = \dfrac{2u - 1}{u + 1},\ u = 8x^2 + 3$
20. $y = \dfrac{3u - 2}{(u + 1)^4},\ u = 10x + 7$

5.3 IMPLICIT DIFFERENTIATION AND RELATED RATES

We have noted the usefulness of the chain rule for differentiating composite functions $f(g(x))$. Sometimes, as variables are related by a function, (as in the composite function $f(g(x))$), they may also be related by an equation as in $f(x)g(y)$, say, where y is a function of x. **Implicit differentiation** arises when variables are (or can be) related.

Using the chain rule, one can determine dy/dx by differentiating **implicitly**.

Example 5.3.1 Direct Differentiation and the Chain Rule

Determine the derivative of $y = (5x^2 + 3x + 1)^5$ directly and using implicit differentiation.

Solution:
Taking the derivative directly and using the power rule yields

$$5(5x^2 + 3x + 1)^4 \frac{d}{dx}(5x^2 + 3x + 1) = 5(5x^2 + 3x + 1)^4(10x + 3)$$

Implicitly, replace $(5x^2 + 3x + 1)$ by u so that $y = u^5$. Using the chain rule,

$$\frac{dy}{dx} = \frac{dy}{du} \cdot \frac{du}{dx} = 5u^4(10x + 3) = 5(5x^2 + 3x + 1)^4(10x + 3)$$

In general, a rule for using the power rule in an implicit differentiation is:

The Implicit Differentiation Rule

$$\frac{d}{dx}(y^r) = ry^{r-1}\frac{dy}{dx}$$

Example 5.3.2 Implicit Differentiation

Use implicit differentiation to calculate $\frac{dy}{dx}$ *of* $x^3 + y^4 = 5x + 4$.

Solution:

Each term on both sides of the equation is differentiated with respect to x *to yield* $3x^2 + 4y^3\left(\frac{dy}{dx}\right) = 5$.

Next, solve for $\frac{dy}{dx}$. *Therefore,* $\frac{dy}{dx} = \frac{5 - 3x^2}{4y^3}$.

Example 5.3.3 More on Implicit Differentiation

Use implicit differentiation to obtain $\frac{dy}{dx}$ *for* $9x^2 + 2y^3 = 5y$.

Solution:

Each term on both sides of the equation is differentiated to yield $18x + 6y^2\frac{dy}{dx} = 5\frac{dy}{dx}$.

Next, solve for $\frac{dy}{dx}$. *Therefore,*

$$18x = 5\frac{dy}{dx} - 6y^2\frac{dy}{dx}$$

$$18x = (5 - 6y^2)\frac{dy}{dx}$$

$$\frac{dy}{dx} = \frac{18x}{5 - 6y^2}$$

Note that the previous two examples were relatively simple in that the variables did not appear as products or quotients. The following examples illustrate more complex situations.

Example 5.3.4 Implicit Differentiation and the Product Rule

Use implicit differentiation to calculate $\frac{dy}{dx}$ *of* $5x^2y^7 = 8x^2 + 7x + 3$.

Solution:
Each term on both sides of the equation is differentiated to yield

$$5x^2\left[7y^6\frac{dy}{dx}\right] + y^7[10x] = 16x + 7$$

Note $\frac{dy}{dx}$ *doesn't appear in the second term,* $y^7[10x]$, *as the derivative is with respect to* x.
Next, solve for $\frac{dy}{dx}$.

$$(35x^2y^6)\frac{dy}{dx} = -10xy^7 + 16x + 7$$

$$\frac{dy}{dx} = \frac{-10xy^7 + 16x + 7}{35x^2y^6}$$

Economics Application – Cobb–Douglas Function

Economists use a *Cobb–Douglas Production Function* to model, say, a manufacturing enterprise. It takes the form $Ax^{\alpha}y^{\beta} = q$, where x and y are units of capital and labor, respectively; A is a constant, and q represents the quantity of goods produced. When $\alpha + \beta = 1$, the function is said to exhibit "*constant returns to scale*". This means that doubling inputs (x and y), doubles outputs (q); tripling inputs, triples outputs, and so on (also see page 221).

Example 5.3.5 Cobb–Douglas Production Function

Calculate $\frac{dy}{dx}$ *using implicit differentiation of the production function* $80x^{1/3}y^{2/3} = 960$.

Solution:
Differentiating implicitly and using the chain rule yield

$$80x^{1/3}\left[\frac{2}{3}y^{-1/3}\frac{dy}{dx}\right] + y^{2/3}\left[\frac{80}{3}x^{-2/3}\right] = 0$$

Next, solve for $\frac{dy}{dx}$ *as*

$$80x^{1/3}\left[\frac{2}{3}y^{-1/3}\frac{dy}{dx}\right] = -y^{2/3}\left[\frac{80}{3}x^{-2/3}\right]$$

$$\frac{dy}{dx} = \frac{-y^{2/3}\left[\frac{80}{3}x^{-2/3}\right]}{\frac{160}{3}x^{1/3}y^{-1/3}} = -\frac{y}{2x}.$$

⟶

↑ *The absolute value of* $\frac{dy}{dx}$ *is called the marginal rate of substitution (MRS) of x for y. Economists regard the MRS as the rate at which one is willing to substitute one good for another without loss or gain.*

Related Rates

Many everyday quantities of interest change with time! Interesting situations arise when two or more rates of change are related to each other. Examples abound!

- The top of a ladder propped against a wall falls at one rate as the foot of the ladder moves away from the wall at a different rate.
- An airplane in flight moves at one speed while its ground shadow recedes at another rate.
- Sales of a product increase at one rate while the related advertising expenditure increases at still another rate.
- A chemist stirs at one rate while a substance dissolves in a solvent at another rate, and so on.

In each of these instances, a mathematical relation – sometimes geometric – connects the two rates. Differentiation with respect to time connects the related rates.

A series of examples helps you gain useful skills.

Example 5.3.6 Related Rates

A 25-foot ladder rests on a horizontal surface and leans against a vertical wall. If the bottom of the ladder slides from the wall at a rate of 3 ft/s, how fast is the ladder sliding down the wall when the top of the ladder is 15 ft from the ground?

Solution:

Let x be the ladder's distance from the base of the wall and y its height on the wall. Note that both x and y are actually functions of time. The length of the ladder being fixed, the variables x and y are related geometrically by a right triangle as

$$x^2 + y^2 = 25^2 = 625$$

As we are interested in a time-related rate, we differentiate with respect to time, t, even as it does not appear explicitly.

$$2x\frac{dx}{dt} + 2y\frac{dy}{dt} = 0 \quad \text{or equivalently,}$$

$$x\frac{dx}{dt} + y\frac{dy}{dt} = 0$$

⟶

At an instant when y is 15 ft, x is 20 ft. At this instant, the rate at which the ladder moves from the wall is $\frac{dx}{dt} = 3ft/\text{s}$. *To solve for* dy/dt, *the rate at which it moves down the wall substitute,*

$$(20)(3) + (15)\frac{dy}{dt} = 0$$

$$\frac{dy}{dt} = -4$$

At the instant that $x = 20ft$ *and* $y = 15ft$, *the ladder is moving down the side of the wall at a rate of 4 ft/s.*

Example 5.3.7 A Water Tank

A water tank in the shape of a right circular cone has radius 8 ft and height 16 ft. Water is pumped into the tank at a rate of 20 gallons per minute. Clearly, the rate at which water is pumped and its depth in the tank are related. At what rate is the water level rising when it is 2 ft deep ($1\ gal \approx 0.134 ft^3$)*?*

Solution:

Let r be the radius at the water surface when its depth is h. The volume of a right circular cone is $V = 1/3\ \pi r^2 h$. *As* r *and* h *are functions of time, t, we seek* $\frac{dh}{dt}$, *the rate at which the water level is rising.*

For this cone, $r = h/2$ *as the ratio of r to* h *is 8/16. When the depth is* h, *the volume is* $V = \frac{1}{3}\ \pi\left(\frac{h}{2}\right)^2 h = \frac{1}{12}\pi h^3$.

Next, differentiation with respect to time, t, yields $\frac{dV}{dt} = \frac{1}{4}\pi h^2 \left(\frac{dh}{dt}\right)$.

From the data, the change in volume,

$\frac{dV}{dt} = (20\,gals/min)(0.134 ft^3/gal) = 2.68 ft^3/min$ *when* $h = 2ft$.

Solving $\frac{dV}{dt}$ *for* $\frac{dh}{dt}$ *yields*

$$\frac{dh}{dt} = \frac{4}{\pi h^2}\left(\frac{dV}{dt}\right) = \frac{4}{\pi(2ft)^2}(2.68 ft^3/min) = 0.853 ft/min$$

↓ To solve related rate problems, a geometric or other relation between the variables is necessary. Then obtain expressions for the derivatives at an arbitrary time and substitute for the instantaneous values of variables.

EXERCISES 5.3

In Exercises 1–14, suppose that x and y are related as indicated. Calculate dy/dx implicitly.

1. $4x^2 + 9y^2 = 25$
2. $2x^3 - y^2 = 1$
3. $2x^3 + 7y^4 = 3x + 7$
4. $x^3 - 2x^2 = y^2 + 1$
5. $x^2 - 2x + y^3 = 3y + 7$
6. $\dfrac{1}{x^3} + \dfrac{1}{y^2} = 1$
7. $\dfrac{1}{x^2} + \dfrac{3}{y^4} = 14x$
8. $xy = 6$
9. $x^2y^2 = 18x$
10. $x^2y + 3x = 6y$
11. $x^3y^5 + 2x = 6x^2$
12. $3x^2 - 2xy + 4y^3 = x^3$
13. $4xy^3 - x^2y + x^3 + 3 = 9x$
14. $5x - y = 3y^2 + 2y^3 + 7x^5$

In Exercises 15–19, obtain the indicated slope by differentiating implicitly.

15. $xy + 15 = 0$ at $(-3, 5)$
16. $x^2 + y^2 = 169$ at $(12, 5)$
17. $y^4 + 2y - 3x^3 = 2x + 8$ at $(-1, 1)$
18. $\sqrt[3]{x} + \sqrt{y} = 3$ at $(8, 1)$
19. $xy + y^3 = 18$ at $(5, 2)$
20. Determine the tangent to $x^4y^2 = 4$ at $(1, 2)$ and $(1, -2)$
21. Determine the tangent to $xy^4 = 48$ at $(3, 2)$

In Exercises 22–25, x and y are differentiable functions of time, t, and are related by the given equation. Use implicit differentiation to find $\dfrac{dy}{dt}$ in terms of x, y, and $\dfrac{dx}{dt}$.

22. $x^2 + y^2 = 25$
23. $x^4 + y^4 = 17$
24. $5x^2y - 3y = 2x + 5$
25. $x^2 + 4xy = 7x + y^2$
26. A 13-foot ladder rests against a wall. If the bottom of the ladder slides horizontally from the wall at 2 ft/s, how fast is the top of the ladder sliding down the wall when it is 12 ft above the ground?
27. Air is pumped into a spherical balloon at a rate of $3\,\text{ft}^3/\text{min}$. Find the rate at which the radius is changing when its diameter is 12 inches, assuming constant pressure.
28. A kite is flying at a height of 60 ft. Wind carries the kite horizontally away from an observer at a rate of 2 ft/s. At what rate must the string be released when the kite is 100 ft away?

29. The current, I, (in amperes) in an electrical circuit is $I = 100/R$ where R is the resistance (in ohms). Find the rate of change of I with respect to R when the resistance is 5 ohms.

30. Suppose that the price p (in dollars) and the monthly sales x (in thousands of units) of a commodity satisfy $3p^3 + 2x^2 = 3800$. Determine the rate at which sales are changing at a time when $x = 20$, $p = 10$ and the price is falling at a rate of \$1 per month.

31. Consider the Cobb–Douglas Production Function, $20x^{1/4}y^{3/4} = 1080$. Use implicit differentiation to calculate $\frac{dy}{dx}$ when $x = 16$ and $y = 81$.

5.4 FINITE DIFFERENCES AND ANTIDIFFERENCES

Chapter 2 concluded with a section on finite differences. Recall that the finite difference of a function $f(x)$, $x = 0, 1, \ldots$ is defined by

$$\Delta f(x) = f(x+1) - f(x) \quad x = 0, 1, \ldots$$

↓ Note that we have defined $f(x)$ only for the non-negative integers $x = 0, 1, \ldots$

This section extends these two related calculi!

◆ The finite calculus, or finite differences as it is sometimes known, was a precursor to the differential calculus (the subject of this text). They differ in that the "step size" in the finite calculus is discrete or finite rather than infinitesimal as in the differential calculus.

The aforementioned definition uses a unit change in x, say x to $x + 1$. In differential calculus, the limit has the step size approaching zero – an instantaneous change.

Differencing

Differencing is the finite calculus analogue of differentiation. We begin with an example.

Example 5.4.1 First Finite Difference

If $f(x) = x^2$, find $\Delta f(x)$.

Solution:

$$\Delta f(x) = (x+1)^2 - x^2 = 2x + 1$$

Similarly, one defines a **second difference**, analogous to a second derivative, as

$$\begin{aligned}\Delta^2 f(x) &= \Delta(\Delta f(x)) = \Delta(f(x+1) - f(x)) \\ &= [f(x+2) - f(x+1)] - [f(x+1) - f(x)] \\ &= f(x+2) - 2f(x+1) + f(x)\end{aligned}$$

♦ Note that the use of Δ as a "difference" differs from its frequent use as "a small change" in many calculus texts. Some texts attempt to overcome possible confusion by indicating a first difference by a capital E as in

$$E(f(x)) = f(x+1) - f(x)$$

However, that choice is not without possible confusion.

Example 5.4.2 Second Finite Difference

If $f(x) = x^2$ *and* $\Delta f(x) = 2x + 1$ *find* $\Delta^2 f(x)$

Solution:

$$\begin{aligned}\Delta^2 f(x) &= \Delta(\Delta f(x)) = [2(x+1) + 1] - (2x + 1) \\ &= 2\end{aligned}$$

Alternately, using the second difference formula directly,

$$\begin{aligned}\Delta^2 f(x) &= f(x+2) - 2f(x+1) + f(x) \\ &= (x+2)^2 - 2(x+1)^2 + x^2 \\ &= x^2 + 4x + 4 - 2x^2 - 4x - 2 + x^2 \\ &= 2\end{aligned}$$

Example 5.4.3 First and Second Finite Differences (cont.)

If $f(x) = a^x$, *find* $\Delta f(x)$ *and* $\Delta^2 f(x)$

Solution:

$$\Delta f(x) = a^{x+1} - a^x = a^x(a - 1)$$

$$\Delta^2 f(x) = \Delta(\Delta f(x)) = \Delta(a^x(a - 1))$$

⟶

$$= a^{x+1}(a-1) - a^x(a-1)$$
$$= a^x(a-1)\ [a-1]$$
$$= a^x(a-1)^2$$

Generally, the $r + 1^{st}$ difference of $f(x)$ is defined as

$$\Delta^{r+1}f(x) = \Delta(\Delta^r f(x)) = \Delta^r[f(x+1) - f(x)] = \Delta^r f(x+1) - \Delta^r f(x)\ r,\ \text{a positive integer}$$

where $\Delta^r f(x)$ is read as "delta to the r of $f(x)$" or the "r^{th} difference of $f(x)$". Note that $\Delta^1 = \Delta$.

It is easily shown that

$$\Delta(f(x) \pm g(x)) = \Delta f(x) \pm \Delta g(x)$$

and

$$\Delta(f(x)g(x)) = f(x+1)g(x+1) - f(x)g(x)$$

Adding and subtracting $f(x)g(x+1)$ to the right hand side above yield

$$f(x+1)g(x+1) - f(x)g(x+1) + f(x)g(x+1) - f(x)g(x)$$
$$= g(x+1)\Delta f(x) + f(x)\Delta g(x)$$

Clearly,

$\Delta(c(f(x))) = c\Delta f(x)$ where c is a constant.

For the difference of the quotient,

$$\Delta\left(\frac{f(x)}{g(x)}\right) = \frac{f(x+1)}{g(x+1)} - \frac{f(x)}{g(x)} = \frac{g(x)f(x+1) - f(x)g(x+1)}{g(x+1)g(x)}$$
$$= \frac{g(x)f(x+1) - g(x)f(x) + g(x)f(x) - f(x)g(x+1)}{g(x+1)g(x)}$$
$$= \frac{g(x)\Delta f(x) - f(x)\Delta g(x)}{g(x+1)g(x)}$$

Note the striking similarity of these results to those in the differential calculus.

Example 5.4.4 Difference Operations

Let $f(x) = x^2 + 3x + 5$ *and* $g(x) = 3^x$.

Calculate $\Delta[f(x) \pm g(x)]$, $\Delta[f(x)g(x)]$, *and* $\Delta[f(x)/g(x)]$.

→

Solution:
Firstly, calculate $\Delta f(x)$ and $\Delta g(x)$.

$$\Delta f(x) = [(x+1)^2 + 3(x+1) + 5] - [x^2 + 3x + 5] = 2x + 4$$
$$\Delta g(x) = 3^{x+1} - 3^x = 3^x(3-1) = 2(3^x)$$

Now, using the expressions in the text,

$$\begin{aligned}
\Delta[f(x) \pm g(x)] &= \Delta[f(x)] \pm \Delta[g(x)] = (2x+4) \pm 2(3^x) \\
\Delta[f(x)g(x)] &= g(x+1)\Delta f(x) + f(x)\Delta g(x) \\
&= [3^{x+1}](2x+4) + (x^2+3x+5)[2(3x)] \\
&= 3^x[3(2x+4) + 2(x^2+3x+5)] \\
&= 3^x[2x^2 + 12x + 22] \\
&= 3^x(2)[x^2 + 6x + 11]
\end{aligned}$$

and

$$\begin{aligned}
\Delta[f(x)/g(x)] &= \frac{g(x)\Delta f(x) - f(x)\Delta g(x)}{g(x+1)g(x)} \\
&= \frac{3^x(2x+4) - (x^2+3x+5)(2)3^x}{3^{x+1}3^x} \\
&= \frac{3^x[(2x+4) - 2(x^2+3x+5)]}{3^{x+1}3^x} \\
&= \frac{-2x^2 - 4x - 6}{3^{x+1}} = \frac{-2(x^2+2x+3)}{3^{x+1}}
\end{aligned}$$

Maxima and Minima

Discrete functions have maxima and minima just as continuous ones. They are the respective largest and smallest values in a region. Locating maxima and minima is usually more arduous in the discrete case, although the basic principle is the same as for differential calculus.

A minimum of $f(x)$ occurs at x^* if

$$f(x^* + 1) > f(x^*) \text{ and } f(x^* - 1) > f(x^*).$$

That is, the function value is larger than $E(x^*)$ on each side of x^*. This can also be expressed as

$$\Delta f(x^* - 1) < 0 < \Delta f(x^*).$$

If $f(x^*) < f(x)$ for all x, then the minimum is local.
If $f(x^*) \leq f(x)$ for all x, then the minimum is absolute or global.

We state a sufficient, but not necessary, condition for a minimum:
$f(x^* - 1) < 0 < f(x^*)$ and $\Delta^2 f(x) \geq 0$ for all x.

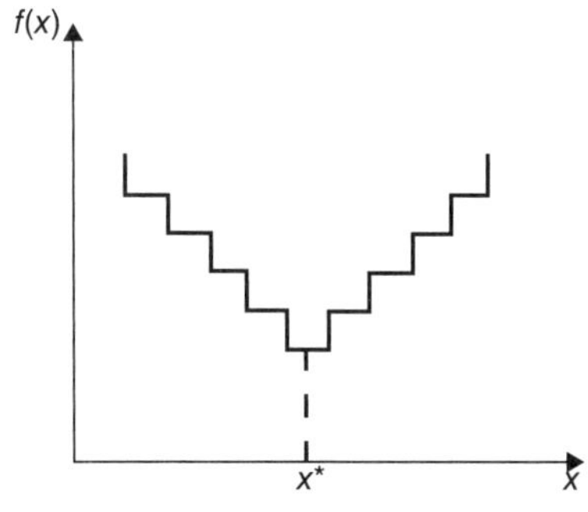

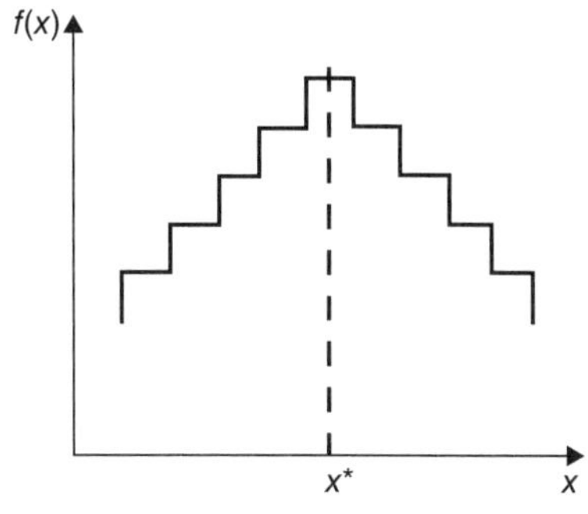

A similar analysis applies for a maximum of $f(x)$. There is a maximum at x^* if

$$f(x^*) > f(x^* + 1) \quad \text{and} \quad f(x^*) > f(x^* - 1)$$

These can be expressed as

$$\Delta f(x^*) < 0 < f(x^* - 1).$$

This is equivalent to a sufficiency condition for a maximum, $\Delta^2 f(x) \leq 0$ for all x. Note the similarity of the second difference to the second derivative as a sufficiency condition.

Example 5.4.5 Finding Maxima and Minima

Examine $f(x) = 3x^2 - 4x + 1$, $x = 0, 1, 2, \ldots$ for maxima and minima.

Solution:
In this case,

$$\begin{aligned}\Delta f(x) &= [3(x+1)^2 - 4(x+1) + 1] - [3x^2 - 4x + 1]\\ &= [3x^2 + 6x + 3 - 4x - 4 + 1] - [3x^2 - 4x + 1]\\ &= 6x - 1\end{aligned}$$

By trial and error,

$$f(0) = 1, \quad f(1) = 0, \quad f(2) = 5, \quad f(3) = 16$$

so

$$\Delta f(0) = -1, \qquad \Delta f(1) = 5, \qquad \text{and} \qquad \Delta f(2) = 11$$

The change of sign in $\Delta f(x)$ is a signal for passing through a maximum or a minimum. In this case, $\Delta f(0) < 0 < \Delta f(1)$ so $x^* = 1$ locates the minimum of $f(x)$, $f(x^*) = f(1) = 0$.

Economic Applications – Inventory Policy

A significant concern, especially in manufacturing enterprises, is the maintenance of economically defensible inventories. Too much stock and costs rise for warehousing, deterioration, idle capital, and obsolescence. Stock too little and costs increase for production delays, lost sales, and idle workers. It is a vital managerial function to specify the trade-off of these "opposing" costs.

The figure illustrates Inventory vs. Time. An initial inventory, Q, is used to satisfy a fixed demand of R items per unit of time, t, ($t = 0, 1, 2, \dots, T$) until depletion at $t = T$ when the cycle repeats.

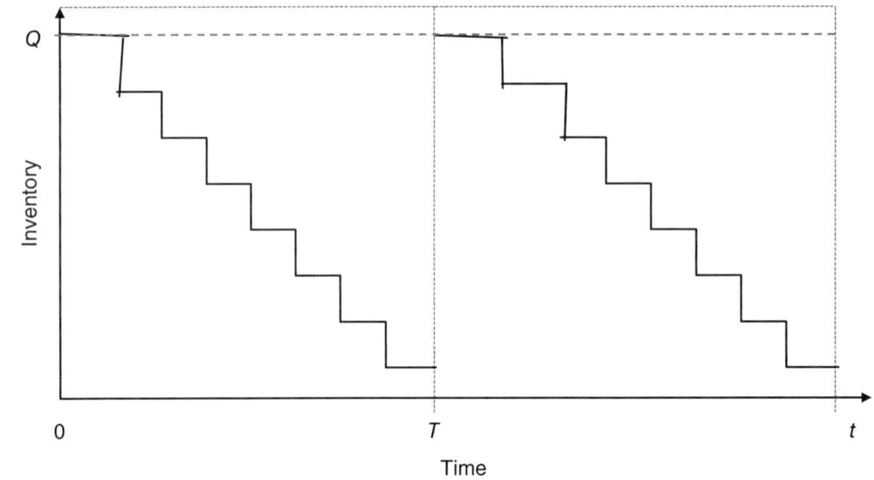

There is a "holding cost" to maintain and store inventory. It usually varies directly with the amount and duration of inventory held. There is also a "set up cost" to initiate a cycle to produce inventory (if there were no setup cost in this example, there would be no need to hold inventory as items could be manufactured to meet demand).

The trade-off is clear! If holding costs are reduced by shorter production runs, the need for more frequent set ups increases and vice versa. Management seeks a balance that minimizes the total cost per unit time.

As the inventory varies uniformly from Q at replenishment to zero on depletion at time T, the average inventory is $Q/2$. Let C_{h} be the holding cost per unit of inventory per unit time. The holding cost in a cycle is $C_{\mathrm{h}}QT/2$. Also, let C_{s} be the setup cost per cycle.

The total cost per cycle is $C_{\mathrm{s}} + C_{\mathrm{h}}QT/2$. The total cost per unit time, Z, is

$$Z = \frac{C_{\mathrm{s}}}{T} + C_{\mathrm{h}}\frac{Q}{2}$$

The total cost per unit time, Z, can also be written as

$$Z = \frac{C_{\mathrm{s}}R}{Q} + \frac{C_{\mathrm{h}}Q}{2}.$$

The quest for a minimum begins with a first difference of Z. That is,

$$\begin{aligned}\Delta Z &= \frac{C_s R}{Q+1} + \frac{C_h}{2}(Q+1) - \frac{C_s R}{Q} - \frac{C_h Q}{2} \\ &= C_s R\left[\frac{Q-(Q+1)}{Q(Q+1)}\right] + \frac{C_h}{2} \\ &= \frac{-C_s R}{Q(Q+1)} + \frac{C_h}{2}\end{aligned}$$

Setting $\Delta Z = 0$ and solving the resulting quadratic yield the roots

$$Q = \frac{-1 \pm \sqrt{1 + \dfrac{8RC_s}{C_h}}}{2}$$

Clearly, the negative root is extraneous.

♦ Inventory policies are among the most important in the profitability of manufacturing enterprises. Large sums are spent in devising better inventory policies and it is the subject of a large and sophisticated literature. There are actually professional societies for inventory-related personnel as a web search will attest. These examples only scratch the surface.

Example 5.4.6 Production Run

Suppose that the cost of initiating a production run is $C_s = \$100$ and the unit holding cost is $C_h = \$1$ item/day. The demand rate is $R = 2$ items/day. Devise an optimal inventory policy.

Solution:

Use the positive root for Q, above, to locate the approximate value,

$$\frac{-1 + \sqrt{1 + \dfrac{(8)(2)(100)}{1}}}{2} = 19.5$$

Next, calculate the value of Z at the nearest integer values.

$$\text{At } Q = 20,\quad \Delta Z = \frac{-100}{(20)(21)} + \frac{1}{2} = 0.262 \text{ and}$$

$$\text{at } Q = 19,\quad \Delta Z = \frac{-100}{(19)(20)} + \frac{1}{2} = -0.263$$

⟶

The change in sign of ΔZ, as Q varies from 19 to 20, signals the passage through a minimum in this case.

$$Z(19) = \frac{100(2)}{19} + \frac{(1)(19)}{2} = 20.03\,\$/day$$

$$Z(20) = \frac{100(2)}{20} + \frac{(1)(19)}{2} = 20.00\,\$/day$$

There is little choice between the two policies, $Q = 19$ and $Q = 20$ in this instance.

Antidifferences

Antidifferences, as the name suggests, is the inverse of differencing. It is analogous to the *integral,* the subject of the next chapter when the study of the "other" calculus resumes. This finite counterpart usefully introduces the next chapter.

The antidifference of $f(x)$, $F(x)$, is defined by $f(x) = \Delta F(x)$. It is also expressed as $F(x) = \Delta^{-1}f(x)$ – the symbol Δ^{-1} denoting an antidifferencing operation. In words, the difference of the antidifference yields the original function.

♦ Caution! The -1 in Δ^{-1} is not an exponent! It is a commonly used mathematical notation signifying an inverse.

Example 5.4.7 Finding Antidifferences

Find the antidifferences of a^x and the permutation ${}_xP_r$.

Hint: ${}_xP_r = \dfrac{x!}{(x-r)!}$

Solution:

We have already noted that $\Delta a^x = (a-1)a^x$ is the first finite difference of a^x so $\Delta^{-1}a^x = \dfrac{a^x}{(a-1)}$.

For the permutation $f(x) = {}_xP_r = x(x-1)\cdots(x-(r-1))$

$$\begin{aligned}\Delta_x P_r &= [(x+1)(x)\cdots(x+1-(r-1))] - [x(x-1)\cdots(x-(r-1))]\\ &= [(x+1)-(x-r+1)]((x)(x-1)\cdots(x+1-(r-1))\\ &= r({}_xP_{r-1})\end{aligned}$$

Therefore $\Delta_x P_{r+1} = (r+1)_x P_r$ *so*

$$\Delta^{-1}{}_xP_r = \frac{{}_xP_{r+1}}{r+1}$$

Antidifferences are particularly useful when summing series and summations. Consider the sum $\sum_{x=a}^{b} f(x) = f(a) + f(a+1) + f(a+2) + \ldots + f(b-1) + f(b)$. Substituting, the antidifference, $F(x)$ yields

$$\begin{aligned}\sum_{x=a}^{b} f(x) &= \sum_{x=a}^{b} [F(x+1) - F(x)] \\ &= F(a+1) + F(a+2) + \ldots + F(b) + F(b+1) - F(a) \\ &\quad - F(a+1) - \ldots - F(b-1) - F(b) \\ &= F(b+1) - F(a)\end{aligned}$$

Again, you will note the similarity to the integral calculus in the next chapter.

Example 5.4.8 Sum of the First n *Positive Integers*

Find the sum of the first n *positive integers.*

Solution:

We seek $\sum_{x=1}^{n} x$. *To evaluate the summation,* $F(x)$, *by antidifferences, seek an antidifference* $F(x)$ *such that* $\Delta F(x) = x$. *As* $F(x) = x$ *is a polynomial, try* $F(x) = ax^2 + bx + c$. *The coefficients* a, b, c *are determined so that* $\Delta F(x) = x$. *That is*

$$\Delta F(x) = [a(x+1)^2 + b(x+1) + c] - [ax^2 + bx + c] = 2ax + a + b$$

For $\Delta F(x)$ *to equal* x *requires* $2ax = x$ *and* $a = 1/2$. *With* $c = 0, a + b = 0$ *and* $a = -b = 1/2$.

Therefore,

$$\sum_{x=1}^{n} x = \sum_{x=1}^{n} \Delta\left(\frac{x^2 - x}{2}\right) = \frac{(n+1)^2 - (n+1)}{2} - \frac{1-1}{2} = \frac{n(n+1)}{2}$$

You may recall this well-known algebraic result.

Summation by Parts

In many applications, challenging summations arise. Evaluating them can sometimes be simplified using a **summation by parts**.

Consider evaluation of a sum

$$\sum_{x=a}^{b} f(x)h(x)$$

The "trick" is to know an antidifference of either $f(x)$ or $h(x)$. Let us suppose that $h(x) = \Delta g(x)$ where $g(x)$ is the antidifference of $h(x)$. Now, we seek to evaluate

$$\sum_{x=a}^{b} f(x)\Delta g(x)$$

Earlier we had

$$\sum_{x=a}^{b} \Delta[f(x)g(x)] = \sum_{x=a}^{b} f(x)\Delta g(x) + \sum_{x=a}^{b} g(x+1)\Delta f(x)$$

Also, it can be shown (Exercise 5.4.12) that

$$\sum_{x=a}^{b} \Delta[f(x)g(x)] = f(b+1)g(b+1) - f(a)g(a)$$

Substituting on the left side, rearranging, and solving for the term of interest

$$\sum_{x=a}^{b} f(x)\Delta g(x) = f(b+1)g(b+1) - f(a)g(a) - \sum_{x=a}^{b} g(x+1)\Delta f(x)$$

Now, the first two terms require no summation and are easily obtained by simple substitution. The last term does require a summation, and the central purpose in using a summation by parts is that this remaining summation is easier than the original one.

You will encounter the same idea in the integral calculus (see Chapter 7) where it is known as **integration by parts**.

Example 5.4.9 Summation by Parts

Evaluate $\sum_{x=1}^{3} x^2 \Delta x^3$.

Solution:

Using the expression for a summation by parts:

$$\sum_{x=1}^{3} x^2 \Delta x^3 = (3+1)^2(3+1)^3 - (1)^2(1)^3 - \sum_{x=1}^{3} \{(x+1)^3[(x+1)^2 - x^2]\}$$
$$= 1024 - 1 - [8(3) + 27(5) + 64(7)] = 416$$

Enumerating, as a check,

$$\sum_{x=1}^{3} x^2 \Delta x^3 = \sum_{x=1}^{3} \{x^2[(x+1)^3 - x^3]\} = \sum_{x=1}^{3} [3x^4 + 3x^3 + x^2]$$
$$= (3+3+1) + (48+24+4) + (243+81+9) = 416$$

Example 5.4.10 Another Summation by Parts

Evaluate $\sum_{x=1}^{k} xa^x$.

Solution:

The summation by parts formula does not apply directly. In an earlier example, we found that $\Delta a^x = (a-1)a^x$ *so* $\Delta(a^x/(a-1)) = a^x$. *Substituting for* a^x.

$$\sum_{x=1}^{k} xa^x = \sum_{x=1}^{k} x\left(\frac{\Delta a^x}{a-1}\right) = \left(\frac{1}{a-1}\right)\sum_{x=1}^{k} x\Delta a^x$$

Set $f(x) = x$ *and* $g(x) = a^x$ *and the summation by parts expression applies!*

$$\left(\frac{1}{a-1}\right)\sum_{x=1}^{k} x\Delta a^x = \left(\frac{1}{a-1}\right)\left[f(k+1)\,g(k+1) - f(1)g(1) - \sum_{x=1}^{k} g(x+1)\Delta f(x)\right]$$

$$= \left(\frac{1}{a-1}\right)\left[(k+1)\,a^{k+1} - a - \sum_{x=1}^{k} g(x+1)\Delta f(x)\right]$$

$$= \left(\frac{1}{a-1}\right)\left[(k+1)\,a^{k+1} - a - \sum_{x=1}^{k} a^{x+1}\right]$$

$$= \left(\frac{1}{a-1}\right)\left[(k+1)\,a^{k+1} - a - a\left(\frac{a\left(1-a^k\right)}{1-a}\right)\right]$$

Note that $\Delta f(x) = \Delta x = (x+1) - x = 1$ *and that the last summation was evaluated as a geometric sum.*

EXERCISES 5.4

1. If $f(x) = x^3$ find $\Delta f(x)$ and $\Delta^2 f(x)$.
2. Verify that $\Delta\left(\frac{a^x}{a-1}\right) = a^x$.
3. Repeat Example 5.4.4 using an algebraic operation before differencing.
4. Find $\Delta(1/(f(x)))$.
5. For the permutation ${}_nP_r = n(n-1)\ \ldots\ (n-r+1)$, a product of r factors, show that $\Delta_n P_r = (r)[{}_nP_{r-1}]$ for fixed r.

6. Show that

a) $\Delta\{x\} = 1$ b) $\Delta \binom{n}{r} = \binom{n}{r-1}$ c) $\Delta(a^n) = a^n(a-1)$

7. Show that the condition for a maximum at x^* is

$$\Delta f(x^*) \le 0 \le \Delta f(x^* - 1).$$

8. Repeat Example 5.4.9 by direct differencing.

9. *Summation by parts*

Verify the expression in the text when $a = 1$ and $b = 3$ starting with the differencing of $\sum_{x=a}^{b} f(x)g(x)$ for arbitrary $f(x)$ and $g(x)$.

10. Find the antidifferences of

a) 8^x b) ${}_nP_3$ c) x^4 d) $\binom{n}{r-1}$

11. Use antidifferences and summations to evaluate

a) $\sum_{x=1}^{n} x^2$ b) $\sum_{x=1}^{n} a^x$ c) $\sum_{x=1}^{n} x^3$

12. Prove that $\sum_{x=a}^{b} \Delta[f(x)g(x)] = f(b+1)g(b+1) - f(a)g(a)$. Hint: carry out and expand the first difference.

CHAPTER 5 SUPPLEMENTARY EXERCISES

In Exercises 1–12, differentiate the given functionsc

1. $f(x) = (5x^3 - 2x + 1)^4(2x + 3)$
2. $f(x) = (2x^5 - 3x^2 + 10x + 10)^4(3x^4 + 2)^5$
3. $f(x) = (2x^4 - 3x^2 + e^{3x})^2(2x + 1)^3$
4. $f(x) = (2x^5 - 3x)^7(2 + \ln 3x)^4$
5. $f(x) = \left(\dfrac{5x+1}{3x-5}\right)^{10}$
6. $f(x) = \dfrac{(5e^{7x+1} + 3)^4}{(2x^5 - 7x + 1)^3}$
7. $f(x) = \dfrac{(2x + 3 + \ln 4x)^8}{(25x + 3)^5}$
8. $f(x) = \dfrac{x^2e^{4x}}{(3x + 1)^7}$

9. $f(x) = \dfrac{(x^3 + 5x + 1)^3 \ln 3x}{(4x^2 - 11x - 9)^6}$

10. $f(x) = \left(\dfrac{x^5 \ln 2x}{(5x + 3)} \right)^{50}$

11. $f(x) = (3x + 1)^4 \ln(5x^3 - 9x^2 + 2x + 1)$

12. $f(x) = \dfrac{e^{x^3+5x+1}}{(9x^7 + 3x^2 + 4)^5}$

13. Determine the equation of the tangent line to $f(x) = (3x - 5)^7(5x + 1)$ at $x = 2$.

14. Determine the equation of the tangent line to $f(x) = (6x - 5)^{5/2}e^{x-1}$ at $x = 1$.

15. Determine the equation of the tangent line to $f(x) = (x + 1)^5 e^{3x}$ at $x = 0$.

16. If $y = u^5$ and $u = x^3 + 2x + 3$, find dy/dx using a chain rule.

17. If $y = 2u^7$ and $u = 3x^5 - 9x^2 + 5x + 10$, find dy/dx using a chain rule.

18. If $y = 3u^4$ and $u = 5x^2 + 9x + 3$, find dy/dx using a chain rule.

19. Suppose $f(x) = x^{10}$ and $g(x) = x^7 + 2x^5 + x^2 + 1$
    a) Determine $f(g(x))$.
    b) Use the chain rule to determine the derivative of $f(g(x))$.
    c) Verify by differentiating part (a) directly, the result is the same.

20. Use implicit differentiation to calculate dy/dx when y is related to x by

$$3x^9y^3 + 3x^4 = 6y^2 + 11x$$

21. Use implicit differentiation to calculate dy/dx when y is related to x by

$$3x^4y^3 = 5y^2 + 6x^2 + 3$$

22. Suppose that x and y are differentiable functions of time, t, and related by $2y^3 - x^2 = 15$, solve for dy/dt in terms of x, y, and dx/dt.

23. Boyle's law for an ideal gas at constant temperature is $pv = c$ where p is the pressure; v the volume; and c a constant. When the volume is 75 in^3, the pressure is 30 lbs/in^2 and the volume is increasing at a rate of 5 in^3/min. At what rate is the pressure decreasing at that instant?

24. A cubical block of ice melts at a constant rate. If its edge changed from 12″ to 8″ in one-half hour, how fast is the ice melting?

25. If $f(x) = x^4$ find $\Delta f(x)$ and $\Delta^2 f(x)$.

6 *Integral Calculus*

INTRODUCTION

The calculus is composed of two main parts: the **differential calculus**, the subject of prior chapters, and the **integral calculus**, the subject of this chapter.

The mathematical operation of **integration** arises in two contexts. One, as the opposite of differentiation – called **antidifferentiation** or **antiderivatives**. The other, as the area under a graph of a function – "an area under a curve!". We briefly illustrate both as a preliminary to their study.

Antidifferentiation is essentially synonymous with integration. If $3x^2$ is the derivative of x^3, for example, then x^3 is the antiderivative of $3x^2$. More precisely, $x^3 + C$, C a constant, is the antiderivative of $3x^2$. This follows as the derivative of a constant is zero.

Fundamentals of Calculus, First Edition. Carla C. Morris and Robert M. Stark.

Companion Website: http://www.wiley.com/go/morris/calculus

↓ This is an important point! If the derivative of a function is, say, $3x^2$, clearly the function must have x^3 as its principal part. However, it could have been $f(x) = x^3 + 5$ (or any other constant than 5) as the derivative of any constant is zero. Without other information, we must allow for the constant in the antiderivative.

Introductory Example   Antiderivative of 2x

What is the antiderivative of 2x?

Solution:

Recall, $f(x)=2x$ is the derivative of x^2, so x^2+C is the antiderivative of 2x, C being an arbitrary constant.

Not all antiderivatives are as easily arrived at as $3x^2$ and $2x$, as shown previously. Systematic techniques for antidifferentiation (integration) are developed in this and subsequent chapters.

Antiderivative as an Area Under a Curve

Next, we show the relation of the antiderivative to an area under a curve. The following figure is a plot of $f(x) = 3x^2$ for $0 \leq x \leq 1$ divided into two sections $0 \leq x \leq 1/2$ and $1/2 \leq x \leq 1$.

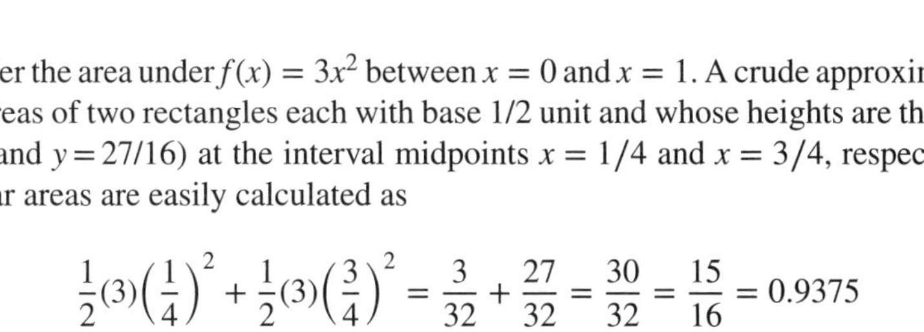

Consider the area under $f(x) = 3x^2$ between $x = 0$ and $x = 1$. A crude approximation is to sum the areas of two rectangles each with base 1/2 unit and whose heights are the ordinates ($y=3/16$ and $y=27/16$) at the interval midpoints $x = 1/4$ and $x = 3/4$, respectively. The rectangular areas are easily calculated as

$$\frac{1}{2}(3)\left(\frac{1}{4}\right)^2 + \frac{1}{2}(3)\left(\frac{3}{4}\right)^2 = \frac{3}{32} + \frac{27}{32} = \frac{30}{32} = \frac{15}{16} = 0.9375$$

↑ This crude approximation is about 94% correct as the exact area is one square unit. As the antiderivative of $3x^2$ is x^3, it is the area under the curve of $f'(x) = 3x^2$.

Note that the estimate of the area can be improved by increasing the number of approximating rectangles. This is an important insight! The following figure approximates the area using four such rectangles.

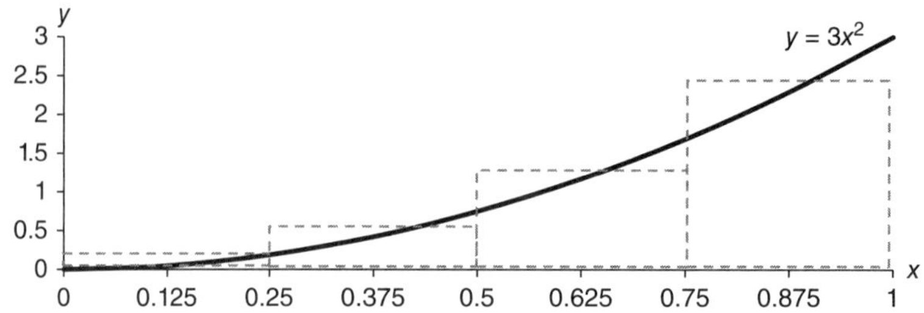

A similar calculation for the sum of the rectangular areas is $\frac{63}{64} \approx 0.9844$ – more than 98% of the actual area of 1 square unit. The estimate improves with increasing numbers of rectangles to approximate the area under the curve. These rectangular approximations, known as **Riemann Sums**, arise in Section 6.2.

It is a remarkable fact that the actual area is the antiderivative. In this case, the antiderivative of $3x^2$ is x^3. The evaluation between $x = 0$ and $x = 1$ is expressed as $x^3\big|_0^1 = (1)^3 - (0)^3 = 1$. It is the exact area under the curve $y = 3x^2$ (and above the x-axis) between $x = 0$ and $x = 1$.

6.1 INDEFINITE INTEGRALS

Antidifferentiation or, synonymously, **integration** is a reversal operation, as it returns the original function. That is, if $f'(x)$ is the derivative, $f(x)$ is its antiderivative.

For instance, if a derivative is $3x^2$, its antiderivative (or integral), is written as

$$\int 3x^2 dx = x^3 + C$$

The elongated "S" being the famous integral sign signifying an **integration** operation. The result has an arbitrary constant, C, and "dx" is an infinitesimal element of x.

The derivative of x^3, $x^3 - 10$, $x^3 + 3$, $x^3 - 7$, for examples, are $3x^2$. This follows because the derivative of a constant vanishes. It follows in seeking the antiderivative of $3x^2$ to allow for the possibility of a constant in the original function. That is

$$\int 3x^2 dx = x^3 + C$$

More generally,

$$\int f(x)dx = F(x) + C$$

where $F(x)$ is the antiderivative of $f(x)$.

The function $f(x)$ is called the **integrand**, C the constant of integration, $F(x)$ the integration result, and dx an element of the integration variable. This integration is called **indefinite integration** because the value of C is arbitrary without additional information.

↓ We follow frequent custom and use the letter "x" as the integration variable. Other letters are common, as examples, $\int 3v^2 dv = v^3 + C$, $\int 3t^2 dv = t^3 + C$, and so on.

It is easy to verify whether a resulting integration is correct: its derivative must match the integrand! That is,

$$\frac{d}{dx}(F(x) + C) = \frac{d}{dx}\left[\int f(x)\,dx\right] = f(x)$$

In words, the derivative of an integral returns the integrand. As a simple instance, integrate $2x$ as

$$\int 2xdx = x^2 + C$$

As a check, hardly necessary in this instance, the derivative of $x^2 + C$ is $2x$; matching the integrand.

↓ The derivative has the effect of "reversing" the integration.

Some basic rules of integration are tabulated as follows. You can check the results by differentiation.

Some Integration Rules

1. $\int rdx = rx + C$
2. $\int x^r dx = \frac{x^{r+1}}{r+1} + C \quad r \neq -1$
3. $\int \frac{1}{x} dx = \ln|x| + C$
4. $\int e^{rx} dx = \frac{e^{rx}}{r} + C$
5. $\int rf(x)dx = r\int f(x)dx$
6. $\int (f+g)(x)dx = \int f(x)dx + \int g(x)dx$

r and C are constants

To help integrate expressions with radicals or a variable in the denominator, rewrite them with fractional or negative exponents to differentiate, as in the following example.

Example 6.1.1 Integration

Integrate the following:

a) $\int 5dx$ b) $\int e^{3x}dx$ c) $\int \sqrt{x}dx$ d) $\int (1/x^4)dx$ e) $\int 7x^2dx$

Solution:

a) *Using Rule 1,* $\int 5dx = 5x + C$.
b) *Using Rule 4,* $\int e^{3x}dx = (e^{3x}/3) + C$.

⟶

c) Rewriting and using Rule 2 yields, $\int x^{1/2}dx = \frac{x^{3/2}}{3/2} + C$ or $\frac{2}{3}x^{3/2} + C$.

d) Rewriting and using Rule 2 yields, $\int x^{-4}dx = \frac{x^{3}}{-3} + C$ or $-\frac{1}{3}x^{-3} + C$.

e) Using Rules 2 and 5 yields, $\int 7x^2dx = 7\int x^2dx = \frac{7x^3}{3} + C$.

↑ An extensive table of integrals appears in *CRC Standard Mathematical Tables and Formulae*, D. Zwillinger (CRC Press). Many Internet sites compute integrals. One such site is http://integrals.wolfram.com/index.jsp.

◆ A common "sport" among calculus students is to challenge one another with a difficult integral. The evaluation of many integrals depends on a clever variable transformation that reduces a complicated integral to a simpler one.

Example 6.1.2 More Integrations

Integrate the following:

a) $\int (5x^4 + 3x^2 + 2x + 4)dx$

b) $\int (\sqrt[3]{x^2} + 7x)dx$

c) $\int (4x^3 + \frac{1}{x^2} + \frac{5}{x})dx$

Solution:

As these integrands have more terms, use Rule 6.

a)

$$\begin{aligned}\int (5x^4 + 3x^2 + 2x + 4)dx &= \int 5x^4dx + \int 3x^2dx + \int 2xdx + \int 4dx \\ &= \frac{5x^5}{5} + \frac{3x^3}{3} + \frac{2x^2}{2} + 4x + C \\ &= x^5 + x^3 + x^2 + 4x + C\end{aligned}$$

b) Rewriting and integrating yield

$$\begin{aligned}\int (x^{2/3} + 7x)dx &= \frac{x^{5/3}}{5/3} + \frac{7x^2}{2} + C \\ &= \frac{3x^{5/3}}{5} + \frac{7x^2}{2} + C\end{aligned}$$

⟶

c) Rewriting and integrating yield

$$\int (4x^3 + x^{-2} + \frac{5}{x})dx = \frac{4x^4}{4} + \frac{x^{-1}}{-1} + 5\,ln\,|x| + C$$
$$= x^4 - \frac{1}{x} + 5\,ln\,|x| + C$$

Initial Conditions

Indefinite integration results in a family of functions that only differ by a constant called the constant of integration. In many applications, there are **initial conditions** that determine the integration constant. This is illustrated graphically as follows.

Family of $y = x^2 + C$

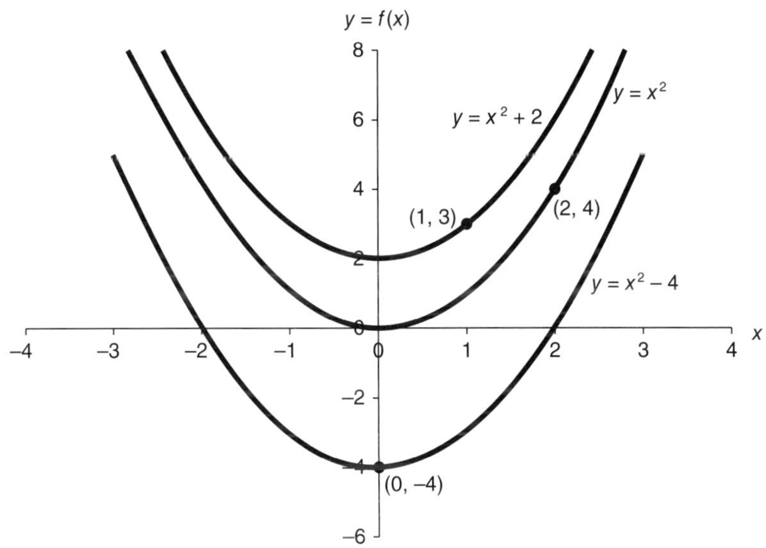

Three members of the family $y = x^2 + C$ are sketched corresponding to $C = -4$, 0, and 2. When $f(0) = -4$, that family member must be $y = f(x) = x^2 - 4$ (as labeled). Similarly, if $f(1) = 3$, one sets $3 = (1)^2 + C$ to determine $C = 2$ and the function is $y = x^2 + 2$ (as shown). Finally, when $f(2) = 4$, then $y = x^2$ as $C = 0$.

Example 6.1.3 Integrals with Initial Conditions

a) Determine f(x) when: $f'(x) = 3x^2 + 2x + 1$ *and* $f(1) = 5$.

b) Determine f(x) when: $f'(x) = \sqrt{x}$ *and* $f(4) = 9$.

→

Solution:

a) *Firstly, integration yields* $\int(3x^2 + 2x + 1)dx = x^3 + x^2 + x + C$. *The initial condition means (1, 5) must satisfy the function. So,* $5 = (1)^3 + (1)^2 + 1 + C$ *yields* $C = 2$. *Therefore, the required function is* $f(x) = x^3 + x^2 + x + 2$.

b) *Rewriting and integrating yield* $\int x^{1/2}dx = \frac{2}{3}x^{3/2} + C$. *Next, the initial condition, (4, 9), must satisfy the integration. So,* $9 = \frac{2}{3}(4)^{3/2} + C$ *and* $C = \frac{11}{3}$ *so* $f(x) = \frac{2}{3}x^{3/2} + \frac{11}{3}$.

Example 6.1.4 Marginal Revenue and Marginal Cost with Initial Conditions

Section 3.5 introduced marginal analysis. For Marginal Revenue and Marginal Cost functions $MR(x) = -x + 150$ *and* $MC(x) = C'(x) = \frac{x^2}{10} - 4x + 110$ *with* $C(30) = 4000$, *determine revenue, cost, and profit functions.*

Solution:

Firstly, integration of Marginal Revenue yields $R(x) = \int(-x + 150)dx = \frac{-x^2}{2} + 150x + C$. *As R(0) must be zero, we conclude that* $R(x) = \frac{-x^2}{2} + 150x$.

Next, integration of Marginal Cost yields
$C(x) = \int(\frac{x^2}{10} - 4x + 110)dx = \frac{x^3}{30} - 2x^2 + 110x + C$.

Using the initial condition $C(30) = 4000$, *we have* $\frac{(30)^3}{30} - 2(30)^2 + 110(30) + C = 4000$ *and, therefore,* $C = 1600$. *The Cost function is* $C(x) = \frac{x^3}{30} - 2x^2 + 110x + 1600$.

These agree with results in Example 3.5.3.

Next, using
$Pr(x) = R(x) - C(x)$, *we have a Profit function* $Pr(x) = -\frac{x^3}{30} + \frac{3x^2}{2} + 40x - 1600$.

EXERCISES 6.1

In Exercises 1–8, carry out the integration. Check your result by differentiation.

1. $\int 7dx$
2. $\int -\frac{2}{3}dz$
3. $\int 5xdx$
4. $\int 9pdp$

5. $\int 3x^{-5}dx$

6. $\int x^{3/5}dx$

7. $\int 2tdt$

8. $\int 5v^4dv$

In Exercises 9–26, carry out the integration.

9. $\int \sqrt{x}dx$

10. $\int \sqrt[3]{x}dx$

11. $\int \sqrt[3]{x^2}dx$

12. $\int \frac{1}{3x^4}dx$

13. $\int \frac{1}{2x^3}dx$

14. $\int 2x^{-5}dx$

15. $\int (x^{3/5} + x^{-2/3})dx$

16. $\int (2x^{-3} + x^{-2})dx$

17. $\int e^{7t}dt$

18. $\int \frac{3}{x}dx$

19. $\int (4x^3 + 3x^2 + 2x + 9)dx$

20. $\int (3x^5 + (9/x))dx$

21. $\int (e^{4x} + 1)dx$

22. $\int (3x^{2/3} - 2x^{1/3} + \frac{1}{3}x^{-2/3})dx$

23. $\int (4 - \frac{2}{x})dx$

24. $\int (\frac{4}{e^{2x}} + 3e^{5x} + 6)dx$

25. $\int ((3/\sqrt{t}) - 2\sqrt{t})dt$

26. $\int (e^{3t} + 3t^3 + 3t^2 + 3t + 3)dt$

In Exercises 27–30, obtain f(x) satisfying the initial conditions.

27. $f'(x) = 3x^2$ and $f(-1) = 6$

28. $f'(x) = 6x^{1/2}$ and $f(4) = 32$

29. $f'(x) = 3x^2 - 2x + 4$ and $f(2) = 10$

30. $f'(x) = \frac{2}{x}$ and $f(1) = 8$

In Exercises 31 and 32, use differentiation to choose the correct integration result.

31. $\displaystyle\int 5xe^{x^2}\,dx$

a) $\dfrac{5}{2}e^{x^2} + C$ b) $5xe^{x^2} + C$

32. $\displaystyle\int x\sqrt{x+1}\,dx$

a) $\dfrac{2}{5}x(x+1)^{5/2} - \dfrac{2}{3}(x+1)^{3/2} + C$

b) $\dfrac{2}{3}x(x+1)^{3/2} - \dfrac{4}{15}(x+1)^{5/2} + C$

6.2 RIEMANN SUMS

Empiric graphs arise often in applications. In such instances, an explicit function is not known, as in the following figure. The area under the curve can be estimated by approximating rectangles whose total area approximates the desired area. Increasing numbers of rectangles of decreasing widths improve the approximation, as noted in the previous section.

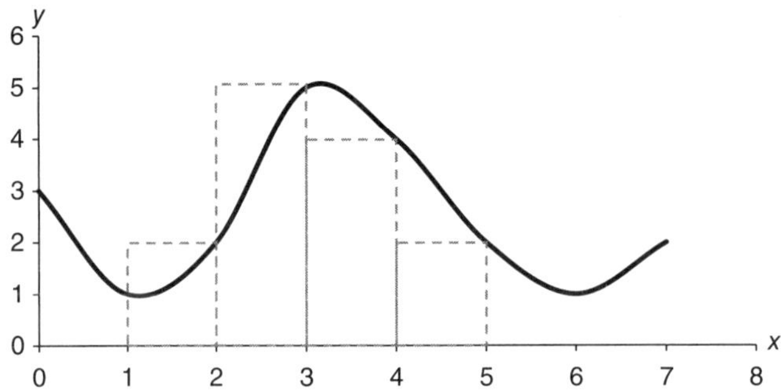

The use of rectangles to approximate an area under a curve known as **Riemann Sums** honors the famous mathematician Georg Riemann (Historical Notes).

Example 6.2.1 Determining Specified Points

The interval $[a,b] = [1,9]$ is divided into $n = 4$ subintervals. What are the coordinates of the four left endpoints, the four right endpoints, and the four midpoints?

Solution:

The interval width is $9 - 1 = 8$ units. It is divided into four subintervals, each of width $\dfrac{9-1}{4} = 2$.

The subintervals are [1, 3], [3, 5], [5, 7], and [7, 9].

⟶

The four left interval endpoints are at $x = 1, 3, 5,$ and 7.
The four right interval endpoints are at $x = 3, 5, 7,$ and 9.
The four interval midpoints are at $x = 2, 4, 6,$ and 8.

One can use left endpoints, right endpoints, or midpoints of the approximating rectangles to estimate areas. The formula that follows indicates how to determine the area under the curve once a choice is made as to whether the x_i correspond to left endpoints, right endpoints, or midpoints. Each choice will give a different approximation with no choice clearly better than another in every case.

Riemann Sums

Using n approximating rectangles, each of width $\frac{b-a}{n}$ on $[a, b]$, the approximate area under the graph of $f(x) \geq 0$ is

$$\textbf{Approximate Area} = \frac{b-a}{n}\sum_{i=1}^{n} f(x_i)\ \textbf{Riemann Sum}$$

Interval coordinates $(x_i, f(x_i))$ refer to the location and height of approximating rectangles. Here, n is a positive integer.

If $f(x)$ is a continuous non-negative function, say x^2, on the interval $1 \leq x \leq 4$, then the area under the graph of $f(x)$ from $x = 1$ to $x = 4$ is depicted in the following figure.

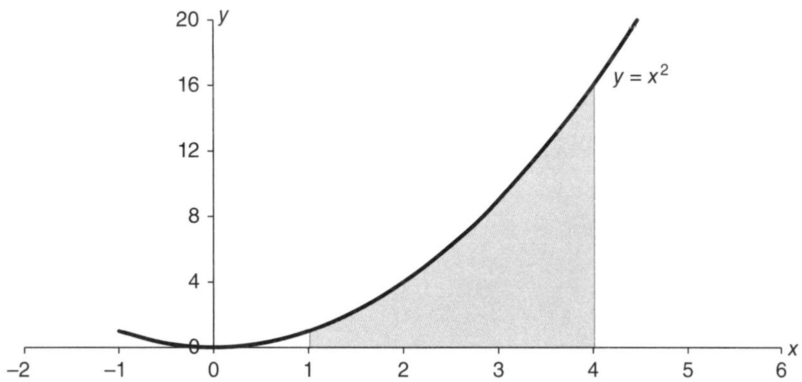

The following example illustrates a Riemann Sum to estimate the area depicted.

Example 6.2.2 Area Under the Curve for $f(x) = x^2$

Approximate the area under $f(x) = x^2$ on [1, 4] using heights of rectangles at left endpoints with $n = 6$.

Solution:

In this case, $a = 1$ and $b = 4$. The interval is to be subdivided into six equal widths, $\frac{4-1}{6} = (1/2)$. The six partitions are at [1, 1.5], [1.5, 2], [2, 2.5], [2.5, 3], [3, 3.5], and [3.5, 4]. The six left endpoints are at $x = 1, 1.5, 2, 2.5, 3,$ and 3.5. Note that left endpoints begin with $x = a$. We seek

$$\frac{1}{2}[f(1) + f(1.5) + f(2) + f(2.5) + f(3) + f(3.5)]$$

Substituting into $f(x) = x^2$, the approximate area is

$$= \frac{1}{2}[(1)^2 + (1.5)^2 + (2)^2 + (2.5)^2 + (3)^2 + (3.5)^2] = 17.375 \ \ sq \ \ units$$

Hint: remember, for Riemann Sums multiply rectangular heights by widths $\frac{b-a}{n}$.

Example 6.2.3 Area Under a Graph

Estimate the area under f(x) between [1, 5]. Use right endpoints and $n = 4$ (figure).

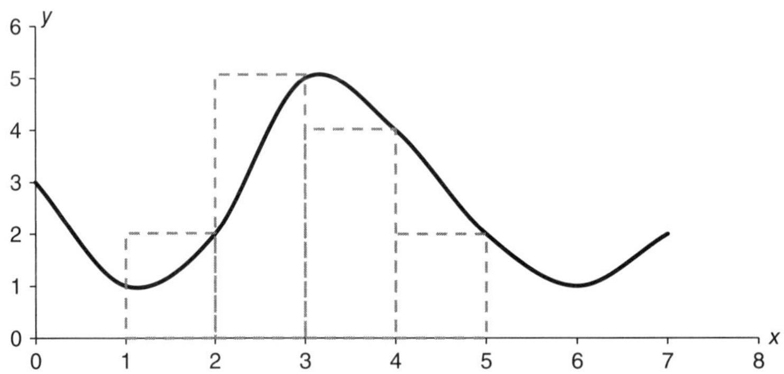

Solution:

In this case, the four right endpoints are at $x = 2, 3, 4,$ and 5 (see graph). Here, f(x) is not known explicitly. Therefore, the graph itself is used to estimate f(x).

The required area is approximated by $\frac{5-1}{4}[f(2) + f(3) + f(4) + f(5)]$.

Reading values of f(x) from the graph, the approximated area is $1[2 + 5 + 4 + 2] = 13$.

♦ Trapezoids, parabolas, and other polygons are also used to approximate areas.

EXERCISES 6.2

In Exercises 1–4, determine the subinterval widths and left endpoints of the partitions.

1. $0 \leq x \leq 4, \quad n = 4$
2. $2 \leq x \leq 12, n = 5$
3. $0 \leq x \leq 3, n = 4$
4. $1 \leq x \leq 7, n = 2$

In Exercises 5–8, determine the subinterval widths and right endpoints of the partitions.

5. $1 \leq x \leq 15, \quad n = 7$
6. $2 \leq x \leq 10, n = 4$
7. $3 \leq x \leq 27, n = 6$
8. $11 \leq x \leq 21, n = 5$

In Exercises 9–12, determine the subinterval widths and midpoints of the partitions.

9. $0 \leq x \leq 4c, \quad n = 8$
10. $1 \leq x \leq 7, \quad n = 3$
11. $5 \leq x \leq 21, \quad n = 4$
12. $10 \leq x \leq 25, \quad n = 5$

In Exercises 13–24, use Riemann Sums to approximate areas under the graph of f(x) on the given intervals.

13. $f(x) = x^2 + 1$ on [1, 9], $n = 4$, using right endpoints.
14. $f(x) = x^2 + 5x + 3$ on [2, 14], $n = 6$, using right endpoints.
15. $f(x) = x^3$ on [0, 20], $n = 5$, using right endpoints.
16. $f(x) = e^{4x}$ on [1, 3], $n = 4$, using right endpoints.
17. $f(x) = 5x + 4$ on [3, 23], $n = 4$, using left endpoints.
18. $f(x) = \ln x$ on [3, 9], $n = 6$, using left endpoints.
19. $f(x) = \dfrac{1}{x}$ on [1, 9], $n = 4$, using left endpoints.
20. $f(x) = x^2 + 5x + 3$ on [2, 14], $n = 6$, using left endpoints.
21. $f(x) = x^2 + 1$ on [1, 9], $n = 4$, using midpoints.
22. $f(x) = x^2 + 5x + 3$ on [2, 14], $n = 6$, using midpoints.
23. $f(x) = e^{4x}$ on [1, 3], $n = 4$, using midpoints.
24. $f(x) = 5x + 4$ on [3, 23], $n = 4$, using midpoints.

In Exercises 25–30, use a Riemann Sum to approximate areas under the graph on the given interval.

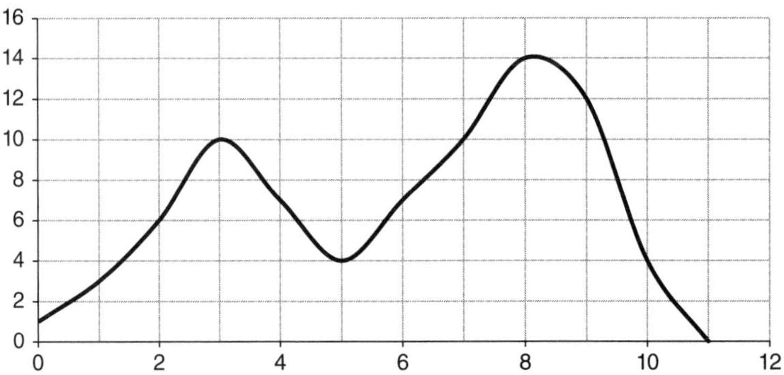

25. Use right endpoints, [1, 7] with $n = 6$.
26. Use right endpoints, [3, 7], with $n = 4$.
27. Use left endpoints, [2, 10] with $n = 8$.
28. Use left endpoints, [0, 5], with $n = 5$.
29. Use midpoints, [1, 9] with $n = 4$.
30. Use midpoints, [4, 8], with $n = 2$.
31. Calculate the area under the curve $f(x) = 4x + 1$ on [0, 4]
    a) geometrically.
    b) from a Riemann Sum with $n = 4$ and using right endpoints.
    c) by integration.
32. The graph of $f(x) = \sqrt{9 - x^2}$ on $[-3, 3]$ is a semicircle of area $\frac{9\pi}{2}$. Use Riemann Sums with $n = 6$ and left endpoints, right endpoints, and midpoints to approximate the indicated area. Compare with each other and the actual area. What are your conclusions about relative merits?

6.3 INTEGRAL CALCULUS – THE FUNDAMENTAL THEOREM

In the previous section, areas under a curve, $f(x)$, are approximated using sums of rectangular areas – Riemann Sums. Smaller widths of approximating rectangles virtually always yield better approximations to areas. Carried to an extreme, as decreasing widths of the rectangles approach zero, the approximating area approaches the actual area. However, there is an alternate and elegant way to determine areas!

An integration is an amazing shortcut alternative to limiting Riemann Sums. If $f(x)$ is a non-negative function on an interval, then its integral on that interval is the actual area under its graph; it is a **definite integral**.

More generally, removing the non-negative restriction on $f(x)$, the definite integral is defined as:

Definite Integral

The definite integral of $f(x)$ on $[a, b]$ is given by

$$\int_a^b f(x)dx = \lim_{\Delta x \to 0} \sum_{i=1}^{n} f(x_i)\Delta x$$

a and b are *limits of integration*, $\Delta x = \frac{b-a}{n}$, $f(x)$ is continuous and n is the number of approximating rectangles.

Earlier, we noted that the antiderivative of an indefinite integral is its integrand. For a definite integral there is the **Fundamental Theorem of Integral Calculus.**

Fundamental Theorem of Integral Calculus

The definite integral of $f(x)$ on $[a, b]$

$$\int_a^b f(x)dx = F(x)|_a^b = F(b) - F(a)$$

exists when $F(x)$ is continuous and $F'(x) = f(x)$ for all x on [a, b].

Example 6.3.1 Riemann Sums and Definite Integrals

Using the Fundamental Theorem evaluate $\int_1^4 x^2 dx$. *Compare the result with the estimate in Example 6.2.2.*

Solution:

The area using a Riemann Sum was estimated at 17.375. Using the Fundamental Theorem,

$$\int_1^4 x^2 dx = \left.\frac{x^3}{3}\right|_1^4 = \frac{64}{3} - \frac{1}{3} = \frac{63}{3} = 21$$

The actual area exceeds the Riemann estimate. The reason is that f(x) is increasing on [1, 4] and using left endpoints underestimates the actual area. Similarly, right endpoints overestimate the area in this case.

Five useful properties of definite integrals are summarized here.

Properties of Definite Integrals

1. $\int_a^b [f(x) \pm g(x)]dx = \int_a^b f(x)dx \pm \int_a^b g(x)dx.$
2. $\int_a^b rf(x)dx = r\int_a^b f(x)dx,$ **where r is a constant.**
3. $\int_a^a f(x)dx = 0.$
4. $\int_a^b f(x)dx = -\int_b^a f(x)dx.$
5. $\int_a^b f(x)dx = \int_a^c f(x)dx + \int_c^b f(x)dx.$

Example 6.3.2 Evaluating Definite Integrals

Evaluate

a) $\int_3^5 5xdx$ *b)* $\int_0^4 e^{3x}dx$ *c)* $\int_1^3 (3x^2 + 2x + 6)dx$

Solution:

a) Using Property 2,

$$\int_3^5 5x\ dx = 5\int_3^5 5x\ dx = 5\left(\frac{x^2}{2}\right)\Bigg|_3^5 = 5\left[\frac{(5)^2}{2} - \frac{(3)^2}{2}\right] = 5\left[\frac{16}{2}\right] = 40.$$

b) $\int_0^4 e^{3x}dx = \frac{e^{3x}}{3}\Big|_0^4 = \frac{e^{12}}{3} - \frac{1}{3}.$

c) Using Property 1,

$$\int_1^3 (3x^2 + 2x + 6)dx = (x^3 + x^2 + 6x)|_1^3 = (27 + 9 + 18) - (1 + 1 + 6) = 46.$$

Example 6.3.3 More on Evaluating Definite Integrals

Evaluate

a) $\int_1^4 (7x^6 + 5x^4 + 3x^2 + 6)dx$

b) $\int_8^{27} \sqrt[3]{x^2}dx$

c) $\int_1^3 \left(\frac{4}{x^2} + \frac{3}{x} + 2x\right) dx$

Solution:

a) $\int_1^4 (7x^6 + 5x^4 + 3x^2 + 6)dx = (x^7 + x^5 + x^3 + 6x)|_1^4$

$= (16,384 + 1024 + 64 + 24) - (1 + 1 + 1 + 6) = 17,487$

b) *Use a fractional exponent for the radical and integrate. So,*

$$\int_8^{27} \sqrt[3]{x^2}dx = \int_8^{27} x^{2/3}dx = \frac{3}{5}x^{5/3}\Big|_8^{27} = \frac{729}{5} - \frac{96}{5} = \frac{633}{5}$$

c) *Rewrite the first term with a negative exponent for integration. So,*

$$\int_1^3 \left(\frac{4}{x^2} + \frac{3}{x} + 2x\right) dx = \int_1^3 \left(4x^{-2} + \frac{3}{x} + 2x\right)dx$$

$$= \left(\frac{-4}{x} + 3\ln|x| + x^2\right)\Big|_1^3 = \left(\frac{-4}{3} + 3\ln 3 + 9\right) - (-4 + 3\ln 1 + 1)$$

$$= \frac{32}{3} + 3\ln 3 = \frac{32}{3} + \ln 27$$

Geometrically, $\int_a^b f(x)dx$ is an area bounded by $y = f(x)$ on $[a, b]$ and the x-axis. A positive value corresponds to a net area above the axis, while a negative value indicates a net area below the axis as the following example illustrates. A *net area* refers to the differences of areas above and below the x-axis.

Example 6.3.4 Definite Integrals and Areas

Evaluate the integrals and sketch the corresponding areas.

a) $\int_{-2}^2 x^3dx$

b) $\int_{-2}^5 (3x + 1)dx$

⟶

Solution:

a) *A graph of $f(x) = x^3$ from $x = -2$ to $x = 2$ (shaded area). By symmetry, the two shaded areas are equal and oppositely signed. The net area is zero, although the actual area is $2 \times 4 = 8$ sq units.*

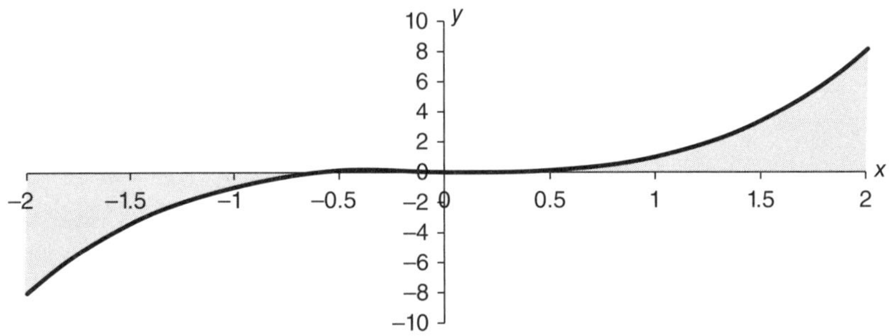

Formally, the definite integral is

$$\int_{-2}^{2} x^3 dx = \left(\frac{x^4}{4}\right)\Bigg|_{-2}^{2} = 4 - 4 = 0$$

The integration does indeed yield zero as expected because the areas above and below the x-axis are equal.

b) *A graph of $f(x) = 3x + 1$ from $x = -2$ to $x = 5$ is shown and the appropriate areas shaded. The area of the triangle below the x-axis is $\left(\frac{1}{2}\right)\left(\frac{5}{3}\right)(5) = \frac{25}{6}$. The area of the triangle above the x-axis is $\left(\frac{1}{2}\right)\left(\frac{16}{3}\right)(16) = \frac{256}{6}$. The net area is $\frac{256}{6} - \frac{25}{6} = \frac{231}{6} = \frac{77}{2}$.*

In many applications, it is the absolute area that is of interest. In that instance, one would add 256/6 to 25/6 to obtain 281/6.

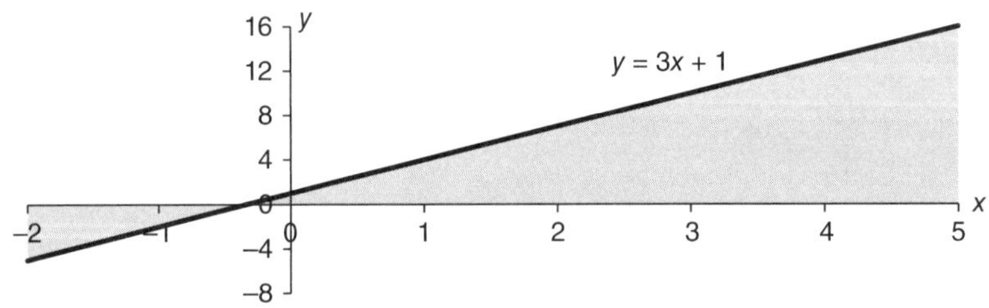

Performing the definite integration yields

$$\int_{-2}^{5} (3x + 1)dx = \left(\frac{3x^2}{2} + x\right)\Bigg|_{-2}^{5} = \left(\frac{75}{2} + 5\right) - (6 - 2) = \frac{77}{2}.$$

The integration agrees with the geometric result. Again, if the absolute area is sought, one uses two definite integrals corresponding to areas above and below the x-axis.

EXERCISES 6.3

In Exercises 1–24, evaluate the definite integral.

1. $\int_4^9 dx$
2. $\int_3^6 dt$
3. $\int_1^4 3dx$
4. $\int_2^{10} 5dv$
5. $\int_2^5 (2x+3)dx$
6. $\int_0^6 (7t+1)dt$
7. $\int_1^2 (4x^3+2x+5)dx$
8. $\int_2^5 (3x^2+9x+4)dx$
9. $\int_0^{15} (e^{3p})dp$
10. $\int_1^5 (e^{4x})dx$
11. $\int_3^6 \frac{1}{t}dt$
12. $\int_1^6 (6t^2+\frac{1}{t})dt$
13. $\int_1^3 \frac{4}{t^2}dt$
14. $\int_2^7 (-\frac{3}{t^4})dt$
15. $\int_{-1}^4 (8x^3+5x+2)dx$
16. $\int_1^2 (5x^4+4x^3)dx$
17. $\int_1^9 \sqrt{t}dt$
18. $\int_{27}^{125} \sqrt[3]{x}dx$
19. $\int_1^8 \sqrt[3]{x^2}dx$
20. $\int_0^{32} \sqrt[5]{x^3}dx$
21. $\int_1^8 \left(x^3+3x^2+\sqrt[3]{x}\right)dx$
22. $\int_1^2 (5x^4+3x+2+\frac{4}{x^2})dx$
23. $\int_0^3 (4-\frac{x^2}{4})dx$
24. $\int_1^4 (4x-2\sqrt{x}+\frac{16}{x^3})dx$

In Exercises 25–28, sketch the area represented by the definite integral. Calculate the area geometrically and verify consistency with the integration.

25. $\int_2^7 6dx$
26. $\int_0^4 2xdx$
27. $\int_1^3 3xdx$
28. $\int_{-1}^5 (2x+3)dx$

6.4 AREA BETWEEN INTERSECTING CURVES

Let $f(x)$ and $g(x)$ be two non-negative continuous valued functions on the interval $[a, b]$ whose intersecting graphs form a *bounded area*. Suppose that $f(x) \geq g(x)$ on $[a, b]$. Then, the area of the region bounded by them is the definite integral $\int_a^b [f(x) - g(x)]dx$.

To see this, consider $f(x) - g(x)$ as the height of an approximating rectangle and dx as the width (base), so its area is $[f(x) - g(x)]dx$. The integral effectively and elegantly sums the elemental rectangular areas.

Area Between Two Curves

The intersection of functions $f(x)$ and $g(x)$ on $[a, b]$ is a bounded region, A

$$A = \int_a^b [f(x) - g(x)]dx$$

where $f(x)$ and $g(x)$ are continuous on $[a, b]$, $f(x) \geq g(x)$.

Roughly,

$$A = \int_a^b [\text{upper curve} - \text{lower curve}]dx$$

Example 6.4.1 Area Bounded by Curves

Determine the area bounded by $f(x) = x^2 - 4x + 1$ and $g(x) = x - 3$.

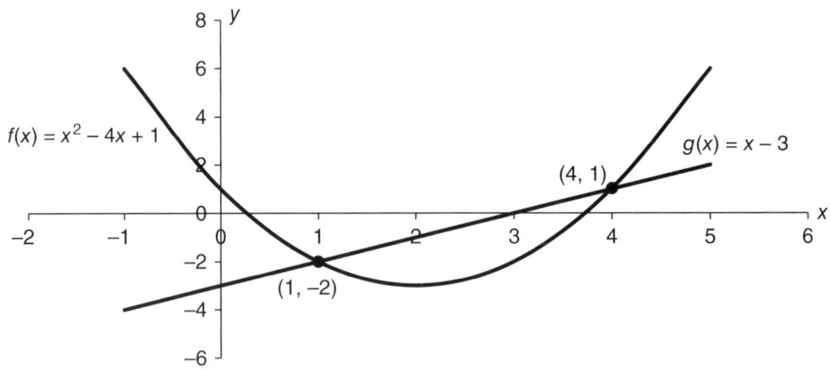

Solution:

At their intersections $f(x) = g(x)$ and the values of x there are the integration limits. Equating $f(x)$ and $g(x)$, and solving for their intersections, yields $x = 1$ and $x = 4$ as shown in the figure. Here, the linear function $g(x) = x - 3$, exceeds $f(x)$ on $(1, 4)$. The area bounded by

⟶

the two curves is

$$\int_1^4 [(x-3)-(x^2-4x+1)]dx = \int_1^4 [-x^2+5x-4]dx = \left(\frac{-x^3}{3}+\frac{5x^2}{2}-4x\right)\Big|_{x=1}^{x=4}$$
$$= \left(\frac{-64}{3}+\frac{80}{2}-16\right)-\left(\frac{-1}{3}+\frac{5}{2}-4\right)=\frac{9}{2}$$

Although points of intersection may be found graphically, it is usually more convenient to equate the two functions, as $x^2 - 4x + 1 = x - 3$ and solve. Simplifying, x satisfies $x^2 - 5x + 4 = 0$. Factored as $(x - 4)(x - 1) = 0$, it is clear that the intersections are at $x = 1$ and $x = 4$. These are the integration limits.

Next, to identify the larger of the two functions on the interval, choose an arbitrary point (within the interval), say, $x = 2$. At $x = 2$, $f(2) = (2)^2 - 4(2) + 1 = -3$. Similarly, $g(2) = 2 - 3 = -1$, so $g(x)$ is the larger value (upper curve). Therefore, express the integrand as $g(x) - f(x) = (x - 3) - (x^2 - 4x + 1)$ in this instance. However, be alert! Sometimes, which function is the larger depends on the choice of interval. This arises in the following example.

Example 6.4.2 More on Area Bounded by Curves

Determine the area enclosed by $f(x) = x^3 - x^2 + 3$ and $g(x) = 2x + 3$.

Solution:
Set $f(x) = g(x)$ for intersections at $x = -1, x = 0$, and $x = 2$. That is, set $x^3 - x^2 + 3 = 2x + 3$ or $x^3 - x^2 - 2x = x(x - 2)(x + 1) = 0$. Here, $f(x)$ is larger than $g(x)$ on $(-1, 0)$, while $g(x)$ is larger on $(0, 2)$. The area bounded by the curves is

$$\int_{-1}^0 [(x^3-x^2+3)-(2x+3)]dx + \int_0^2 [(2x+3)-(x^3-x^2+3)]dx$$
$$= \int_{-1}^0 [x^3-x^2-2x]dx + \int_0^2 [-x^3+x^2-2x]dx$$
$$= \left(\frac{x^4}{4}-\frac{x^3}{3}-x^2\right)\Big|_{-1}^0 + \left(\frac{-x^4}{4}+\frac{x^3}{3}+x^2\right)\Big|_0^2$$
$$= \frac{5}{12}+\frac{8}{3}=\frac{37}{12}$$

A graph can be an aid to obtain areas. On graph paper, counted squares in the bounded region multiplied by the area of each square can roughly approximate the area.

More Applications

Average Value of a Function A function, $f(x)$, generally varies with x on an interval, so one can speak of the **average value of the function** in much the manner of the average of some numbers. The average of $f(x)$, a continuous function on $[a, b]$, written $\overline{f(x)}$, is defined by

$$\overline{f(x)} = \frac{1}{b-a}\int_a^b f(x)\ dx$$

↓ For instance, for $f(x) = x$ on $[0, 1]$, the average is 1/2. Using the formula

$$\frac{1}{1-0}\int_0^1 xdx = 1\left(\frac{x^2}{2}\bigg|_0^1\right) = 1\left(\frac{1}{2}-0\right) = \frac{1}{2}; \text{ as expected!}$$

> ***Average Value of a Function***
>
> **The average value of $f(x)$ on $[a, b]$ is defined as**
>
> $$\textbf{Average value} = \overline{f(x)} = \frac{1}{b-a}\int_a^b f(x)dx.$$
>
> **where $f(x)$ is continuous on $[a, b]$.**

A rectangle of height $\overline{f(x)}$ and base $(b - a)$ is the geometrical equivalent of the definite integral $\int_a^b f(x)dx$ when $f(x)$ is non-negative. Dividing by the width of the base, $(b - a)$, gives the average value of $f(x)$ on $[a, b]$.

♦ You may have noted bank checking account statements that report "average daily balance." These are calculated by multiplying a balance by the number of days it was maintained, repeating for each change in balance, and adding them together in much the manner of a Riemann Sum. To obtain the daily average, one divides the sum by the number of days in that month.

Example 6.4.3 Average Value

Compute the average value of $f(x) = x^2 + 2$ on $[0, 3]$

Solution:

The average value is between $f(0) = 2$ and $f(3) = 11$. Using the formula, the average value is

$$\frac{1}{3-0}\int_0^3 (x^2+2)dx = \frac{1}{3}\left(\frac{x^3}{3}+2x\right)\bigg|_0^3 = \frac{1}{3}\left[\left(\frac{27}{3}+6\right)-(0+0)\right] = 5$$

Applications to Economics

Example 6.4.4 EOQ – Economic Order Quantity

The figure illustrates a fundamental model for an inventory at time t, I(t). A starting inventory, Q, is drawn upon at a constant rate until dissipated at time T. The cycle repeats. What is the average inventory over a cycle?

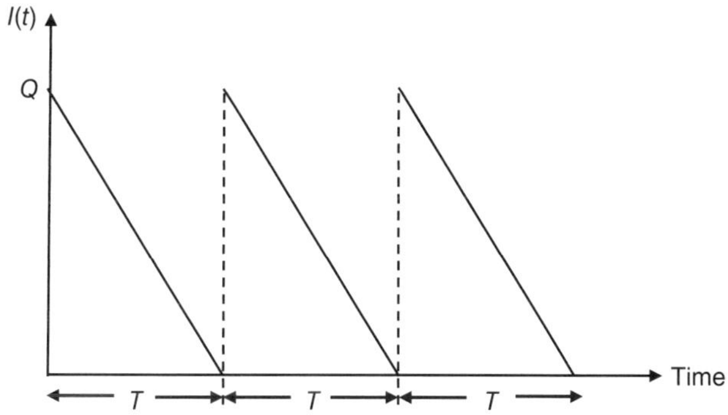

Solution:

Firstly, establish a relation for I(t). As $I(0) = Q$ and $I(T) = 0$, it follows that $I(t) = \frac{-Q}{T}t + Q$ expresses the inventory at any time t, $0 \leq t \leq T$.

The average value on [0, T], $\overline{I(t)}$, using $a = 0$, $b = T$, and $f(x) = I(t)$ in the definition is

$$\overline{I(t)} = \frac{\int_0^T \left(\frac{-Qt}{T} + Q\right) dt}{\int_0^T dt} = \frac{\left(\frac{-Qt^2}{2T} + Qt\right)\Big|_0^T}{T} = \frac{\frac{-QT}{2} + QT}{T} = \frac{Q}{2}$$

Therefore, the average inventory is $\frac{Q}{2}$ over a cycle.

Likely, the reader surmised the result by inspecting the figure. For a linear "draw down," the average inventory is the average of the starting and ending value. That is, $\frac{Q+0}{2} = \frac{Q}{2}$.

♦ Inventory analysis is a very important industrial consideration. Try a web search!

Supply and Demand Curves Supply and demand curves are a staple of economics. Roughly, in a plot of price versus quantity for a supply curve, price rises as the quantity demanded increases while for the demand curve, price decreases as the quantity increases (figure).

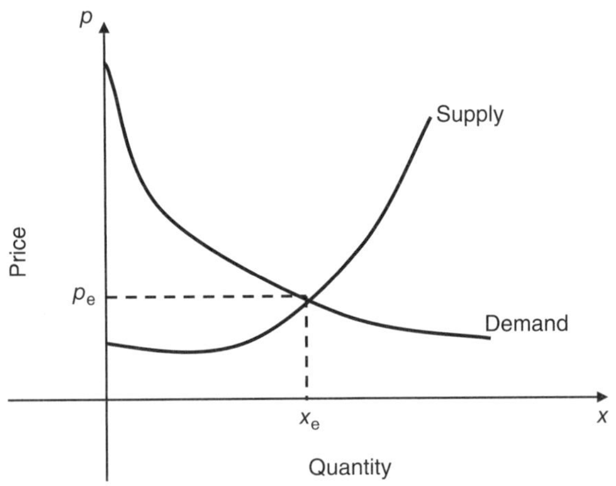

The intersection of the two curves at $x = x_e$ represents a market equilibrium in which *supply equals demand at price* p_e.

Sometimes, consumers are willing to pay more than market equilibrium price, and sometimes, producers sell for less than this price. These savings are known as **consumers' surplus** and **producers' surplus**, respectively. Areas under the supply and demand curves are used to calculate such surpluses. The following figure depicts areas of consumer surplus and producer surplus.

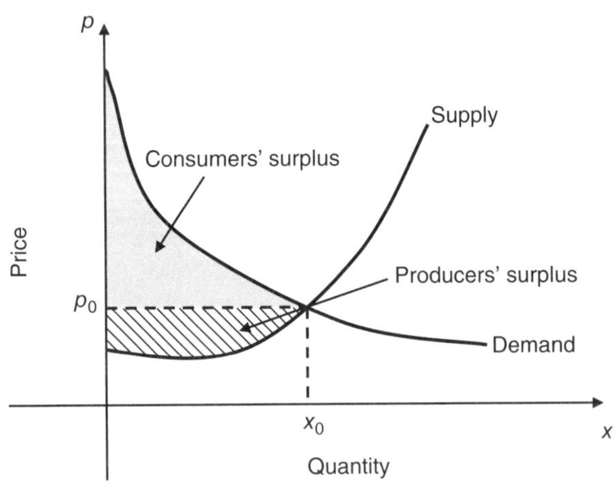

Consumers' Surplus

The consumers' surplus for a commodity with a demand curve $p = p(x)$ is

$$CS = \int_0^{x_0} [p(x) - p_0]dx$$

where x_0 is the sales level (quantity demanded) and $p_0 = p(x_0)$

Example 6.4.5 Consumers' Surplus

A commodity has the demand function $p(x) = \dfrac{100}{x+1}$. What is the consumers' surplus at a sales level of $x = 4$?

Solution:

In this case, $x_0 = 4$ and $p_0 = \dfrac{100}{4+1} = 20$. Therefore,

$$CS = \int_0^4 \left[\frac{100}{x+1} - 20\right]dx = [100\ln|x+1| - 20x]|_0^4$$

$$= [100\ln 5 - 80] - [100\ln 1 - 0] = 100\ln 5 - 80 = \$80.94$$

The consumers' surplus is \$80.94.

EXERCISES 6.4

In Exercises 1–10, determine the area bounded by the intersections of the given functions.

1. $f(x) = 8 - x^2,\ g(x) = -2x + 5.$
2. $f(x) = x + 7, g(x) = x^2 - 3x + 10.$
3. $f(x) = x^2, g(x) = \sqrt{x}.$
4. $f(x) = x^2,\ g(x) = 9x.$
5. $f(x) = x^2 + 3,\ g(x) = 4x + 3.$
6. $f(x) = x^3 - x + 5, g(x) = 5.$
7. $f(x) = x^2 - 3x + 1, g(x) = -x + 4.$
8. $f(x) = x, g(x) = -x^2 + 6.$
9. $f(x) = x^2 - 4x + 3, g(x) = -x^2 + x + 3.$

10. $f(x) = x + 1, g(x) = x^2 - 4x + 5$.
11. Determine the area between $y = x^2$ and the x-axis for $x = -2$ to $x = 3$.
12. Determine the area between $y = x^2 + 4$ and $y = x$ for $x = -1$ to $x = 4$.
13. Determine the area enclosed by $y = x^3 - x^2 + 1$ and $y = 2x + 1$.
14. Determine the area bounded by $y = x^2$ and $y = 9$.

In Exercises 15–18, determine the average value of the function on the given interval.

15. $f(x) = 4x + 3$ on $[1, 4]$.
16. $f(x) = e^{2x}$ on $[0, 3]$.
17. $f(x) = 1/x$ on $[2, 6]$.
18. $f(x) = x^3 + 2$ on $[0, 4]$.
19. Determine the average value of $f(x) = \sqrt[3]{x}$ on [0, 8].
20. An investment of \$2500 earns 6% interest compounded continuously. Determine the average value of the investment over the first 3 years.
21. Consider the cost function $C(x) = \frac{1}{2}x^2 + x + 100$. Find its average value on [6, 20].
22. The demand function for a commodity is $p = 30 - x$. Find the consumers' surplus at a sales level of 5 units.
23. The demand function for a commodity is $p = -x^2 + 34$. Find the consumers' surplus at a sales level of 5 units.
24. The demand function for a commodity is $p = 100 - 4x$. Find the consumers' surplus at a sales level of 15 units.

HISTORICAL NOTES

Georg Friedrich Bernhard Riemann (1826–1866) — was born in Hanover, Germany; he died in Italy. In 1846, Riemann enrolled at the University of Göttingen as a theology student. He became interested in mathematics and studied under mathematicians Moritz Stern and Karl Gauss. The next year, he transferred to Berlin University to study under such luminaries as Steiner, Jacobi, Dirichlet, and Eisenstein. There, Riemann devised a general theory of complex variables, perhaps his most important work. He returned to Göttingen to complete his doctorate with Gauss in 1851.

CHAPTER 6 SUPPLEMENTARY EXERCISES

1. Integrate $\int 4t^3 dt$ and check your result by differentiation.
2. Integrate $\int (4x^3 + 3x^2 + \sqrt{x} + \frac{2}{x})dx$.

3. Integrate $\int (e^{3x} + 6x + 5)dx$.
4. Find all functions $f(x)$ with the following properties $f'(x) = 5x^4 + 2$ and $f(1) = 10$.
5. Determine the width of the subintervals and the right endpoints by partitioning [3, 21] into six subintervals.
6. Determine the width of the subintervals and the midpoints by partitioning [3, 5] into four subintervals.
7. Determine the fourth Riemann Sum of $f(x) = x^2$ on [3, 11] using left endpoints.
8. Determine the sixth Riemann Sum of $f(x) = x^3 + 2x + 3$ on [3, 27] using right endpoints.
9. Evaluate $\int_1^4 (5 + 2e^{3x})dx$.
10. Evaluate $\int_1^3 (5x^4 + 3x^2 - \frac{4}{x^2})dx$.
11. Evaluate $\int_1^4 x^{3/2} dx$.
12. Evaluate $\int_{-1}^3 (4x^3 - 6x + 2)dx$.
13. Determine the area bounded by $f(x) = x + 1$ and $g(x) = x^3 + 1$.
14. Determine the area bounded by $f(x) = x^2 - 2x$, the x-axis and the lines $x = 1$ and $x = 3$.
15. Determine the area bounded by $f(x) = x^2 - 7$ and $g(x) = -x^2 + 2x + 5$.
16. Determine the area bounded by $f(x) = 2x^2 - 4x - 5$ and $y = 11$.
17. Determine the average value of $f(x) = 4 - x^2$ on [0, 2].
18. The demand function for a commodity is given by $p = -x^2 + 91$ and the supply function by $p = x^2 + 4x + 21$. Find the equilibrium quantity and the consumers' surplus there.

7 Techniques of Integration

Differentiation of a function is often a direct operation. Integration, a reverse operation, is less direct and usually requires more skill. Basically, one seeks variable substitutions, (sometimes ingenious!), which transform a complicated integral into a simpler one.

Fundamentals of Calculus, First Edition. Carla C. Morris and Robert M. Stark.

Companion Website: http://www.wiley.com/go/morris/calculus

7.1 INTEGRATION BY SUBSTITUTION

The following are three basic integral types:

$$\int e^u du = e^u + C$$

$$\int \frac{1}{u} du = \ln|u| + C$$

$$\int u^r du = \frac{u^{r+1}}{r+1} + C$$

Recall these from Chapter 6.

Substitution is the foremost means of rendering more complicated integrals less so! The following example is illustrative.

Example 7.1.1 Using Substitution

Evaluate $\int e^{5x^2+5x}(2x+1)dx$. *Hint: seek a simplifying substitution.*

Solution:

The derivative of the exponent, $g(x) = 5x^2 + 5x$, *is usefully,* $g'(x) = 10x + 5 = 5(2x + 1)$, *and this coordinates with the integrand factor* $2x + 1$.

Start by setting a new variable $u = 5x^2 + 5x$, *so the integral appears as* $\int e^u[(2x+1)dx]$.

Next, consider the bracketed term ("leftover portion")

As $u = 5x^2 + 5x$, $\frac{du}{dx} = 10x + 5$ *or* $du = (10x + 5)dx = 5(2x + 1)dx$.

Solving for $[(2x+1)]dx$ *yields* $\frac{1}{5}du = (2x + 1)dx$.

Substituting into the integral yields,

$\int e^u[(2x+1)dx] = \int e^u \frac{1}{5}(du) = \frac{1}{5}\int e^u du.$

This is easily integrated as $\frac{1}{5}\int e^u du = \frac{1}{5}e^u + C$.

Finally, replace u by $5x^2 + 5x$ *to yield* $\frac{e^{5x^2+5x}}{5} + C$.

Example 7.1.2 Algebraic Substitution

Simplify and integrate $\int 6x^2\sqrt{x^3+8}dx$.

Solution:

Rewrite as $\int 6x^2(x^3+8)^{1/2}dx$. *Use the substitution:* $u = x^3 + 8$ *to obtain* $\int (u)^{1/2}[6x^2dx]$.

Next, focus on the bracketed portion ("leftover portion").

If $u = (x^3 + 8)$, *then* $\frac{du}{dx} = 3x^2$ *so that* $2du = 6x^2dx$, *matching the bracketed term.*

The integration becomes $\int (u)^{1/2}[6x^2]dx = \int (u)^{1/2}[2du] = 2\int u^{1/2}du$.

⟶

This is the basic integration $2\int u^{1/2}du = \frac{4}{3}u^{3/2} + C.$

Finally, replace u by $x^3 + 8$ *to yield* $\int 6x^2\sqrt{x^3+8}\,dx = \frac{4}{3}(x^3+8)^{3/2} + C.$

Unlike differentiation, integration often requires trial and error substitutions. When a function is a fraction, its denominator is a candidate for substitution.

Example 7.1.3 Substitution for a Fraction

Integrate $\int \left(\frac{x^3}{x^4+7}\right) dx.$

Solution:

The fractional integrand suggests substitution for the denominator $u = x^4 + 7$.

Rewrite the integral as $\int \left(\frac{1}{x^4+7}\right)[x^3dx] = \int \frac{1}{u}[x^3dx].$

Usefully, the bracketed term is related to $\frac{du}{dx} = 4x^3$ *or* $du = 4x^3dx$. *So* $\frac{1}{4}du = x^3dx.$

Now, $\int \frac{1}{u}[x^3dx]$ *becomes* $\int \left(\frac{1}{u}\right)\left(\frac{1}{4}du\right)$ *or* $\frac{1}{4}\int\left(\frac{1}{u}\right)du$*: a basic integral.*

Therefore, $\frac{1}{4}\int\left(\frac{1}{u}\right)du = \frac{1}{4}\ln|u| + C.$

Finally, replace u by $x^4 + 7$ *to yield* $\frac{1}{4}\ln|x^4+7| + C$ *as the desired integration.*

Substitution can reduce a challenging integral to any of three basic ones – it is well to remember them!

A new variable, *u*, may replace an entire expression, appropriately exponentiated, whose derivative relates to the "left-over portion."

Expressions with *e* raised to a power may benefit by the substitution of *u* for the exponent.

If the integrand has a denominator, try replacing it by *u*, and so on.

Example 7.1.4 Multiple Terms

Simplify and integrate $\int (16xe^{8x^2} + e^{4x} + 4x^3 + 2)dx.$

Solution:

Firstly, note that the last three integrand terms are of a basic type, so no substitution is needed. As the integral of a sum is the sum of its integrals, form

$\int (16xe^{8x^2})dx + \int e^{4x}dx + \int 4x^3dx + \int 2dx.$ *Next, integrate each in turn.*

Now, for the first integral, let $u = 8x^2$ *so* $du = 16xdx$. *This yields*

$\int (e^{8x^2})[16xdx] = \int e^u du = e^u + C = e^{8x^2} + C.$

⟶

Executing the other integrations in turn and combining yield the desired result.

$$\left(\int 16xe^{8x^2} + e^{4x} + 4x^3 + 2\right) dx = e^{8x^2} + \frac{e^{4x}}{4} + x^4 + 2x + C.$$

Integration by Substitution

1. **Seek a new variable $u = g(x)$ to form a simpler integral in u.**
2. **Substitute for its differential du.**
3. **Complete the integration in the variable u.**
4. **Replace u by $g(x)$.**

Practice with many of the exercises to gain skill and confidence.

EXERCISES 7.1

In Exercises 1–20, integrate by substitution.

1. $\int (5x + 3)^{-3/4} dx$
2. $\int (7x + 1)^{10} dx$
3. $\int x(x^2 + 5)^4 dx$
4. $\int (4x^3 + 6x^2)(2x^4 + 4x^3 + 9)^{25} dx$
5. $\int (8x^3 + 4x)(x^4 + x^2 + 11)^9 dx$
6. $\int (6x^2 + 18)(x^3 + 9x + 1)^{12} dx$
7. $\int 8xe^{x^2} dx$
8. $\int x^2 e^{x^3+5} dx$
9. $\int \frac{2x + 3}{x^2 + 3x + 5} dx$
10. $\int \frac{x^4 + 2}{x^5 + 10x + 3} dx$
11. $\int \frac{x^2 - 2x}{x^3 - 3x^2 + 1} dx$
12. $\int \frac{3e^{3x} - 3e^{-3x}}{e^{3x} + e^{-3x}} dx$
13. $\int \frac{x^4}{x^5 + 1} dx$
14. $\int \frac{\ln 3x}{x} dx$
15. $\int (x/\sqrt{x^2 + 2}) dx$
16. $\int x\sqrt{100 - x^2} dx$
17. $\int 6x\sqrt{x^2 + 9} dx$
18. $\int (6x^2 e^{x^3} - 2x + 8) dx$
19. $\int (6x^3 + 5x)(3x^4 + 5x^2 + 8)^{10} dx$
20. $\int (2xe^{x^2} - 6x^2 + 2x + 4) dx$

7.2 INTEGRATION BY PARTS

Integration by Parts is another useful technique for evaluating integrals. It is a sophisticated form of substitution that relies on the differential of a product of functions to replace a more difficult integration by a simpler one.

The differential of the product $f(x)g(x)$ can be written as:

$$d((f(x)g(x)) = f(x)g'(x)dx + g(x)f'(x)dx$$

Next, consider its integral.

$$\int d(f(x)g(x)) = \int f(x)g'(x)dx + \int g(x)f'(x)dx$$

$$\text{so} \quad f(x)g(x) = \int f(x)g'(x)dx + \int g(x)f'(x)dx$$

Note that while one of the right hand integrals is the original, the other is simpler if a useful selection has been made.

Example 7.2.1 An Integration by Parts

Integrate $\int xe^{4x}dx$.

Solution:

Firstly, note that the integrand is not suited for a simple algebraic substitution (as in the previous section), as a derivative of the exponent cannot be absorbed in the integrand. A new approach is required.

Consider the integrand as a product of $f(x) = x$ *and* $g'(x) = e^{4x}$.

With $f'(x) = 1$ *and the integral of* $g'(x)$, $g(x) = \frac{e^{4x}}{4}$ *an integration by parts yields*

$$xe^{4x} = \int xe^{4x}dx + \int \frac{e^{4x}}{4}(1)dx$$

Rearranging terms for convenience, $\int xe^{4x}dx = x\frac{e^{4x}}{4} - \int \frac{e^{4x}}{4}dx.$

The desired result follows after executing the basic integral on the right.

$$\frac{xe^{4x}}{4} - \frac{e^{4x}}{16} + C$$

The example makes clear that integration by parts is also a substitution of sorts requiring a subtle consideration of the product of functions.

↑ What may confuse is that both integration and differentiation are utilized in an integration by parts.

Integration by Parts

$$\int f(x)g'(x)dx = f(x)g(x) - \int g(x)f'(x)dx$$

Notice that $f(x)$, $g(x)$, $f'(x)$, and $g'(x)$ appear in an integration by parts. As $f(x)$ appears in the integral on the left, it requires differentiation to yield $f'(x)$, which appears in the right hand integral. The term $g'(x)$ on the left side requires integration to determine $g(x)$ for the right hand integral. So integration by parts involves both integration and differentiation. The following are helpful guidelines!

In integration by parts, the left hand side can represent the original integration. However, the integral on the right side with $g(x)f'(x)$ needs to be a simpler integration. A judicious choice of $f(x)$ is the key to a successful integration by parts.

Example 7.2.2 Another Integration by Parts

Integrate $\int (x^{10} \ln 3x)dx$.

Solution:

A direct substitution is not apparent. Often, a mixed product, an algebraic factor (x^{10}) *and a transcendental factor as* $(\ln 3x)$*, suggest trying an integration by parts.*

Try $f(x) = \ln 3x$ *and* $g'(x) = x^{10}$*, as they lead easily to* $f'(x) = 1/x$ *and* $g(x) = \frac{x^{11}}{11}$.

Using an integration by parts with $f(x)g(x) = x^{11}/11 \ln 3x$*, we have*

$$\int (x^{10} \ln 3x)dx = \frac{x^{11}}{11} \ln 3x - \int \left(\frac{x^{11}}{11}\right)\left(\frac{1}{x}\right)dx.$$

The last integrand simplifies as $\frac{x^{10}}{11}$.

The desired integration is $\int (x^{10} \ln 3x)dx = \frac{x^{11}}{11} \ln 3x - \frac{x^{11}}{121} + C$.

In general, in any integration by parts, choose $f(x)$ to be the function with the "simpler" derivative. For example, the derivative of $\ln x$ is $1/x$ – an algebraic term no longer involving a logarithm. This means $\ln x$ is a good choice for $f(x)$. On the other hand, e^x or more complicated exponential functions have derivatives that are still exponential and more complex than the original function. This suggests exponential functions as good choices for $g'(x)$.

If a "wrong" initial choice of $f(x)$ and $g'(x)$ is made, the resulting right hand integral of the "formula" may be harder to integrate than the original. If this happens, start again with the choices for $f(x)$ and $g'(x)$ reversed.

Example 7.2.3 Integration of ln x

Integrate $\int \ln x dx$.

Solution:
Again, a simple substitution will not help. We try integration by parts. At first glance, this integral does not appear to be the product of two functions. However, setting $f(x) = \ln x$ *and* $g'(x) = dx$ *yields the product* $f(x)g'(x) = \ln x dx$.

Now, setting $f(x) = \ln x$ *and* $g'(x) = dx$ *yields* $f'(x) = \frac{1}{x}$ *and* $g(x) = x$.

The result is $\int \ln x\, dx = x \ln x - \int x\left(\frac{1}{x}\right)dx = x \ln x - x + C$.

Example 7.2.4 Still Another Integration by Parts

Integrate $\int x\sqrt[3]{x+3}dx$.

Solution:
Firstly, replace the radical as: $\int x(x+3)^{1/3}dx$.
Again, a simple substitution is not evident. Choose the function with the "simpler" derivative as f(x). So, set $f(x) = x$ *and* $g'(x) = (x+3)^{1/3}$.

Therefore, $f'(x) = 1$ *and* $g(x) = \frac{3}{4}(x+3)^{4/3}$. *The integration by parts yields*

$$\begin{aligned}\int f(x)g'(x)dx = \int x\sqrt[3]{x+3}dx &= \int (x)(x+3)^{1/3}dx \\ &= \frac{3}{4}x(x+3)^{4/3} - \frac{3}{4}\int (x+3)^{4/3}dx \\ &= \frac{3}{4}x(x+3)^{4/3} - \frac{9}{28}(x+3)^{7/3} + C.\end{aligned}$$

EXERCISES 7.2

Complete the integrations in Exercises 1–20.

1. $\int xe^{9x}dx$
2. $\int 4xe^{8x}dx$
3. $\int xe^{-x}dx$
4. $\int (3x+7)e^{2x}dx$
5. $\int (xe^{7x} + 4x + 3)dx$
6. $\int \left(xe^{4x} + 3x^2 + \frac{1}{x}\right)dx$

7. $\int (x^3 \ln 5x)dx$

8. $\int (x^{15} \ln 7x)dx$

9. $\int (6x^5 \ln 9x)dx$

10. $\int (3x^4 \ln 6x)dx$

11. $\int (x^8 \ln 3x + e^{2x} + 6)dx$

12. $\int (4x^9 \ln 2x + \sqrt{x})dx$

13. $\int \frac{5x}{\sqrt{x+2}}dx$

14. $\int \frac{3x}{\sqrt{3x+5}}dx$

15. $\int x(x+4)^{-2}\, dx$

16. $\int \frac{x}{(x+2)^5}\, dx$

17. $\int \ln x^3 dx$

18. $\int x^2 e^{5x} dx$ *(Hint: use integration by parts twice)*

19. $\int (3xe^x + 6xe^{x^2} + e^{5x})dx$

20. $\int (x^5 \ln 7x + 6x^2 e^{x^3} + 5x + 1)dx$

7.3 EVALUATION OF DEFINITE INTEGRALS

Recall, from Chapter 6, the Fundamental Theorem of Integral Calculus.

> ***Fundamental Theorem of Integral Calculus***
>
> **The definite integral of $f(x)$ is**
>
> $$\int_a^b f(x)dx = F(b) - F(a)$$
>
> **where $f(x)$ is continuous on $[a, b]$ and $F(x)$ is such that $F'(x) = f(x)$ for all x on $[a, b]$.**

When definite integration is by substitution or by parts, limits of definite integrals need alteration when variables are substituted. Limits of integration are changed to accord with the actual integration variable, as the following example illustrates.

Example 7.3.1 Limit Substitutions for a Definite Integral

Evaluate $\int_0^4 2x\sqrt{x^2+9}dx$.

Solution:

Substitute $u = x^2 + 9$; *so* $du = 2xdx$ *and, on substitution, the integral becomes* $\int u^{1/2}du$.

⟶

Now, consider the integral limits.

At the lower limit $x = 0$, so $u = 0^2 + 9 = 9$ and at the upper limit, $x = 4$ and $u = 4^2 + 9 = 25$.

Therefore, the integral with u as the integration variable and limits reflecting the variable change becomes

$$\int_9^{25} u^{1/2}du = \frac{2}{3}u^{3/2}\Big|_9^{25} = \frac{2}{3}(25)^{3/2} - \frac{2}{3}(9)^{3/2} = \frac{196}{3}$$

↓ *The result mirrors x as the integration variable with the original limits.*

That is, $\int u^{1/2}du = \frac{2}{3}(x^2+9)^{3/2}$. *Evaluating* $\frac{2}{3}\left(x^2+9\right)^{3/2}\Big|_0^4 = \frac{250}{3} - \frac{54}{3} = \frac{196}{3}$ *as previously.*

Example 7.3.2 Integration by Parts with a Definite Integral

Evaluate $\int_0^3 \frac{x}{\sqrt{x+1}}dx.$

Solution:

As a substitution is not evident, try an integration by parts with the integrand $x(x+1)^{-1/2}$. Set $f(x) = x$ and $g'(x) = (x+1)^{-1/2}dx$.

Next, $f'(x)dx = 1dx$ and $g(x) = 2(x+1)^{1/2}$.

Using an integration by parts yields

$$\int_0^3 \frac{x}{\sqrt{x+1}}dx = 2x(x+1)^{1/2}\Big|_0^3 - \int_0^3 2(x+1)^{1/2}dx$$

$$= 2x(x+1)^{1/2}\Big|_0^3 - \frac{4}{3}(x+1)^{3/2}\Big|_0^3$$

$$= [6(4)^{1/2} - (0)(1)^{1/2}] - \left[\frac{4}{3}(4)^{3/2} - \frac{4}{3}(1)^{3/2}\right]$$

$$= 12 - \frac{28}{3} = \frac{8}{3}.$$

↓ *The product differential is $d(f(x)g(x)) = d(x(2(x+1)^{1/2}) = x(x+1)^{-1/2} + 2(x+1)^{1/2}$*

Example 7.3.3 Another Definite Integral

Evaluate $\int_1^2 6x^2e^{x^3+1}dx$

⟶

Solution:
Substitution is an option in this case. Set $u = x^3 + 1$ *so* $du = 3x^2dx$ *and* $2du = 6x^2dx$ *(the "leftover portion"). On substitution, the integral becomes* $2\int e^u du$*, a basic integral.*

Now, consider the limits of integration.

At the lower limit $x = 1$ *so* $u = 1^3 + 1 = 2$*, and at the upper limit* $x = 2$ *so* $u = 2^3 + 1 = 9$*.*

Therefore, the integral with u as the integration variable and limits reflecting the variable change becomes

$$2\int_2^9 e^u du = 2e^u|_2^9 = 2(e^9 - e^2)$$

Note that the substitution is effective in this case because the algebraic portion of the integrand is related to the derivative of the exponential's exponent.

EXERCISES 7.3

Evaluate these definite integrals.

1. $\int_2^7 \left(\frac{x}{\sqrt{x+2}}\right)dx$
2. $\int_2^5 \left(\frac{x}{\sqrt{x-1}}\right)dx$
3. $\int_1^4 3x^2e^{x^3}dx$
4. $\int_0^4 2x\sqrt{x^2+9}dx$
5. $\int_1^2 8x^3e^{x^4}dx$
6. $\int_0^3 \left(\frac{x}{\sqrt{x+1}}\right)dx$
7. $\int_0^5 8x\sqrt{x^2+144}dx$
8. $\int_3^5 x\sqrt{x^2-9}dx$
9. $\int_1^2 (2xe^{x^2} + 4x + 3)dx$
10. $\int_1^2 6x^2\sqrt{x^3+8}dx$
11. $\int_0^1 (9x^2 + 12x)(x^3 + 2x^2 + 1)^3dx$
12. $\int_1^2 2xe^{3x}dx$
13. $\int_1^e x^4 \ln x dx$
14. $\int_1^e \frac{\ln x}{x}dx$
15. $\int_1^5 6x(x+3)^{-3}dx$
16. $\int_0^2 3xe^{x^2+3}dx$

7.4 PARTIAL FRACTIONS

Partial Fractions is another important aid in transforming difficult integrations of rational functions into simpler ones. Recall, $\int \frac{p(x)}{q(x)}dx$ as an integral with rational functions where

$p(x)$ and $q(x)$ are polynomials. To use partial fractions, the degree of the numerator polynomial is less than the degree of the denominator polynomial. (If it is not, use a long division to reduce the degree of the numerator.)

To begin, factor the denominator polynomial when possible. For example, suppose that the denominator is the quadratic $q(x) = 2x^2 - x - 15$. This factors as $(2x + 5)(x - 3)$. The fraction can be written as

$$\frac{1}{q(x)} = \frac{1}{2x^2 - x - 15} = \frac{A}{2x + 5} + \frac{B}{x - 3}$$

where A and B are to be determined. In general, for a factorable quadratic denominator form fractions $\frac{\mathbf{A}}{\mathbf{LF1}}$ and $\frac{\mathbf{B}}{\mathbf{LF2}}$ where **LF1** and **LF2** represent linear factors. We illustrate its usage in the following section.

Example 7.4.1 Partial Fractions

Integrate $\int \frac{2x+3}{x^2-9}dx$ *using partial fractions.*

Solution:

A long division isn't needed as the numerator is of lesser degree than the denominator.

The fraction decomposes into partial fractions with LF1 and LF2 as $x - 3$ *and* $x + 3$*, respectively.*

Set,

$$\frac{2x+3}{(x-3)(x+3)} = \frac{A}{x-3} + \frac{B}{x+3} \text{ and, forming a common denominator, it equals}$$
$$= \frac{A(x+3) + B(x-3)}{(x-3)(x+3)}$$

So, equating numerators, $2x + 3 = A(x + 3) + B(x - 3)$*. Using the critical points (zeros for the denominator of the fraction), in this case* ± 3*, yields:*

If $x = -3$*, the equation becomes* $2(-3) + 3 = -6B$ *so* $B = 1/2$*.*

If $x = 3$*, the equation becomes* $2(3) + 3 = 6A$ *so* $A = 3/2$*.*

So,

$$\int \frac{2x+3}{x^2-9}dx = \int \left(\frac{3}{2\,(x-3)} + \frac{1}{2(x+3)} \right) dx$$
$$= \frac{3}{2} \ln|x-3| + \frac{1}{2} \ln|x+3| + C$$
$$= \ln (x-3)^{3/2}(x+3)^{1/2} + C$$

Note, A and B could have been evaluated by equating the coefficients of like powers of the two polynomials. In this example, the numerator $2x + 3$ *is equated to corresponding coefficients so* $3A - 3B = 3$ *(constants) and* $A + B = 2$ *(coefficient of x). Then, solve the two equations simultaneously for A and B.*

The following is another example of partial fractions.

Example 7.4.2 Partial Fractions and Long Division

Integrate $\int \frac{x^2}{x^2-4}dx$ *using partial fractions.*

Solution:

In this case, a long division is needed as the degree of the numerator is not less than that of the denominator. That is,

$$\begin{array}{r} 1 \\ x^2-4\,\overline{)\,x^2+0x+0} \\ -\quad \underline{x^2 \qquad -4} \\ 4 \end{array}$$

$$\text{So, } \frac{x^2}{x^2-4} = 1 + \frac{4}{x^2-4} = 1 + \frac{4}{(x-2)(x+2)}$$

and the integral is rewritten as

$$\int \frac{x^2}{x^2-4}dx = \int \left(1 + \frac{4}{x^2-4}\right) dx = \int 1dx + 4\int \frac{dx}{(x-2)(x+2)}$$

Next, decompose the last integrand into partial fractions. That is, set

$$\frac{4}{(x-2)(x+2)} = \frac{A}{x-2} + \frac{B}{x+2} = \frac{A(x+2)+B(x-2)}{(x-2)(x+2)}$$

Using the forms with a common denominator, equate numerators to solve for A and B as $4 = A(x+2) + B(x-2)$ *to yield* $A = 1$ *and* $B = -1$.

The integral we seek to evaluate using the partial fraction technique is

$$\int \frac{x^2}{x^2-4}dx = \int \left(1 + \frac{1}{x-2} - \frac{1}{x+2}\right)dx = \int 1dx + \int \frac{dx}{x-2} - \int \frac{dx}{x+2}$$

These are integrations of basic types. It follows that

$$\int \left(1 + \frac{1}{x-2} - \frac{1}{x+2}\right)dx = x + ln\,|x-2| - ln\,|x+2| + C$$

$$= x + ln\left|\frac{x-2}{x+2}\right| + C$$

This completes the required integration.

When a denominator factor is a quadratic, say **QF1** and **LF2**, then form the partial fractions as $\frac{Ax+B}{QF1}, \frac{C}{\text{LF2}}$. The technique generalizes.

If the factorization yields a linear factor raised to a power then use $\frac{A}{\text{LF}}, \frac{B}{(\text{LF})^2}, \frac{C}{(\text{LF})^3}$ and so on, until the highest power term is reached.

The integration of many of these types of problems involves logarithms. The properties of logarithms (see Chapter 4) can simplify the results.

↓ Note: often integrating partial fractions can involve trigonometric functions.

Example 7.4.3 Partial Fractions – Repeated Root

Integrate $\int \frac{8x^2 + 11x - 9}{(x-1)^2(x+4)}dx$ *using partial fractions.*

Solution:

Firstly, note that the numerator is of lower degree than the denominator.

Next, express the integrand as partial fractions. The factored denominator, having three linear factors ($x - 1$ appearing twice) results in three partial fractions.

Note the repeated factor, $(x-1)^2$, as it requires partial fraction terms of $\frac{A}{x-1} + \frac{B}{(x-1)^2}$ *for the decomposition. So,*

$$\frac{8x^2 + 11x - 9}{(x-1)^2(x+4)} = \frac{A}{x-1} + \frac{B}{(x-1)^2} + \frac{C}{x+4} = \frac{A(x-1)(x+4) + B(x+4) + C(x-1)^2}{(x-1)^2(x+4)}$$

To determine A, B, and C simplify and compare numerator terms of the left and right hand sides.

$$8x^2 + 11x - 9 = A(x^2 + 3x - 4) + B(x + 4) + C(x^2 - 2x + 1)$$

If $x = 1$, then $8(1)^2 + 11(1) - 9 = B(1 + 4)$ or $10 = 5B$. Therefore, $B = 2$.

If $x = -4$, then $8(-4)^2 + 11(-4) - 9 = C((-4)^2 - 2(-4) + 1)$ or $75 = 25C$. Therefore, $C = 3$.

Next, using $(A + C)x^2 + (3A + B - 2C)x + (-4A + 4B + C) = 8x^2 + 11x - 9$ results in $B = 2, C = 3, A = 5$.

Integration yields:

$$\int \frac{8x^2 + 11x - 9}{(x-1)^2(x+4)}dx = \int \left(\frac{5}{(x-1)} + \frac{2}{(x-1)^2} + \frac{3}{(x+4)}\right)dx$$

$$= 5\,ln|x-1| - \frac{2}{x-1} + 3\,ln|x+4| + C$$

$$= ln\,(x-1)^5(x+4)^3 - \frac{2}{x-1} + C$$

EXERCISES 7.4

Use partial fractions to integrate in Exercises 1–10.

1. $\int \left(\frac{2}{x^2-1}\right)dx$
2. $\int \left(\frac{x^2}{x^2+x-6}\right)dx$
3. $\int \left(\frac{x^3+3x^2-2}{x^2-1}\right)dx$
4. $\int \left(\frac{2x+3}{x^3-x}\right)dx$
5. $\int \frac{dx}{x^2-9}$
6. $\int \left(\frac{2x+1}{x^2+x}\right)dx$
7. $\int \left(\frac{x^2+2}{x^3-3x^2+2x}\right)dx$
8. $\int \left(\frac{4x^2+13x-9}{x^3+2x^2-3x}\right)dx$
9. $\int \left(\frac{6x^2+7x-4}{x^3+x^2-2x}\right)dx$
10. $\int \left(\frac{4x^2+12x-18}{x^3-9x}\right)dx$

7.5 APPROXIMATING SUMS

Recall the earlier discussion of Riemann Sums (see Chapter 6) where widths $\Delta x = \frac{b-a}{n}$ and heights of rectangles and left endpoints, right endpoints, or midpoints of intervals were used to approximate areas as estimates of definite integrals.

Midpoint Rule

The midpoint rule for approximating the area under a curve was indirectly used in Section 6.2 on Riemann Sums. The following is a variation of the formula used there.

Midpoint Rule

An estimate of the definite integral $\int_a^b f(x)dx$ is

$$[f(m_1)+f(m_2)+\cdots+f(m_n)]\Delta x$$

where $f(m_i)$ is the height of an approximating rectangle at the midpoint, m_i, of the ith interval of width $\Delta x = \frac{b-a}{n}$, $i = 1, \ldots, n$ and n, the number of approximating rectangles. This approximating Riemann Sum is an estimate of the area.

Example 7.5.1 A Midpoint Rule Approximation

Use the midpoint rule to approximate $\int_{1}^{4}(x^2 + x + 1)dx$ *using* $n = 6$.

Solution:
Firstly, note that $\Delta x = \frac{4-1}{6} = \frac{1}{2}$ *and that the six midpoints are at* $x = 1.25, 1.75, 2.25, 2.75, 3.25,$ *and* 3.75.

Therefore,

$$\int_{1}^{4}(x^2 + x + 1)dx \approx [f(1.25) + f(1.75) + f(2.25) + f(2.75) + f(3.25) + f(3.75)]\left(\frac{1}{2}\right)$$

$$= [3.8125 + 5.8125 + 8.3125 + 11.3125 + 14.8125 + 18.8125]\left(\frac{1}{2}\right)$$

$$= 31.4375$$

The actual area is the integration result $\left(\frac{x^3}{3} + \frac{x^2}{2} + x\right)\Bigg|_{1}^{4} = 31.5.$

Trapezoidal Rule

Trapezoids can be used to approximate areas. Consider trapezoids with two parallel sides of height $f(x)$ at any point x.

To illustrate the method, we estimate the area under the curve below from $x = 1$ to $x = 4$ using three trapezoids of equal width $\Delta x = 1$ as shown. There are three subintervals of interest whose endpoints are $e_0 = 1, e_1 = 2, e_2 = 3$, and $e_n = 4$. The integral estimate is

$$\int_{1}^{4} f(x)dx \approx [(f(e_0) + f(e_1)] + [(f(e_1) + f(e_2)](1) + [f(e_2) + f(e_n)](1)$$

$$= [f(e_0) + 2f(e_1) + 2f(e_2) + f(e_n)](1)$$

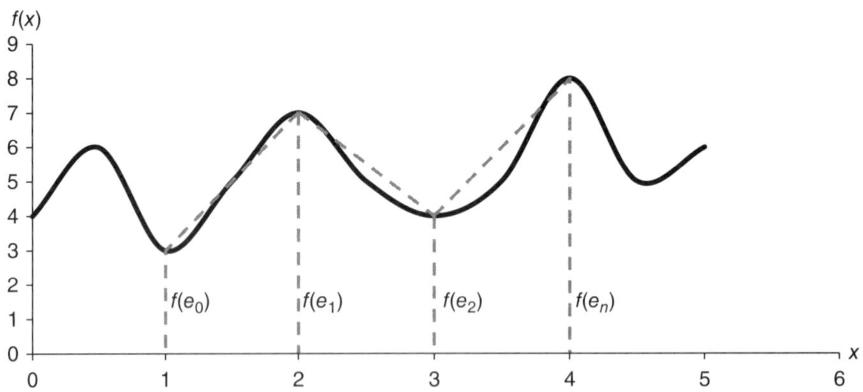

Note that $\Delta x = h = \dfrac{4-1}{3} = 1$ and that the first and last boundary sides are used only once but the interior ones are each used twice. Recall that the area of the trapezoid below is $\left(\dfrac{a+b}{2}h\right)$.

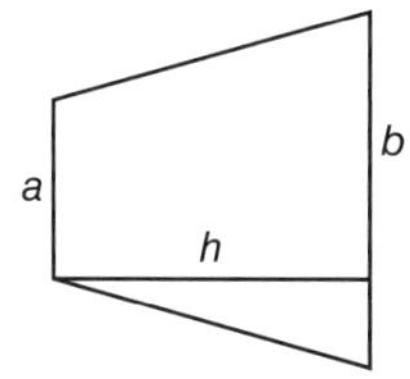

This generalizes to the next formula for the trapezoidal rule.

Trapezoidal Rule

The area under a non-negative function $f(x)$ on $[a, b]$ can be determined by using trapezoids of width $\Delta x = \dfrac{b-a}{n}$. Endpoints of each interval (denoted by e_i) are used to determine the heights of the sides of the trapezoid $f(e_i)$.

$$\int_a^b f(x)dx \approx [f(e_0) + 2f(e_1) + 2f(e_2) + \cdots + 2f(e_{n-1}) + f(e_n)]\left(\frac{\Delta x}{2}\right)$$

where n is the number of trapezoids.

Example 7.5.2 A Trapezoidal Approximation

Use the trapezoidal rule to approximate $\int_1^4 (x^2 + x + 1)dx$ *for* $n = 6$.

Solution:
Firstly, $\Delta x = \dfrac{4-1}{6} = \dfrac{1}{2}$ *and the endpoints including* $e_0 = a$ *and* $e_n = b$ *are, respectively,*

$$x = 1, 1.5, 2, 2.5, 3, 3.5, \text{ and } 4$$

Therefore,

$$\int_1^4 (x^2 + x + 1)dx \approx [f(1) + 2f(1.5) + 2f(2) + 2f(2.5) + 2f(3) + 2f(3.5) + f(4)]\left(\frac{1}{4}\right)$$

$$= [3 + 9.5 + 14 + 19.5 + 26 + 33.5 + 21]\left(\frac{1}{4}\right) = 31.625$$

Numerical approximation techniques, such as the trapezoidal rule, are of importance for digital computation and for functions in a tabular format.

In general, the error from the midpoint rule using rectangles is about half the error of using the trapezoidal rule. These errors are usually in opposite directions, that is, one overestimates actual areas while the other underestimates them. In the two previous examples, the midpoint rule underestimated the actual area by 0.0625(31.4375 – 31.5000), while the trapezoidal rule overestimated the actual area by 0.1250(31.6250 – 31.5000).

Simpson's Rule

Simpson's Rule is another approximation using an even number of subintervals, n, and credited to British mathematician Thomas Simpson (Historical Notes). It is a weighted average of the midpoint and trapezoidal rules.

Simpson's Rule

The area under a non-negative function $f(x)$ on $[a, b]$ can be determined by using both endpoints and midpoints of intervals (denoted by m_i and e_i) as

$$\int_a^b f(x)dx \approx [f(e_0) + 4f(m_1) + 2f(e_1) + \cdots + 2f(e_{n-1}) + 4f(m_n) + f(e_n)]\left(\frac{\Delta x}{6}\right)$$

where n is even and $\Delta x = \dfrac{b-a}{n}$.

Example 7.5.3 A Simpson's Rule Approximation

Use Simpson's rule to approximate $\int_1^4 (x^2 + x + 1)dx$ *when* $n = 6$.

Solution:

Firstly, $\Delta x = \dfrac{4-1}{6} = \dfrac{1}{2}$ *and the six midpoints at* $x = 1.25, 1.75, 2.25, 2.75, 3.25,$ *and* 3.75.

The seven endpoints are at $x = 1, 1.5, 2, 2.5, 3, 3.5, 4$.

Therefore,

$$\int_1^4 (x^2 + x + 1)dx \approx [f(1) + 4f(1.25) + 2f(1.5) + 4f(1.75) + 2f(2) + 4f(2.25)$$
$$+ 2f(2.5) + 4f(2.75) + 2f(3) + 4f(3.25) + 2f(3.5) + 4f(3.75) + f(4)]\left(\frac{1}{12}\right)$$
$$= [3 + 15.25 + 9.5 + 23.25 + 14 + 33.25 + 19.5 + 45.25 + 26 + 59.25 + 33.5$$
$$+ 75.25 + 21]\left(\frac{1}{12}\right) = 31.5$$

EXERCISES 7.5

Divide the designated intervals into the indicated number of subintervals. Record the value of Δx and midpoints.

1. $2 \le x \le 5, \quad n = 6$
2. $1 \le x \le 9, \quad n = 8$

Divide the designated intervals into the indicated number of subintervals. Record the value of Δx and endpoints.

3. $2 \le x \le 5, \quad n = 6$
4. $1 \le x \le 9, \quad n = 8$

In Exercises 5–7, approximate using the midpoint rule.

5. $\int_0^4 (x^2 + 5)dx \quad n = 4$
6. $\int_0^3 (x^3 + 5x + 4)dx \quad n = 6$
7. $\int_1^9 (x^2 + 9x + 8)dx \quad n = 8$

In Exercises 8–10, approximate using the trapezoidal rule.

8. $\int_0^4 (x^2 + 5)dx \quad n = 4$
9. $\int_0^3 (x^3 + 5x + 4)dx \quad n = 6$
10. $\int_1^9 (x^2 + 9x + 8)dx \quad n = 8$

In Exercises 11–13, approximate using Simpson's Rule.

11. $\int_0^4 (x^2 + 5)dx \quad n = 4$
12. $\int_0^3 (x^3 + 5x + 4)dx \quad n = 6$
13. $\int_1^9 (x^2 + 9x + 8)dx \quad n = 8$

In Exercises 14 and 15, the table defines f(x)

x	2.0	2.5	3.0	3.5	4.0	4.5	5.0	5.5	6.0
$f(x)$	4.1	5.3	5.7	6.1	7.4	3.6	4.5	3.9	7.8

14. Approximate $\int_2^5 f(x)dx$ using the trapezoidal rule with $n = 6$.

15. Approximate $\int_2^6 f(x)dx$ using the trapezoidal rule with $n = 4$.

7.6 IMPROPER INTEGRALS

Improper integrals are definite integrals with infinite limit(s) such as

$$\int_a^\infty f(x)dx = \lim_{t\to\infty}\int_a^t f(x)dx$$

$$\text{and}\quad \int_{-\infty}^a f(x)dx = \lim_{t\to-\infty}\int_t^a f(x)dx$$

$$\text{or}\quad \int_{-\infty}^\infty f(x)dx = \lim_{t\to-\infty}\int_t^0 f(x)dx + \lim_{t\to\infty}\int_0^t f(x)dx$$

The integrals are said to **converge** if the limits exist as $|t|$ increases without bound. The integrals are said to **diverge** when, as $|t|$ increases, the limits do not exist.

Example 7.6.1 A Divergent Improper Integral

Evaluate the following integral $\int_3^\infty \frac{1}{x-1}dx$, *if it exists.*

Solution:

Firstly, rewrite the integral as $\int_3^\infty \frac{1}{x-1}dx = \lim_{t\to\infty}\int_3^t \frac{1}{x-1}dx$. *Next, evaluate the integral:*

$$\int_3^t \frac{1}{x-1}dx = ln|x-1||_3^t = ln|t-1| - ln\,2\,.$$

Now, in the limit as $t \to \infty$

$\lim_{t\to\infty}[ln|t-1| - ln2]$ *is infinite so the integral diverges.*

Example 7.6.2 A Convergent Improper Integral

Evaluate the integral $\int_{-\infty}^{2} e^{3x}dx$, *if it exists.*

Solution:

Firstly, rewrite the integral as $\int_{-\infty}^{2} e^{3x}dx = \lim_{t\to-\infty}\int_{t}^{2} e^{3x}dx.$
Next, evaluate the last integral.

$$\int_{t}^{2} e^{3x}dx = \left.\frac{e^{3x}}{3}\right|_{t}^{2} = \frac{e^{6}}{3} - \frac{e^{3t}}{3}$$

Now, taking limits

$\lim_{t\to-\infty}\left[\frac{e^{6}}{3} - \frac{e^{3t}}{3}\right]$ *converges to* $\frac{e^{6}}{3}$ *as* $\frac{e^{3t}}{3}$ *tends to zero as* $t \to -\infty$.

When calculating improper integrations, as $|t|$ approaches infinity, either the limit doesn't exist (approaches infinity) so the integral diverges, or the limit tends to zero and the integral converges.

Example 7.6.3 Improper Integrals and Substitution

Evaluate the integral $\int_{2}^{\infty} 4x^{3}e^{-x^{4}}dx$, *if it exists.*

Solution:

Firstly, rewrite the integral as $\int_{2}^{\infty} 4x^{3}e^{-x^{4}}dx = \lim_{t\to\infty}\int_{2}^{t} e^{x^{4}}(4x^{3}dx)$. *Next, evaluate the integral on the right. Use a substitution with* $u = x^{4}$, $du = 4x^{3}dx$. *When* $x = 2$, $u = 16$, *and when* $x = t$, $u = t^{4}$. *Thus,*

$$\int_{16}^{t^{4}} e^{-u}du = -e^{-u}|_{16}^{t^{4}} = -e^{-t^{4}} + e^{-16}$$

Now, taking the limit

$\lim_{t\to\infty}[-e^{-t^{4}} + e^{-16}] = e^{-16}$ *since* $-e^{-t^{4}}$ *tends to zero.*
The integral converges.

EXERCISES 7.6

In Exercises 1–16, evaluate the integrals, if they exist.

1. $\int_{5}^{\infty} \frac{1}{x-3}dx$

2. $\int_{6}^{\infty} \frac{1}{(x-4)^{2}}dx$

3. $\int_1^\infty \frac{2}{x^4}dx$

4. $\int_8^\infty \frac{3}{x^{4/3}}dx$

5. $\int_{-\infty}^4 e^{4x}dx$

6. $\int_{-\infty}^1 e^{-5x}dx$

7. $\int_1^\infty e^{3x+1}dx$

8. $\int_4^\infty e^{-x}dx$

9. $\int_1^\infty 2xe^{-x^2}dx$

10. $\int_1^\infty 6x^2e^{-x^3}dx$

11. $\int_2^\infty \frac{1}{x\ln x}dx$

12. $\int_3^\infty \frac{\ln x}{x}dx$

13. $\int_2^\infty \frac{x^2}{\sqrt{x^3-4}}dx$

14. $\int_0^\infty \frac{4x}{(x^2+1)^3}dx$

15. $\int_{-\infty}^2 \frac{2}{(4-x)^3}dx$

16. $\int_{-\infty}^1 \frac{1}{(x-2)^2}dx$

HISTORICAL NOTES

Thomas Simpson (1710–1761) – British mathematician at the Royal Military Academy in Woolwich, England. A solar eclipse in 1724 fostered his early mathematical interests.

Simpson authored texts on algebra (1745), geometry (1747), and trigonometry (1748). In the latter, he introduced abbreviations for trigonometric functions still in use today. However, he is usually remembered for his work on numerical integration; curiously he did not originate "Simpson's Rule".

CHAPTER 7 SUPPLEMENTARY EXERCISES

In Exercises 1–10, integrate the integrals using any of the techniques studied thus far.

1. $\int 8x^2\sqrt{x^3+9}dx$

2. $\int 2x^3e^{x^4}dx$

3. $\int \frac{4x^3+2x}{x^4+x^2+5}dx$

4. $\int (3x^2+5x)(2x^3+5x^2+3)^8dx$

5. $\int \frac{5x}{\sqrt[3]{5x+4}}dx$

6. $\int (5xe^x+8xe^{x^2}+e^{4x}+5)dx$

7. $\int (5x+1)e^{3x}dx$

8. $\int (5x^6\ln 9x)dx$

9. $\int_1^2 20xe^{5x^2}dx$

10. $\int_2^5 \left(\frac{x}{\sqrt{x-1}}\right)dx$

11. Use partial fractions to integrate $\int \frac{2x^2-25x-33}{(x+1)^2(x-5)}dx$.

12. Approximate, using the midpoint rule, trapezoidal rule, and Simpson's rule.

$$\int_{1}^{13} (2x^2 + 5x - 3)dx \quad n = 6$$

13. Approximate using the midpoint rule, trapezoidal rule, and Simpson's rule.

$$\int_{2}^{6} e^x dx \quad n = 4$$

In Exercises 14–16, evaluate the improper integrals, if they exist.

14. $\int_{2}^{\infty} \frac{3}{x^2} dx$

15. $\int_{-\infty}^{3} e^{2x+1} dx$

16. $\int_{4}^{\infty} (2xe^{-x^2} + 4x - 1)dx$

8 *Functions of Several Variables*

To here, functions, $f(x)$, have been of a single variable, x. Many applications, if not most, use functions of more than a single independent variable to adequately describe phenomena of interest.

Fundamentals of Calculus, First Edition. Carla C. Morris and Robert M. Stark.

Companion Website: http://www.wiley.com/go/morris/calculus

↓ You can easily imagine examples: net profit depends on variables as price, cost, quantity on hand, advertising, replenishment, and deterioration in inventory. Life spans of animals are a function of nutrition, prenatal experience, environment, genetics, and so on.

Let, $f(x, y)$, for example, represent a function of two independent variables, x and y. Generally, $f(x_1, x_2, \ldots, x_n)$ represents a function of n independent variables $x_1, x_2, \ldots, x_n$, $n = 1, 2, \ldots$.

This chapter extends the calculus of a single variable to a calculus of several variables. Although it suggests a formidable effort, it is a surprisingly easy generalization.

8.1 FUNCTIONS OF SEVERAL VARIABLES

Functions such as $f(x) = x^2 + 5x + 4$ are of a single variable. Functions of two variables, $f(x, y)$, (also known as **bivariate**), are a rule that assigns a number to each pair of values of the variables. The function $f(x, y) = x^2 + 3y + 2x + 1$ is an example. In this case, $f(1, 2)$ means that $x = 1$ and $y = 2$. Therefore, $f(1, 2) = (1)^2 + 3(2) + 2(1) + 1 = 10$.

Functions may have three or more variables. For the **trivariate** function $f(x, y, z) = x^2 + y^3 + e^z$, for example, $f(1, 2, 0)$ has $x = 1$, $y = 2$, $z = 0$, and a value of 10.

Example 8.1.1 Evaluate a Bivariate Function

For $f(x, y) = 3x^2 + 2y^3 + 4y + 2$, evaluate $f(1, 4)$, $f(2, 3)$, and $f(1 + h, 2)$.

Solution:

$f(1, 4) = 3(1)^2 + 2(4)^3 + 4(4) + 2 = 3 + 128 + 16 + 2 = 149.$
$f(2, 3) = 3(2)^2 + 2(3)^3 + 4(3) + 2 = 12 + 54 + 12 + 2 = 80.$
$f(1 + h, 2) = 3(1 + h)^2 + 2(2)^3 + 4(2) + 2 = 3h^2 + 6h + 3 + 16 + 8 + 2 = 3h^2 + 6h + 29.$

Example 8.1.2 Evaluate a Trivariate Function

For $f(x, y, z) = 2x^3 + 3y + z^2$, evaluate $f(1, 2, 3)$, and $f(2a, b, 3c)$.

Solution:

$f(1, 2, 3) = 2(1)^3 + 3(2) + (3)^2 = 2 + 6 + 9 = 17.$
$f(2a, b, 3c) = 2(2a)^3 + 3(b) + (3c)^2 = 16a^3 + 3b + 9c^2.$

Level Curves

Graphs of functions are familiar for a single variable in the two-dimensional Cartesian plane. For instance, the graph of $f(x) = x^2 - 3x + 2$ is an opening upward parabola.

A bivariate function, such as $f(x, y) = x^2 + y^2$, has a three-dimensional graph. However, if $f(x, y)$ is assigned a constant value, it can be graphed in two dimensions as a **level curve**.

> ***Level Curve of a Bivariate Function***
>
> **Let $f(x, y)$ be a bivariate function. The set of all ordered pairs (x, y) for which $f(x, y) = c$, a fixed value, is a level curve of $f(x, y)$.**

Example 8.1.3 Level Curves

Sketch level curves for $f(x, y) = x^2 + y^2$ when $c = 9$ and $c = 25$.

Solution:
We seek graphs of $x^2 + y^2 = 9$ and $x^2 + y^2 = 25$ on a common coordinate axes. The concentric circles have radii 3 and 5, respectively.

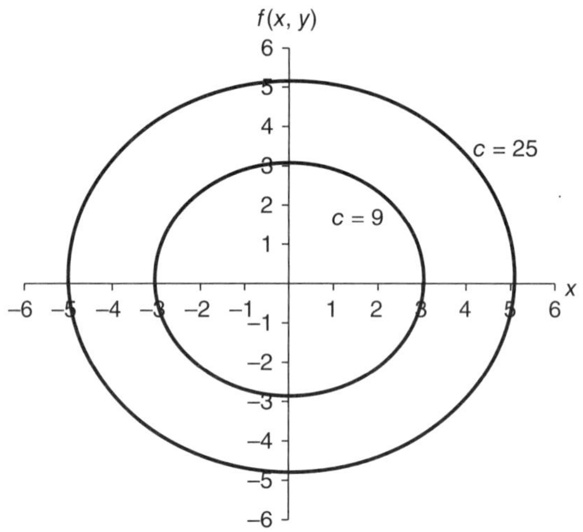

The family of level curves of $f(x, y) = x^2 + y^2$ is concentric circles centered at the origin.

EXERCISE 8.1

Evaluate $f(x, y)$ at the indicated points.

1. $f(x, y) = 2x + 3y$ at $f(2, 5), f(3, -1)$, and $f(4, -3)$.
2. $f(x, y) = x^2 + 5y$ at $f(0, 4), f(1, 3)$, and $f(2.5, 4.2)$.

3. $f(x, y) = 4x + 3y^2$ at $f(1, 0), f(2, -1)$, and $f(2, 1)$.
4. $f(x, y) = x^2 + y^2$ at $f(3, 4), f(5, 0)$, and $f(\sqrt{2}, \sqrt{5})$.
5. $f(x, y) = x^2 + 3x + y^3 + 2y + 5$ at $f(1, 2), f(0, 1)$, and $f(-1, -2)$.
6. $f(x, y) = x^3e^x + 3y^2 + 2y$ at $f(1, 2), f(0, 1)$, and $f(0, 4)$.
7. $(x, y) = x + 2y^3 + e^y$ at $f(2, 0), f(0, 1)$, and $f(3, 0)$.
8. $f(x, y) = 2xy^2 + 3x^4y$ at $f(1, -2), f(0, 3)$, and $f(-1, 2)$.
9. $f(x, y, z) = x^2 + 3y + z^3$ at $f(0, 1, 2), f(0.5, 1, 1)$, and $f(-1, 0, 2)$.
10. $f(x, y, z) = x + 3y + z^2 + 1$ at $f(1, 2, 3)$ and $f(3, -2, 1)$.
11. $f(x, y) = x^2 + y^2$. Show that $f(3 + h, 4) - f(3, 4) = h^2 + 6h$.
12. $f(x, y) = x^2 + e^y$. Show that $f(x + h, y) - f(x, y) = 2xh + h^2$.
13. For $f(x, y) = 7x^{2/3}y^{1/3}$ show that $f(2a, 2b) = 2f(a, b)$.
14. For $f(x, y) = 50x^{3/4}y^{1/4}$ show that $f(5a, 5b) = 5f(a, b)$.

In Exercises 15–18, draw level curves when $f(x, y) = 1, 4$, and 9.

15. $f(x, y) = x + y$
16. $f(x, y) = x^2 + y$
17. $f(x, y) = xy$
18. $f(x, y) = 2x - y$

8.2 PARTIAL DERIVATIVES

Let $z = f(x, y)$ be a function of two independent variables, x and y. Now, two derivatives are possible: one with respect to x and a second with respect to y. Called **partial derivatives** of x and y, they are obtained by differentiating one of the variables, imagining the other variable to be momentarily constant.

The partial derivative of $f(x, y)$ with respect to x, (y held constant!) is denoted by f_x. Likewise, the partial derivative of $f(x,y)$ with respect to y (x held constant!) is denoted by f_y. An alternate and popular notation is $\frac{\partial f}{\partial x} = f_x$ and $\frac{\partial f}{\partial y} = f_y$. For instance, when $z = f(x, y) = x^2y, f_x = \frac{\partial f}{\partial x} = 2xy$ and $f_y = \frac{\partial f}{\partial y} = x^2$. Note that there is a (partial) derivative for each variable.

◆ The symbol for a partial derivative was devised as a "bent back" letter *d* to distinguish it from an ordinary derivative. The term "partial" implies that more than one variable is subject to differentiation. It is necessary to indicate to which variable the derivative applies.

A formal definition of partial derivatives derived from limits is:

Partial Derivatives

The partial derivative of $f(x, y)$ with respect to x at the point (x, y) is

$$f_x = \frac{\partial f}{\partial x} = \lim_{h \to 0} \frac{f(x+h, y) - f(x, y)}{h}$$

Similarly, the partial derivative of $f(x, y)$ with respect to y at (x, y) is

$$f_y = \frac{\partial f}{\partial y} = \lim_{h \to 0} \frac{f(x, y+h) - f(x, y)}{h}$$

Example 8.2.1 Limits and Partial Derivatives

Use limits to determine partial derivatives f_x and f_y of $f(x, y) = 3x^2 + 2y^3$.

Solution:

$$f_x = \lim_{h \to 0} \frac{f(x+h, y) - f(x, y)}{h} = \lim_{h \to 0} \frac{[3(x+h)^2 + 2y^3] - [3x^2 + 2y^3]}{h}$$

$$= \lim_{h \to 0} \frac{[3x^2 + 6xh + 3h^2 + 2y^3] - [3x^2 + 2y^3]}{h}$$

$$= \lim_{h \to 0} \frac{h(6x + 3h)}{h} = 6x$$

$$f_y = \lim_{h \to 0} \frac{f(x, y+h) - f(x, y)}{h} = \lim_{h \to 0} \frac{[3x^2 + 2(y+h)^3] - [3x^2 + 2y^3]}{h}$$

$$= \lim_{h \to 0} \frac{[3x^2 + 2y^3 + 6y^2h + 6yh^2 + 2h^3] - [3x^2 + 2y^3]}{h} = \lim_{h \to 0} \frac{h(6y^2 + 6yh + 2h^2)}{h} = 6y^2$$

The partial derivatives obtained directly are $f_x = 6x$ and $f_y = 6y^2$.

↓ The previous example is to remind you that a limit is at the core of the definition of the concept of a derivative.

Using limits is impractical to obtain derivatives, including partial derivatives. Usually, partial differentiation applies to a single variable, the object of the differentiation, treating other variables as constants. The next example is illustrative.

Example 8.2.2 Partial Derivatives

Obtain partial derivatives of $f(x, y) = 3x^2 + 2y^3$.

⟶

Solution:
To obtain f_x, the partial derivative with respect to x, differentiate with respect to x with y fixed. (Only terms with x have nonzero derivatives). Therefore, the derivative with respect to x of $f(x, y)$ is $\frac{\partial f}{\partial x} = f_x = 6x$, (the $2y^3$ regarded as constant; its derivative is zero). Similarly, to obtain f_y, differentiate with respect to y while with x fixed. (Only terms with y have nonzero derivatives). Therefore, the derivative with respect to y of $f(x, y)$ is $\frac{\partial f}{\partial y} = f_y = 6y^2$ (the derivative of $3x^2$ with respect to y is zero).

Example 8.2.3 More Partial Derivatives

Obtain partial derivatives of $f(x, y) = 5x^3y^4 + 3xy^2$.

Solution:
For f_x, view $f(x, y)$ as $[5y^4]x^3 + [3y^2]x$, where the brackets enclose constants (as differentiation is with respect to x).

$$\frac{\partial f}{\partial x} = f_x = [5y^4](3x^2) + 3y^2(1)$$
$$= 15x^2y^4 + 3y^2$$

For f_y, view $f(x, y)$ as $[5x^3]y^4 + [3x]y^2$ so

$$\frac{\partial f}{\partial y} = f_y = [5x^3](4y^3) + [3x](2y)$$
$$= 20x^3y^3 + 6xy$$

Note, again, that only terms with the variable of differentiation have nonzero derivatives.

Example 8.2.4 The Power Law

Obtain partial derivatives of $f(x, y) = (5x^2 + 2y^3 + 7)^{10}$.

Solution:
For either partial derivative, the function is first viewed as a power law. In taking derivatives, the exponent becomes the coefficient and is decreased by a unit. Then, take the partial derivative with respect to the appropriate variable in the base (within parentheses). That is,

$$f_x = 10(5x^2 + 2y^3 + 7)^9(10x) = 100x(5x^2 + 2y^3 + 7)^9$$

and

$$f_y = 10(5x^2 + 2y^3 + 7)^9(6y^2) = 60y^2(5x^2 + 2y^3 + 7)^9$$

Example 8.2.5 Partial Derivatives with an Exponential

Obtain partial derivatives of $f(x, y) = e^{x^2y^3} + 2x + 3y^2$.

Solution:
For the first term, recall and adapt $d/dx(e^{g(x)}) = e^{g(x)}\, g'(x)$.

To obtain f_x: $e^{x^2y^3}\left[\frac{\partial}{\partial x}(x^2y^3)\right] + 2 + 0 = 2xy^3e^{x^2y^3} + 2$.

To obtain f_y: $e^{x^2y^3}\left[\frac{\partial}{\partial y}(x^2y^3)\right] + 0 + 6y = 3x^2y^2e^{x^2y^3} + 6y$.

Partial derivatives, generally of several variables, are evaluated as other functions. For example, $f_x(a, b)$ is $f_x(x, y)$ evaluated at $x = a$ and $y = b$. Similarly, $f_y(a, b)$ is $f_y(x, y)$ evaluated at $x = a$ and $y = b$.

Example 8.2.6 Evaluating Partial Derivatives at (x, y)

Evaluate partial derivatives of $f(x, y) = 6x^3 + 3y^2 + 5x^2y$ *at* $x = 1$ *and* $y = 0$.

Solution:
Firstly, obtain the partial derivatives. Next, substitute values for x and y to yield

$$f_x = 18x^2 + 10xy \qquad f_x(1, 0) = 18$$
$$f_y = 6y + 5x^2 \qquad f_y(1, 0) = 5$$

The partial derivative of $f(x, y)$ with respect to x is simply an ordinary derivative while treating y as a constant. So, f_x is the rate of change of $f(x, y)$ with respect to x, while y is fixed.

Increasing x by a small amount produces a change in $f(x, y)$ that is approximately f_x with y fixed. In the previous example, $f_x(1, 0) = 18$. We interpret this as $f(x, y)$ changing at a rate of 18 (tiny) units for each (tiny) unit that x changes about $x = 1$ when y is kept at zero.

↓ The reason to qualify change as "tiny" is so $f(x, y)$ and its derivatives change continuously with changes in x and in y.

That is, for a small change, h,

$$f(1 + h, 0) - f(1, 0) \approx 18\, h$$
$$f(a + h, b) - f(a, b) \approx f_x(a, b) \cdot h$$
$$f(a, b + k) - f(a, b) \approx f_y(a, b) \cdot k$$

Similarly for f_y.

A function may have any number of variables. Partial derivatives of functions of many variables are always with respect to a single variable at a time, other variables held constant.

Example 8.2.7 Partial Derivatives of Several Variables

Consider $f(x, y, z) = x^4 + 3x^3y^4 + 5x^2y^3z^4 + z^5$, *a function of three variables. Find the three partial derivatives,* f_x, f_y, *and* f_z.

Solution:

To find f_x, *regard the function as*

$$f(x, y, z) = (x^4) + [3y^4](x^3) + [5y^3z^4](x^2) + z^5$$

in which bracketed quantities are regarded as "constant coefficients".

Hence,

$$f_x = (4x^3) + [3y^4](3x^2) + [5y^3z^4](2x) = 4x^3 + 9x^2y^4 + 10xy^3z^4.$$

For f_y, *regard the function as*

$$f(x, y, z) = x^4 + [3x^3](y^4) + [5x^2z^4](y^3) + z^5$$

from which

$$f_y = [3x^3](4y^3) + [5x^2z^4](3y^2) = 12x^3y^3 + 15x^2y^2z^4.$$

To find f_z, *regard the function as*

$$f(x, y, z) = x^4 + 3x^3y^4 + [5x^2y^3](z^4) + (z^5)$$

from which

$$f_z = [5x^2y^3](4z^3) + (5z^4) = 20x^2y^3z^3 + 5z^4.$$

Economics – Cobb–Douglas Production Function

A firm's *output*, Q, for the *Cobb–Douglas Production Function* (or, simply, *production function*) is the bivariate function

$$Q = AK^{\alpha}L^{1-\alpha}$$

Here, A, is a constant; K, denotes capital investment; L, size of the labor force; and $0 < \alpha < 1$.

↓ Cobb–Douglas Production functions were a landmark in macroeconomics. It was a first aggregate economy wide model. An Internet search is rewarding.

The partial derivatives with respect to labor and capital are called *marginal productivities*. If labor is fixed and capital increased by one unit, then output increases by $Q_k = [AL^{1-\alpha}](\alpha K^{\alpha-1}) = \frac{\alpha}{K}Q$. This partial derivative of Q with respect to K is evaluated at relevant values of Q and K.

Example 8.2.8 Marginal Productivity

Consider the Cobb–Douglas Production Function $f(x, y) = 100\, x^{1/4}\, y^{3/4}$. Here, $f(x, y)$ is the quantity of goods produced utilizing x units of capital and y units of labor.

a) Find the marginal productivities of capital and labor.
b) Evaluate and interpret the marginal productivities when $x = 256$ units and $y = 81$ units.

(For convenience, K, L, and Q used earlier are replaced by x, y, and $f(x, y)$, respectively.)

Solution:
a) In this case, $f_x = 100\,((1/4)\,x^{-3/4})\,y^{3/4}$ is the marginal productivity of capital and $f_y = 100\,x^{1/4}((3/4)y^{-1/4})$ the marginal productivity of labor.
b) The evaluation of the partial derivatives at the indicated point (256, 81) is

$$f_x = 100(1/4)(256^{-3/4})(81^{3/4}) = 100(1/4)(1/64)(27) \approx 10.55 \text{ and}$$
$$f_y = 100(256^{1/4})(3/4)(81^{-1/4}) = 100(4)(3/4)(1/3) = 100$$

In words, if labor is constant at 81 units and capital is increased by one unit, then the quantity of goods produced increases by about 10.55 units. Likewise, if capital is held constant at 256 units and labor is increased by one unit, then the quantity of goods produced increases by 100 units.

◆ Economists often use a *ceteris paribus ("all other things being equal")* assumption to avoid undue influences. This enables exploration in a simplified setting using partial derivatives.

EXERCISES 8.2

In Exercises 1–14, determine f_x and f_y.

1. $f(x, y) = 5x^4 - 3y^2 + 2y - 6x + 4$
2. $f(x, y) = 3x^4 - 5x^9y + 2y^3 + 9x - 3y^2 + 50$
3. $f(x, y) = (5x^7 + 4y^5 + 7y + 3)^{25}$
4. $f(x, y) = (10x^3 + 3y^2 + 6y)^{100}$
5. $f(x, y) = e^{x^3y^5} + 9x^4 + 3y + 2$

6. $f(x, y) = e^{5xy} + 9x^2 + 3y^5$

7. $f(x, y) = \dfrac{y^5}{x}$

8. $f(x, y) = \dfrac{x - 3y}{5x - y}$

9. $f(x, y) = \sqrt[3]{x^2 y}$

10. $f(x, y) = \sqrt[3]{x^4 y^2} + 5x + 2e^{3y}$

11. $f(x, y) = x^3 e^x y^8$

12. $f(x, y) = (x^5 \ln 2x) y^8$

13. $f(x, y) = x^2 \sqrt[3]{y - x}$

14. $f(x, y) = \ln xy$

In Exercises 15–18, determine f_x, f_y, and f_z for the functions.

15. $f(x, y, z) = 5x^3 + 2y^2 + 3z^4$

16. $f(x, y, z) = x^3 y^4 + y^2 z^2 + \ln yz$

17. $f(x, y, z) = 5y + x^3 y^2 + 3yz^4 + e^{4xz}$

18. $f(x, y, z) = 5x^4 + xe^x + 8y^7 z^4 + e^{4yz}$

19. Determine f_x and f_y at $x = 4$ and $y = 1$ for $f(x, y) = 3x^2 y^3 + 2x + 3y^6 + 9$.

20. Determine f_x and f_y at $x = 3$ and $y = 2$ for $f(x, y) = 7x^2 y^5 + 2y + 5x^3 + 1$.

21. For the production function $f(x, y) = 4x^{3/4} y^{1/4}$, find the marginal productivities of x and y.

22. A business' productivity is $f(x, y) = 500\, x^{1/3} y^{2/3}$ where x and y represent units of capital and labor, respectively. Compute and interpret the marginal productivities of capital and of labor when $x = 125$ units and $y = 216$ units.

8.3 SECOND-ORDER PARTIAL DERIVATIVES – MAXIMA AND MINIMA

As functions of a single variable may have second-order, third-order, or higher derivatives, so do functions of several variables, that is, multivariate functions. Consider (second) partial derivatives of a (first) partial derivative; it is aptly called the *second partial derivative*.

Clearly, as a differentiable function, the first partial derivatives can be (partially) differentiated again in analogy to the second derivative studied in single variable calculus. Consider $f(x, y)$ and its partial derivatives

$$\frac{\partial f}{\partial x} = f_x(x, y) = f_x \quad \text{and} \quad \frac{\partial f}{\partial y} = f_y(x, y) = f_y$$

Four second partial derivatives are possible depending on the choice of variable of differentiation:

$$\frac{\partial^2 f}{\partial x^2} = \frac{\partial f_x(x, y)}{\partial x} = f_{xx}(x, y) = f_{xx} \qquad \frac{\partial^2 f}{\partial y^2} = \frac{\partial f_y(x, y)}{\partial y} = f_{yy}(x, y) = f_{yy}$$

$$\frac{\partial^2 f}{\partial x \partial y} = \frac{\partial f_y(x, y)}{\partial x} = f_{xy}(x, y) = f_{xy} \qquad \frac{\partial^2 f}{\partial y \partial x} = \frac{\partial f_x(x, y)}{\partial y} = f_{yx}(x, y) = f_{yx}$$

Note the notation! As $f_x = \dfrac{\partial f}{\partial x}$ denoted the first partial derivative with respect to x, so $f_{xx} = \dfrac{\partial^2 f}{\partial x^2}$ denotes the second partial derivative with respect to x; a partial derivative of the first partial derivative with respect to x.

There are four such second-order partial derivatives. Of the four, two are generally the same! If f_{xy} and f_{yx} are continuous functions, the two *mixed derivatives* are equal. That is,

$$\frac{\partial^2 f}{\partial x \partial y} = f_{xy} = f_{yx} = \frac{\partial^2 f}{\partial y \partial x};$$

the order of differentiation being immaterial. This fact can be a double check on a differentiation. The results are summarized as follows:

Second-Order Partial Derivatives

$$\frac{\partial^2 f(x, y)}{\partial x^2} = \frac{\partial}{\partial x}\left(\frac{\partial f(x, y)}{\partial x}\right) = f_{xx}$$

$$\frac{\partial^2 f(x, y)}{\partial y^2} = \frac{\partial}{\partial y}\left(\frac{\partial f(x, y)}{\partial y}\right) = f_{yy}$$

$$f_{xy} = \frac{\partial^2 f(x, y)}{\partial y \partial x} = \frac{\partial}{\partial y}\left(\frac{\partial f(x, y)}{\partial x}\right) = \frac{\partial^2 f(x, y)}{\partial x \partial y} = \frac{\partial}{\partial x}\left(\frac{\partial f(x, y)}{\partial y}\right) = f_{yx}$$

Example 8.3.1 Second-Order Partial Derivatives

For $f(x, y) = x^5y^3 + 2x^3 + 5y^2 + 3$, find the second partial derivatives of the function.

Solution:
To start, the first-order derivatives are

$$f_x = 5x^4y^3 + 6x^2 \quad \text{and} \quad f_y = 3x^5y^2 + 10y$$

Now, the second partial derivatives follow from first-order derivatives as

$$f_{xx} = 20x^3y^3 + 12x \qquad f_{yy} = 6x^5y + 10$$

$$f_{xy} = 15x^4y^2 \qquad f_{yx} = 15x^4y^2$$

Note that $f_{xy} = f_{yx}$.

Example 8.3.2 More on Second-Order Partial Derivatives

If $f(x, y) = 5x^8y^{2/3} + \ln x + e^{xy}$, find second partial derivatives.

Solution:
To start, the first-order derivatives are

$$f_x = 40x^7y^{2/3} + \frac{1}{x} + ye^{xy} \quad \text{and} \quad f_y = \frac{10}{3}x^8y^{-1/3} + xe^{xy}$$

Now, the second-order partial derivatives are determined from the first-order derivatives

$$f_{xx} = 280x^6y^{2/3} - \frac{1}{x^2} + y^2e^{xy} \quad f_{yy} = \frac{-10}{9}x^8y^{-4/3} + x^2e^{xy}$$

$$f_{xy} = \frac{80}{3}x^7y^{-1/3} + yxe^{xy} + e^{xy} \quad f_{yx} = \frac{80}{3}x^7y^{-1/3} + xye^{xy} + e^{xy}$$

(Note: to find the second partial derivatives, the terms ye^{xy} and xe^{xy} utilize the product rule.)

Maxima and Minima

Recall, that for functions of a single variable, first and second derivatives enabled identification of maxima and minima. Similar properties apply to functions of several variables.

For $f(x, y)$, a continuous function with vanishing partial derivatives f_x and f_y at $x = a$, $y = b$, the presence and type of extremum are indicated by the second derivative as:

Second Derivative Test for Functions of Two Variables

For a bivariate function $f(x, y)$ with, $f_x(a, b) = f_y(a, b) = 0$ at (a, b) let

$$D = f_{xx}(a, b)f_{yy}(a, b) - [f_{xy}(a, b)]^2$$

Then for,

$D > 0$ and $f_{xx}(a, b) > 0$, $f(x, y)$ has a local minimum at (a, b).
$D > 0$ and $f_{xx}(a, b) < 0$, $f(x, y)$ has local maximum at (a, b).
$D < 0$, $f(x, y)$ has no extremum at (a, b).
$D = 0$ the test is inconclusive.

The test derivative, D, is known as a **discriminant**.

Example 8.3.3 Locating Extrema

Identify extrema of $f(x, y) = x^2 + 3y^2 + 4x + 6y + 8$.

⟶

Solution:
To locate extrema, set the first partial derivatives to zero. The resulting system of equations is

$$f_x = 2x + 4 = 0$$

$$f_y = 6y + 6 = 0$$

This system is easily solved for the only possible extremum at $(-2, -1)$.
The second-order partial derivatives are

$$f_{xx} = 2 \qquad f_{yy} = 6 \qquad f_{xy} = 0 \qquad f_{yx} = 0.$$

Next, D, is evaluated at $(-2, -1)$. *That is,*

$$D = f_{xx} f_{yy} - (f_{xy})^2 = (2)(6) - [0]^2 = 12$$

As $D > 0$, *an extremum exists, and as* f_{xx} *is positive, the extremum is a local minimum at* $(-2, -1)$. *The minimum value of* $f(x, y)$ *there is* $f(-2, -1) = 1$.

You are familiar with solving systems of linear equations using either substitution or elimination (Chapter 1). However, sometimes, derivatives of multivariate functions are non-linear systems of equations. Such systems are usually more difficult to solve simultaneously.

Example 8.3.4 More Maxima and Minima

Identify extrema of $f(x, y) = 3x^2 - 6xy + y^3 - 9y$

Solution:
Firstly, find all first and second partial derivatives of $f(x, y)$. *They are*

$$f_x = 6x - 6y, \quad f_y = -6x + 3y^2 - 9, \quad f_{xx} = 6, \quad f_{yy} = 6y, \quad f_{xy} = f_{yx} = -6$$

Setting $f_x = 0$ *yields* $x = y$.
Substituting into f_y *yields* $3y^2 - 6y - 9 = 3(y^2 - 2y - 3) = 3(y - 3)(y + 1) = 0$.
This signals possible extrema at $(-1, -1)$ *and* $(3, 3)$ *as* $x = y$.
Evaluate D at $(-1, -1)$ *to obtain*

$$D = (6)(-6) - [-6]^2 = -72 < 0$$

As $D < 0$, *there is no extremum at* $(-1, -1)$.
Now, to investigate the other possibility at $(3, 3)$, *calculate*

$$D = (6)(18) - [-6]^2 = 72 > 0$$

As $D > 0$, *there is an extremum, and as* $f_{xx} > 0$, *it is a minimum.*
Therefore, $f(3, 3) = 3(3)^2 - 6(3)(3) + (3)^3 - 9(3) = -27$ *is a local minimum.*

EXERCISE 8.3

In Exercises 1–10, determine first and second partial derivatives of the functions.

1. $f(x, y) = 5x^3 - 2x^2 + 4y + 5$
2. $f(x, y) = 4x^7 - 2y^3 + 9y^2 - 2y + 1$
3. $f(x, y) = 5x^2 y + 9xy^5$
4. $f(x, y) = 4x^6 + 7xy + y^5$
5. $f(x, y) = x^3y + 9xy^4$
6. $f(x, y) = x^2e^{3x} \ln y$
7. $f(x, y) = \dfrac{y^2}{x}$
8. $f(x, y) = \dfrac{x+y}{x-y}$
9. $f(x, y) = e^{x^3y^4} + 4x^8 + 7x^4y^5$
10. $f(x, y) = e^{x^3y^6} + 7y^5 + 3x^4y^6$

In Exercises 11–14, determine all points where f(x, y) has possible extrema.

11. $f(x, y) = x^2 - y^2$
12. $f(x, y) = x^2 + y^2 - 4x + 5$
13. $f(x, y) = x^2 + 8xy + y^2$
14. $f(x, y) = x^3 + 6xy + y^3$

In Exercises 15–24, determine local extrema of the functions.

15. $f(x, y) = x^2 + y^2 + 2x - 6y$
16. $f(x, y) = x^2 + y^2 - 4x - 10y + 15$
17. $f(x, y) = 3x^2 - 4xy + 3y^2 + 8x - 17y + 30$
18. $f(x, y) = y^3 - x^2 + 8x - 18y + 10$
19. $f(x, y) = 40x + 50y - x^2 - y^2 - xy$
20. $f(x, y) = 100x + 40y - 5x^2 - 2y^2$
21. $f(x, y) = 2x^3 - 2y^2 - 6x + 8y + 15$
22. $f(x, y) = 3x^2 + 12xy + 6y^4 + 10$
23. $f(x, y) = x^3 - 2xy + 3y^4$
24. $f(x, y) = y^3 - 2xy - x^3$

25. A large freezer box with rectangular cross-section is to have a volume of 125 cubic feet. What should its dimensions be to cover the surface with a minimum of insulation?
26. Postal restrictions limit parcel size to a length plus girth not to exceed 108 inches. Find the volume of the largest package with rectangular cross-sectional area that can be mailed.
27. Revenue, $R(x, y)$, from sales of two products with x units of one and y units of the other as $f(x, y) = 3x + 4y - xy - x^2 - 2y^2$.
    What sales target maximizes revenue?

♦ In a competition with x units of one entity versus y units of a second entity, $f(x, y)$ is the value of the competition at (x, y). Examples are competitions among animal species or industrial products. In this model, $f(x, y) = ax + by - cxy - dx^2 - ey^2$ is frequently used. The linear growth terms (ax and by) are diminished by the interaction ($-cxy$) and diminished by continued increase in x and y by the quadratic terms ($-dx^2$ and $-ey^2$), which represent capacity limits.

8.4 METHOD OF LEAST SQUARES

Centuries of plotting data led to quests for their faithful description in mathematical expression – an empirical equation. A descriptive equation, besides compactness, permits interpolation of values and other insights.

For example, a plot of the daily closing price of a common stock for a long period yields a graph with scattered points. Amidst the scattering there may be a discernible trend that can be usefully described algebraically – a line being a common choice.

Called **curve fitting**, many schemes have been proposed over centuries. Probably, the most popular is the **method of least squares** attributed to the mathematical great, Karl Friedrich Gauss about 1850.

♦ Gauss' standing, to this day, is such that he is called the "Prince of Mathematicians."

His method squares the deviation (error) of each of the plotted points from a candidate polynomial whose coefficients are chosen to minimize the sum of the squared deviations.

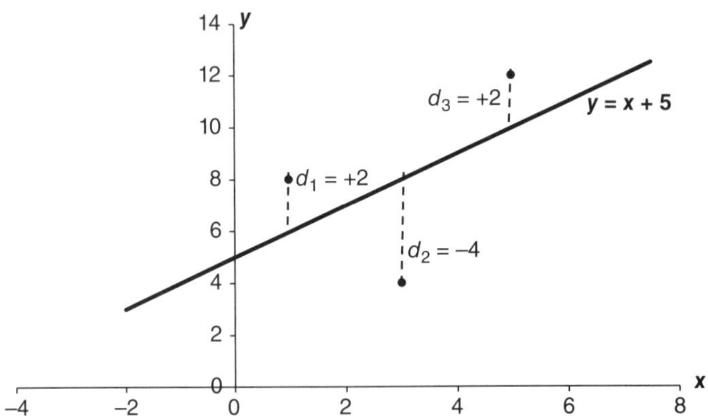

The figure illustrates the deviation, d_i, of the data point (x_i, y_i) from a proposed polynomial – usually a line, $y = \beta_0 + \beta_1 x$. Its coefficients, β_0 and β_1, are determined to minimize the sum of the squared deviations. Hence its name!

Let $y = \beta_0 + \beta_1 x$ be the "fitting" line whose coefficients β_0 and β_1 are sought. The deviation or "error" between a datum (x_i, y_i) and the fitting line is
$d_i = y_i - \beta_0 - \beta_1 x_i$, $i = 1, \ldots, n$. The sum of the squares of these sample errors is to be

minimized. Let E denote their sum so

$$E = d_1^2 + d_2^2 + \cdots + d_n^2 = \sum_{i=1}^{n} (y_i - \beta_0 - \beta_1 x_i)^2$$

Remember, the variables here are the coefficients β_0 and β_1; (ordinarily constants) the x_i and y_i are given data.

A necessary condition for E to be a minimum is for the partial derivatives with respect to β_0 and β_1 to be zero. The two resulting equations to be solved simultaneously are called the *normal equations*.

$$\frac{\partial E}{\partial \beta_1} = 2\sum_{i=1}^{n} [(y_i - \beta_0 - \beta_1 x_i)(-x_i)] = 0$$

$$\frac{\partial E}{\partial \beta_0} = -2\sum_{i=1}^{n} [(y_i - \beta_0 - \beta_1 x_i)] = 0$$

These are to be solved simultaneously for β_0 and β_1.

Example 8.4.1 Least Squares Line

Verify that the least squares fit for the three points (1, 8), (3, 4), and (5, 12) is $y = 5 + x$. (previous figure)

Solution:

Firstly, the square error, E, and derivatives with respect to β_0 and β_1 are set to zero as:

$$E = (8 - \beta_0 - \beta_1)^2 + (4 - \beta_0 - 3\beta_1)^2 + (12 - \beta_0 - 5\beta_1)^2$$

$$\frac{\partial E}{\partial \beta_0} = 2(8 - \beta_0 - \beta_1)(-1) + 2(4 - \beta_0 - 3\beta_1)(-1) + 2(12 - \beta_0 - 5\beta_1)(-1)$$

$$= -48 + 6\beta_0 + 18\beta_1 = 0$$

$$\frac{\partial E}{\partial \beta_1} = 2(8 - \beta_0 - \beta_1)(-1) + 2(4 - \beta_0 - 3\beta_1)(-3) + 2(12 - \beta_0 - 5\beta_1)(-5)$$

$$= -160 + 18\beta_0 + 70\beta_1 = 0$$

The solutions for the resulting system is $\beta_1 = 1$ and $\beta_0 = 5$ so the least squares-line is $\hat{y} = 5 + x$.
The corresponding value of E is $(8 - 5 - 1)^2 + (4 - 5 - 3)^2 + (12 - 5 - 5)^2 = 24$. Note that the deviations sum to zero $[0 = (8 - 5 - 1) + (4 - 5 - 3) + (12 - 5 - 5)]$.
*(Note: the carat ("hat") above $\hat{y}$ denotes it is an **estimate of the true value of y**).*
*Statisticians also refer to least squares lines as **linear regressions**.*

Least square lines are unique. This is not necessarily the case for lines whose deviations sum to zero.

Example 8.4.2 Another Least Squares Line

Determine the least squares line for the data (2, 4), (3, 6), and (4, 9).

Solution:
In this case,

$$E = (4 - \beta_0 - 2\beta_1)^2 + (6 - \beta_0 - 3\beta_1)^2 + (9 - \beta_0 - 4\beta_1)^2$$

Next, set the partial derivatives with respect to β_0 and β_1 to zero.

$$\begin{aligned}\frac{\partial E}{\partial \beta_0} &= 2(4 - \beta_0 - 2\beta_1)(-1) + 2(6 - \beta_0 - 3\beta_1)(-1) + 2(9 - \beta_0 - 4\beta_1)(-1)\\ &= -38 + 6\beta_0 + 18\beta_1 = 0\\ \frac{\partial E}{\partial \beta_1} &= 2(4 - \beta_0 - 2\beta_1)(-2) + 2(6 - \beta_0 - 3\beta_1)(-3) + 2(9 - \beta_0 - 4\beta_1)(-4)\\ &= -124 + 18\beta_0 + 58\beta_1 = 0\end{aligned}$$

Their solution is $\beta_1 = 5/2$ and $\beta_0 = -7/6$ so the (unique) least squares line is $\hat{y} = \frac{-7}{6} + \frac{5}{2}x$.

Example 8.4.3 Least Squares Line for Five Data Points

Fit a least squares line to the data (0, 1.9), (1, 4.8), (2, 8.3), (3, 11.1), and (4, 13.7).

Solution:

$$\begin{aligned}E &= (1.9 - \beta_0)^2 + (4.8 - \beta_0 - \beta_1)^2 + (8.3 - \beta_0 - 2\beta_1)^2\\ &\quad + (11.1 - \beta_0 - 3\beta_1)^2 + (13.7 - \beta_0 - 4\beta_1)^2\\ \frac{\partial E}{\partial \beta_0} &= 2(1.9 - \beta_0)(-1) + 2(4.8 - \beta_0 - \beta_1)(-1) + 2(8.3 - \beta_0 - 2\beta_1)(-1)\\ &\quad + 2(11.1 - \beta_0 - 3\beta_1)(-1) + 2(13.7 - \beta_0 - 4\beta_1)(-1)\\ &= -79.6 + 10\beta_0 + 20\beta_1 = 0\\ \frac{\partial E}{\partial \beta_1} &= 2(1.9 - \beta_0)(0) + 2(4.8 - \beta_0 - \beta_1)(-1) + 2(8.3 - \beta_0 - 2\beta_1)(-2)\\ &\quad + 2(11.1 - \beta_0 - 3\beta_1)(-3) + 2(13.7 - \beta_0 - 4\beta_1)(-4)\\ &= -219 + 20\beta_0 + 60\beta_1 = 0\end{aligned}$$

Next, set the partial derivatives with respect to β_0 and β_1 to zero.
The solution is $\beta_1 = 2.99$ and $\beta_0 = 1.98$, so the least squares line is $\hat{y} = 1.98 + 2.99x$.

In general, the two partial derivatives of $E = \sum_{i=1}^{n} (y_i - \beta_0 - \beta_1 x_i)^2$, with respect to β_0 and β_1, can be solved simultaneously for them. The solutions are

$$\widehat{\beta}_1 = \frac{SS_{xy}}{SS_{xx}} \text{ and } \widehat{\beta}_0 = \overline{y} - \widehat{\beta}_1 \overline{x}$$

where $\overline{x} = \frac{\sum_{i=1}^{n} x_i}{n}$, $\overline{y} = \frac{\sum_{i=1}^{n} y_i}{n}$, $SS_{xy} = \sum x_i y_i - \frac{\left(\sum x_i \sum y_i\right)}{n}$, $SS_{xx} = \sum x_i^2 - \frac{\left(\sum x_i\right)^2}{n}$, and $\widehat{y} = \widehat{\beta}_0 + \widehat{\beta}_1 x$.

These statistical formulas are common in statistics texts.

EXERCISES 8.4

In Exercises 1–12, find the least squares line for the data

1. (3, 6), (4, 9), (5, 15)
2. (2, 10), (4, 6), (6, 5)
3. (4, 6), (5, 8), (6, 4)
4. (3, 8), (5, 14), (7, 12)
5. (0, 5), (4, 11), (8, 18)
6. (6, 4), (8, 12), (12, 14)
7. (1, 12), (2, 11), (3, 9), (4, 6)
8. (1, 25), (2, 42), (3, 60), (4, 85)
9. (3, 10), (4, 27), (5, 52), (6, 102)
10. (3, 5), (4, 10), (5, 18), (7, 25)
11. (2, 9), (3, 7), (4, 6), (5, 10)
12. (2, 8), (5, 4), (6, 5), (8, 6)

13. The points (1, 5), (3, 9), and (4, 11) all lie on the line $y = 2x + 3$. Therefore, it is the least squares line. Verify that this is the least squares equation and the least squares error is zero.

14. a) Use partial derivatives to obtain the least squares line for the data points (3, 2), (4, 5), and (5, 11).

    b) Verify the result using the statistical formulas in the text.

15. A month after time t a commodity price was \$25. It was \$30 four months after t and \$32 five months after t. Use a linear least squares fit to estimate the commodity price 2 months after time t.

8.5 LAGRANGE MULTIPLIERS

We are all familiar with impediments (constraints) that impede attainment of a maximum benefit or minimum of expense. Similarly, in seeking maxima and minima of functions, variables are usually constrained. Such constraints limit values permitted for variables. Examples include spending limits, available labor and materials, a technologic limit such as maximum volume or energy conservation, total consumption of a nutrient, and so on.

For a function of two variables, $z = f(x, y)$, whose maximum (or minimum) is sought, a necessary condition is for its two partial derivatives to vanish.

However, if there is a relationship between variables that must be satisfied, a constraint, simply setting derivatives of $z = f(x, y)$ to zero is insufficient to identify useful extrema. For example, consider a pipe cut into four pieces which form a parallelogram. That their lengths must sum to the pipe length is an equality constraint upon the area maximization.

A simple and ingenious scheme overcomes the obstacle posed by a constraint. Represent a constraint by $g(x, y) = 0$ and introduce a new variable, λ, and a new function, $\boldsymbol{L}$, formed as

$$L = f(x, y) + \lambda\, g(x, y)$$

where λ is called a **Lagrange multiplier** and L the **Lagrangian**.

↓ As $g(x, y) = 0$, L is still, essentially, $f(x, y)$. Addition of the constraint $g(x, y) = 0$ limits possible values of x and y to satisfy it.

The necessary conditions for maxima and minima are found by setting the partial derivatives of L to zero, λ considered a variable. That is,

$$\frac{\partial L}{\partial x} = \frac{\partial f}{\partial x} + \lambda \frac{\partial g}{\partial x} = 0$$
$$\frac{\partial L}{\partial y} = \frac{\partial f}{\partial y} + \lambda \frac{\partial g}{\partial y} = 0$$
$$\frac{\partial L}{\partial \lambda} = g(x, y) = 0$$

Candidates for an optimal solution are obtained by solving these equations simultaneously for x and y. Note that the last equation simply "returns" the constraint $g(x, y) = 0$. Its simultaneous solution with the other necessary conditions ensures that resulting values of x and y satisfy $g(x, y) = 0$.

Example 8.5.1 Cutting a Pipe

A pipe of length d is to be cut into four pieces to form a parallelogram of maximum area. How shall cuts be made?

Solution:

Let x and y represent dimensions of the parallelogram. The enclosed area, z, is proportional to the product xy. That is,

$z = f(x, y) = kxy$ where k is a proportionality constant.

The maximization of z is subject to the pipe length constraint

$$g(x, y) = 2x + 2y - d = 0$$

Forming the Lagrangian and setting derivatives equal to zero yield

$$L = kxy + \lambda\,(2x + 2y - d)$$

⟶

$$\frac{\partial L}{\partial x} = ky + 2\lambda = 0$$

$$\frac{\partial L}{\partial y} = kx + 2\lambda = 0$$

$$\frac{\partial L}{\partial \lambda} = 2x + 2y - d = 0$$

Solving simultaneously yields the solution $x^* = y^* = d/4$ and $\lambda = -kd/8$. As x^* and y^* are equal, the maximizing parallelogram is a square. It follows that $k = 1$ and, substituting, the maximum area is $z^* = d^2/16$. (Asterisks are often used to denote optimality.)

♦ The reader may recognize that the aforementioned problem is solved in basic calculus courses without mention of the Lagrange Multiplier. One simply solves $g(x, y) = 0$ for one of the variables, substitutes into $z = f(x, y)$ so that it becomes a function of a single independent variable and its derivative can be set to zero. That is, indeed, quite correct.

However, there are reasons to favor the more elegant Lagrange Multiplier. For starters, the substitution method can fail when $g(x^*, y^*)$ is a singular point. Also, if there are more than a few constraints and/or variables, eliminations and substitutions can be prohibitive (even with automated computation). Finally, Lagrange Multipliers furnish a stepping stone to more general optimization problems.

In the previous example, one might have suggested simply to avoid two variables by using the constraint to substitute for $y = \frac{d - 2x}{2}$. That would have worked as well. However, that is not always the case as the following example illustrates.

Example 8.5.2 Closest Distance to a Circle

On the circle $x^2 + y^2 = 4$, find the point closest to $(1, 0)$ using substitution and using the Lagrange Multiplier Method.

Solution:

The minimum of $z = f(x, y) = (x - 1)^2 + y^2$ is sought, as it is the squared distance from $(1, 0)$ to an arbitrary point (x, y) on the circle. The circle's equation forms the constraint

$$g(x, y) = x^2 + y^2 - 4 = 0$$

Substituting, replace y^2 in z by $4 - x^2$ to yield

$$z = (x - 1)^2 + (4 - x^2),$$

which, after simplification, becomes $z = -2x + 5$. Clearly, the derivative of z with respect to x cannot be set to zero. Bad luck!

Now try to substitute for x. Solving for x in $g(x, y)$ we have

⟶

$x = \pm\sqrt{4 - y^2}$ so z as a function of y becomes

$$z = (\pm\sqrt{4 - y^2} - 1)^2 + y^2$$

Setting $dz/dy = 0$ yields $x = \pm 2,\ y = 0$.
The solution is $x^* = 2,\ y^* = 0$, and $z^* = 1$ and the closest distance is 1 (Note: remember z^* is the squared distance, which is easier to use than the distance.)
One cannot necessarily predict which variables can be successfully substituted.
The Lagrange Multiplier Method avoids that complication.

$$L = f(x, y) + \lambda g(x, y) = (x - 1)^2 + y^2 + \lambda(x^2 + y^2 - 4)$$

and

$$\frac{\partial L}{\partial x} = 2(x - 1) + 2\lambda x = 0 \tag{8.1}$$

$$\frac{\partial L}{\partial y} = 2y + 2\lambda y = 0 \tag{8.2}$$

$$\frac{\partial L}{\partial \lambda} = x^2 + y^2 - 4 = 0 \tag{8.3}$$

From Equation (8.2), it is clear that $\lambda = -1$ or that $y = 0$.
For $\lambda = -1$, Equation (8.1) cannot be satisfied. As all three equations must be satisfied for a solution, $y = 0$ must be chosen.
With $y = 0$, Equation (8.3) requires $x^2 = 4$ or $x = \pm 2$.
As the shortest distance is required, the optimal solution occurs when $x^* = 2,\ y^* = 0$. There $z^* = 1$ and so the shortest distance is also 1.

Note that there is no proof of optima in the last two examples. A proof requires an examination of second derivatives beyond the present scope.

EXERCISES 8.5

1. Form a Lagrangian to maximize $x^2 - y^2$ subject to the constraint $2x + y = 3$.
2. Form a Lagrangian to minimize $x^2 + y^2$ subject to the constraint $x + y = 10$.
3. Form a Lagrangian to maximize xy subject to the constraint $x + y = 16$.
4. Find the minimum of $x^2 + y^2 - 2xy$ subject to $2x + y = 7$.
5. Find the point on the parabola $y^2 = 4x$ that is closest to (1, 0).
6. What two positive numbers whose product is 100 have the smallest sum?
7. What two positive numbers add to nine and maximize x^2y?
8. To fence a rectangular garden, \$320 is available. The cost of fencing on the north, east, and south sides is \$10 per linear foot, and the fencing for the less visible west side is \$6 per foot. Form a Lagrangian to find the dimensions of the largest garden.

9. To fence a rectangular garden, \$280 is available. The fencing for the north and south sides of the garden costs \$8 per linear foot, and the fencing for the east and west sides \$5 per foot. Form a Lagrangian to find the dimensions of the largest possible garden.

10. Find the dimensions of the rectangle of maximum area that can be inscribed in the unit circle ($x^2 + y^2 = 1$).

11. Find the dimensions of a closed rectangular box with square base and volume 64,000 cubic inches that requires the least material.

8.6 DOUBLE INTEGRALS

Just as an integral may evaluate an area, a **double integral** can evaluate a volume. If $f(x, y)$ is a function of two variables, continuous and non-negative on a closed rectangular region R, then a double integral of $f(x, y)$ over the region R is denoted by

$$\iint_R f(x, y)dx\,dy$$

and corresponds to a volume.

↓ Just as the graph of a function of one variable, $f(x)$, can enclose an area (two dimensional), so the graph of a function of two variables, $f(x, y)$, can enclose a volume (three dimensions).

A double integral is evaluated sequentially and is called an **iterated integral**. Suppose (arbitrarily) that the first integration variable is y. The iterated integral of $f(x, y)$ is

$$\int_a^b \left[\int_{g(x)}^{h(x)} f(x, y)dy \right] dx$$

where a and b are constants. The brackets, usually omitted, appear in this case as a guide to the order of integration.

Note that the limits on the inner integral may, generally, be functions of x, say, $g(x)$ and $h(x)$, as indicated. The inner integral (bracketed) is the first of the two integrations. On completion of the first integration on y, the second integration, on x, follows.

Alternatively, the order of integration could have been reversed. Let $g(y)$ and $h(y)$, functions of y, an iterated integral (on x first, then y) is

$$\int_a^b \left[\int_{g(y)}^{h(y)} f(x, y)dx \right] dy$$

where again, the only purpose of the brackets is to highlight the order of integration.

Double, or iterated integrals, are evaluated outward from the inner integral. Above, the first integration on x (inner integral) is followed by integration on y.

↓ The order of the integration, although theoretically arbitrary, is usually chosen for ease of effort. In many instances, when x and y are independent variables, and the integration limits are constants, the order of integration is immaterial.

Example 8.6.1 Evaluating a Double Integral

Evaluate the double integral $\int_0^2 \int_0^1 (x+y)dx\,dy.$

Solution:

The inner integral is evaluated first. As x is its integration variable, y is regarded as a constant for the moment. That is, extracting and executing the inner integral as,

$$\int_0^1 (x+y)dx = \left(\frac{x^2}{2}+xy\right)\Big|_0^1 = \frac{1}{2}+y.$$

Substituting $\frac{1}{2}+y$ *for the inner integral and carrying out the remaining (outer) integration on y yields*

$$\int_0^2 \left[\int_0^1 (x+y)\,dx\right]dy = \int_0^2 \left(\frac{1}{2}+y\right)dy = \left(\frac{1}{2}y+\frac{y^2}{2}\right)\Big|_0^2 = (1+2)-(0+0) = 3.$$

Example 8.6.2 Another Double Integral

Evaluate the double integral $\int_0^1 \int_0^2 (x+y)dy\,dx.$

Solution:

Although this integrand is the same as in the previous example, the order of integration is reversed. In this case, the inner integration becomes

$$\int_0^2 (x+y)dy = \left(xy+\frac{y^2}{2}\right)\Big|_0^2 = (2x+2)-(0+0) = 2x+2.$$

Substituting $(2x+2)$ *for the inner integral and integrating on x yield*

$$\int_0^1 \int_0^2 [(x+y)dy]dx = \int_0^1 (2x+2)dx = (x^2+2x)\Big|_0^1 = (1+2)-(0+0) = 3.$$

As expected, the result is the same as in the previous example! The order of integration is immaterial in this case because limits are constants.

As noted, the double integral $\iint_R f(x,y)dx\,dy$ represents a volume whose base is the region R and whose height at (x,y) is $f(x,y)$.

Example 8.6.3 Double Integrals and Volume

Calculate the volume formed by $f(x, y) = x^3y^5$ above a region bounded by the graphs of $y = x^2$, and $y = x$ from $x = 0$ to $x = 1$ (see graph).

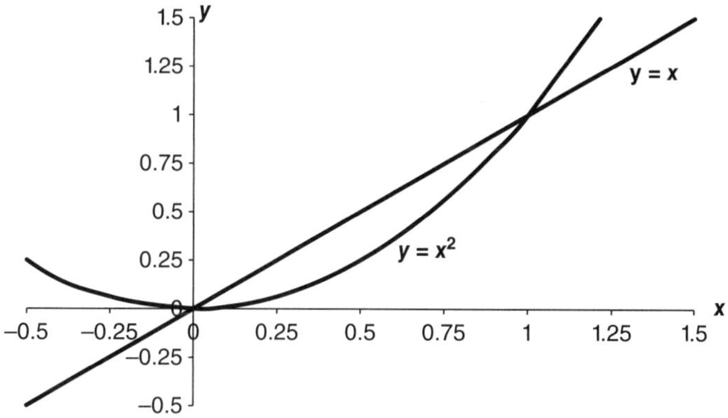

Solution:

As $f(x, y) = x^3y^5$ is the height of the volume above the base R, the value of the double integral is the volume of the solid formed by, $f(x, y)$, over the area of the base R.

The iterated integral has limits on y of x^2 (lower curve) and x (upper curve) for the inner integral with integration variable y. For the outer integral whose integration variable is x, the limits are $x = 0$ and $x = 1$.

The volume is written as

$$\int_0^1 \left[\int_{x^2}^{x} x^3y^5 dy\right] dx$$

The inner bracketed integration is carried out first and evaluated as

$$\left.\frac{x^3y^6}{6}\right|_{x^2}^{x} = \frac{x^9}{6} - \frac{x^{15}}{6}$$

So, the integration becomes $\int_0^1 \left(\frac{x^9}{6} - \frac{x^{15}}{6}\right)dx = \left.\left(\frac{x^{10}}{60} - \frac{x^{16}}{96}\right)\right|_0^1 = \frac{1}{160}.$

It is the desired volume.

Example 8.6.4 Writing an Equivalent Double Integral

Write the double integral of the previous example with order of iteration reversed.

⟶

Solution:

We seek limits for the integral $\int\int x^3 y^5 \, dx \, dy$ *with the integration order reversed from the previous example. The limits of integration for the outside integral are the y values of the intersections of the graphs. In this case, the intersections are at* $y = 0$ *and* $y = 1$*. The inner integration variable is x.*

The two functions forming the base of the volume are $y = x$ *and* $y = x^2$*. So now consider* $x = y$ *and* $x = \sqrt{y}$ *as integration limits.*

Next, consider "right and left" (instead of "top and bottom") curves. The x values are larger for $x = \sqrt{y}$ *so it is the curve to the right. Therefore, the equivalent iterated integral is*

$$\int_0^1 \left[\int_y^{\sqrt{y}} x^3 y^5 \, dx \right] dy$$

EXERCISES 8.6

In Exercises 1–12, evaluate the integrals.

1. $\int_0^3 \left[\int_{-1}^2 (5x - 2y) \, dy \right] dx$
2. $\int_1^3 \left[\int_{-1}^2 (3x^2 - 4y^3) \, dy \right] dx$
3. $\int_1^3 \left[\int_{-2}^3 (6xy^2 - x^4) \, dy \right] dx$
4. $\int_0^3 \left[\int_x^{x^2} (x^3 + 4y) \, dy \right] dx$
5. $\int_1^4 \left[\int_0^2 x^4 y^5 \, dy \right] dx$
6. $\int_1^5 \left[\int_{-1}^3 4x^2 y^3 \, dy \right] dx$
7. $\int_0^2 \left[\int_0^x (x + y + 2) \, dy \right] dx$
8. $\int_1^e \left[\int_1^2 \left(\frac{1}{x} \right) dy \right] dx$
9. $\int_0^3 \left[\int_0^{x/2} (e^{2y-x}) \, dy \right] dx$
10. $\int_1^2 \left[\int_1^4 (2x - 3y) \, dx \right] dy$
11. $\int_0^1 \left[\int_{y^2}^{y-1} (5x - 2y) \, dx \right] dy$
12. $\int_0^1 \left[\int_y^{\sqrt{y}} (5) \, dx \right] dy$

HISTORICAL NOTES

Joseph-Louis Lagrange (1736–1813) – of French ancestry was born and educated in Italy. His interest in mathematics arose from the use of algebra in the study of optics. Largely self-taught, he was appointed a professor of mathematics at the Royal Artillery School in Turin at the age of 19.

He wrote on a variety of topics including the calculus of variations, the calculus of probabilities, number theory, dynamics, fluid mechanics (where he introduced the Lagrangian function), and the stability of the solar system.

CHAPTER 8 SUPPLEMENTARY EXERCISES

1. Find f_x and f_y for $f(x, y) = e^{x^7 y^5} + 9x^4y^3 + 32x + 1$.
2. A business can produce $f(x, y) = 40\sqrt[3]{10x^2 + 20y^2}$ desks by utilizing x units of capital and y units of labor. Calculate the marginal productivities of capital and labor when $x = 50$ and $y = 10$.
3. Verify that a Cobb–Douglas function such as $f(x, y) = kx^\alpha y^\beta$ exhibits constant returns to scale whenever $\alpha + \beta = 1$. This means $f(mx, my) = m(f(x, y))$ or that increasing all inputs by a factor of m also increases outputs by a factor of m.
4. Draw level curves for $c = -6$, $c = 6$, and $c = 12$ when $f(x, y) = 3x + 2y$.
5. For $f(x, y) = (2x^3 - 3x + y^4)^{10}$ obtain f_x and f_y.
6. For $f(x, y) = 5x^3y^4 + 9xy^3 + 3y + 4$ obtain f_x and f_y.
7. For $f(x, y) = x^3 - x^2y + xy^2 - y^3$ use the first and second partial derivatives to show that the mixed derivatives are equal.
8. Find the local extrema of $f(x, y) = x^3 + 3xy^2 + 3y^2 - 15x + 2$.
9. Use partial derivatives to obtain the least squares line for the data (3, 7), (4, 15), (7, 23).
10. Use partial derivatives to obtain the least squares line for the data (1, 8), (2, 7), (3, 5) and (4, 2).
11. Verify that the partial derivative solutions for the least squares line yield the results familiar from statistics given in Section 8.4.
12. Form a Lagrangian to verify that the area of a rectangle with fixed perimeter is maximized when it is a square.
13. The sum of squares of two numbers that add to a fixed number is a minimum when the numbers are equal. Form a Lagrangian to verify this statement.
14. Evaluate $\int_{-2}^{1}\left[\int_{x^2+4x}^{3x+2}(4)\,dy\right]dx$.
15. Evaluate $\int_{1}^{2}\left[\int_{0}^{1}e^{x+y}dx\right]dy$.

9 Series and Summations

Fundamentals of Calculus, First Edition. Carla C. Morris and Robert M. Stark.

Companion Website: http://www.wiley.com/go/morris/calculus

Finite sums and **infinite series** have important roles in mathematics – especially to approximate functions. Often, and surprisingly, the functions approximated are unknown.

We reserve the term summations, or sums when there are a finite number of terms as: $1 + \frac{1}{2} + \frac{1}{3} + \cdots + \frac{1}{n}$, ($n$, a positive integer). We use the term series when the number of terms is infinite as: $1 + \frac{1}{2} + \frac{1}{3} + \cdots + \frac{1}{n} + \cdots$

9.1 POWER SERIES

A series of the form

$$a_0 + a_1 x + a_2 x^2 + \cdots + a_n x^n + \cdots$$

is a **power series** in the variable x with constant coefficients $a_0, a_1, \ldots$.

The special case $a = a_0 = a_1 = \cdots = a_n$ is the well-known **geometric series**

$$a + ar + ar^2 + \cdots + ar^n + \cdots = \sum_{k=0}^{\infty} ar^k \quad \textit{geometric series}$$

An immediate consideration is whether an infinite series converges. That is, whether the infinity of terms sums to a finite number. Such series are said to **converge**. When the sum increases beyond bounds it is said to **diverge**.

To examine the geometric series for convergence and divergence, let S represent its sum

$$S = a + ar + ar^2 + \cdots + ar^n + \cdots$$

Multiplying each side by r yields

$$Sr = ar + ar^2 + ar^3 + \cdots + ar^{n+1} + \cdots$$

Subtracting,

$$S - Sr = a$$

and simplifying

$$S = \frac{a}{1 - r} \qquad |r| < 1$$

Until here, r has not been specified. Clearly, the denominator requires $r \neq 1$. When $|r| > 1$, the terms ar^n, and hence S, are dominated by increasing powers of r as n increases.

So, for $|r| > 1$, the geometric series diverges; it increases beyond bound. Now, for $|r| < 1$, the expression for S is finite and the series converges.

Geometric Series

The geometric series

$$a + ar + ar^2 + \cdots + ar^n + \cdots$$

converges to

$$S = \frac{a}{1-r}$$

when $|r| < 1$ and diverges otherwise.

Example 9.1.1 Geometric Series

Sum the geometric series $2 + 1 + 1/2 + 1/4 + \cdots$

Solution:
The geometric series, $2 + 2(1/2) + 2(1/2)^2 + 2(1/2)^3 + \cdots$, has $a = 2$ and $r = 1/2$.
Its sum is $S = \dfrac{2}{1-(1/2)} = 4$.

Example 9.1.2 Economic Multiplier Effect

Imagine a tax cut that leaves \$150 billion with consumers who in turn spend 90% of their extra income and save 10%. Estimate the total economic multiplier effect of the tax cut on the economy.

Solution:
People spend \$150(0.9) billion of the initial cut. However, there is a ripple effect as those receiving the tax cut dollars in turn spend 90% of this additional income, and so on. Therefore, the initial \$150 billion has the effect of

$$150(0.90) + 150(0.90)^2 + 150(0.90)^3 + \cdots = \frac{150(0.90)}{1-0.90} = \$1350 \text{ billion}$$

This is a geometric series with $a = 150(0.90)$ and $r = 0.90$. Therefore, in addition to the \$150 billion cut, theoretically, \$1.35 trillion in new spending is generated.

Economists can have differing views of the effect of tax cuts. Our examples are motivated solely by mathematics; not economics.

Example 9.1.3 Drug Dosage

People with chronic ailments are often prescribed daily dosages of medications. Physicians know that daily dosages may not be completely absorbed by the body. They are chosen so that drug accumulation over time remains within a therapeutic range.

Suppose that a daily dose is 2 mg and the patient's body eliminates about 92% daily. What is the long run accumulation?

Solution:
At the start, the body has 2 mg. After the first day, the body has a second dose plus the residual from the prior day, $2 + (2)(0.08)$ *mg. After the second day, there is* $2 + (2)(0.08) + (2)(0.08)^2$ *mg, representing the new dose plus the residuals from two prior days. In this manner, the long-run accumulation is* $2 + (2)(0.08) + (2)(0.08)^2 + \cdots$ *mg.*

The sum of the geometric series with $a = 2$ *and* $r = 0.08$ *is*

$$S = \frac{2}{1 - 0.08} = 2.17\,mg$$

If this is not within the therapeutic drug range, the dosage needs adjustment.

In algebra, one learns that repeating decimals can be expressed as rational fractions. For instance, 0.333333... is 1/3. The following example illustrates conversion of repeating decimals to fractions using a geometric series.

Example 9.1.4 A Repeating Decimal

Express the decimal 0.343434... as a rational fraction.

Solution:
The decimal can be expressed as

$$0.34 + 0.0034 + 0.000034 + \cdots \quad or \quad 0.34 + 0.34(0.01) + 0.34(0.01)^2 + \cdots$$

This geometric series has $a = 0.34$ *and* $r = 0.01$ *so*

$$S = \frac{0.34}{1 - 0.01} = \frac{0.34}{0.99} = \frac{34}{99}$$

↓ Consider the decimal 0.343434..., an equivalent alternate algebraic method to determine the rational fraction it represents can be found by noting there are two repeating integers. If x represents the rational fraction as a repeating decimal, then in this instance, $100x = 34.343434\ldots$ still contains a repeating decimal. Subtracting x from $100x$ yields $99x = 34$ or $x = 34/99$. This is the rational fraction whose repeating decimal is 0.343434....

Example 9.1.5 Another Repeating Decimal

Express the decimal 7.352352 … as a rational fraction.

Solution:
In this case, the number is larger than unity, so view the decimal as $7 + 0.352352\ldots$. *Therefore, the number is rewritten as* $7 + 0.352 + 0.00352 + 0.0000352 + \cdots$
$= 7 + 0.352 + 0.352(0.01) + 0.352(0.01)^2 + \cdots$ *where the decimal terms form a geometric series with* $a = 0.35$ *and* $r = 0.001$ *so*

$$S = 7 + \frac{0.352}{1 - 0.001} = 7 + \frac{0.352}{0.999} = 7 + \frac{352}{999} = \frac{7345}{999}$$

Zeno's Paradox

A variety of paradoxes perplexed thinkers dating to antiquity. One of the better known has a race between a tortoise and a hare – called Zeno's Paradox.

On a one meter track, the tortoise and the hare move at constant speed. The confident hare spots the tortoise a 1/2 m lead. At each stride, the hare halves its distance to the tortoise. Can the tortoise elude the hare? If so, by how many strides?

Clearly, the hare's distance is modeled as the geometric series

$$\frac{1}{2} + \frac{1}{4} + \frac{1}{8} + \cdots$$

This series sums to unity in the limit. So, the tortoise can elude the hare for an eternity! The notion that an infinite series could sum to a finite number was a difficult one for early thinkers to grasp (and to prove). The mathematician Augustin-Louis Cauchy deserves considerable credit for clarifying many matters relating to limiting values.

EXERCISES 9.1

In Exercises 1–10, sum the geometric series that converge.

1. $1 + \frac{1}{2} + \frac{1}{2^2} + \frac{1}{2^3} + \cdots$
2. $2 + \frac{2}{3} + \frac{2}{3^2} + \frac{2}{3^3} + \cdots$
3. $\frac{2}{3} + \frac{2}{3^2} + \frac{2}{3^3} + \cdots$
4. $\frac{3}{2^2} + \frac{3}{2^4} + \frac{3}{2^6} + \cdots$
5. $\frac{4}{5} - \frac{4}{5^2} + \frac{4}{5^3} - \cdots$
6. $1 + \frac{9}{7} + \frac{9^2}{7^2} + \frac{9^3}{7^3} + \cdots$
7. $\frac{5}{2} + \frac{5}{2^2} + \frac{5}{2^3} + \cdots$
8. $1 - \frac{2}{3^2} + \frac{2^2}{3^4} - \cdots$
9. $5 - \frac{5}{4} + \frac{5}{4^2} - \cdots$
10. $4 + 1.2 + 0.36 + \cdots$

In Exercises 11–20, use a geometric series to obtain rational fractions.

11. 0.11111...
12. 0.77777...
13. 0.161616...
14. 0.727272...
15. 0.135135...
16. 0.259259...
17. 1.44444 ...
18. 2.55555...
19. 3.4343...
20. 1.147147...

In Exercises 21–24, evaluate:

21. $\sum_{x=3}^{\infty} \left(\frac{1}{3}\right)^x$
22. $\sum_{x=0}^{\infty} (2)^{-x}$
23. $\sum_{x=1}^{\infty} \left(\frac{3}{5}\right)^x$
24. $\sum_{x=0}^{\infty} \left(\frac{5}{7}\right)^x$

25. Imagine a tax cut of \$15 million and people spend 80% of their additional income. Calculate the theoretical multiplier effect of the tax cut on the economy.
26. Verify that 0.9999... is unity.

9.2 MACLAURIN AND TAYLOR POLYNOMIALS

Recall that a finite sum of power terms as

$$p(x) = a_0 + a_1x + a_2x^2 + \cdots + a_nx^n$$

is an nth degree polynomial in the variable x and real coefficients $a_0, a_1, \ldots, a_n$.

Polynomials have an importance as approximations to functions, sometimes of unknown functional form. In general, functions $f(x)$ can be approximated by a polynomial for values of x close to zero.

To form the nth degree polynomial approximation, seek the first n derivatives of $f(x)$. Next, evaluate the derivatives at $x = 0$ to yield:
$f(0) = a_0$, $f'(0) = a_1$, $f''(0) = 2a_2$, $\ldots$, $f^{(n)}(0) = n!a_n$. These derivatives are the coefficients for the polynomial approximation. Replacing the coefficients $a_0, a_1, \ldots, a_n$ by the evaluated derivatives yields

$$p_n(x) = f(0) + f'(0)x + \frac{f''(0)}{2!}x^2 + \cdots + \frac{f^{(n)}(0)}{n!}x^n \approx f(x)$$

This is a **Maclaurin Polynomial**.

Maclaurin Polynomial

$$p_n(x) = f(0) + f'(0)x + \frac{f''(0)}{2!}x^2 + \cdots + \frac{f^{(n)}(0)}{n!}x^n \approx f(x)$$

Coefficients of Maclaurin Polynomials are its derivatives evaluated at $x = 0$.

Example 9.2.1 *Binomial Sum*

Expand $f(x) = (1 + x)^3$ as a Maclaurin Polynomial.

Solution:
Firstly, $f(0) = (1 + 0)^3 = 1$.
Next, the derivatives are evaluated at $x = 0$.

$$f'(x) = 3(1 + x)^2, \quad so \quad f'(0) = 3$$

$$f''(x) = 6(1 + x), \quad so \quad f''(0) = 6$$

$$f'''(x) = 6, \quad so \quad f'''(0) = 6$$

The fourth-order and higher order derivatives are all zero. Therefore, the Maclaurin Polynomial, $p(x)$, is

$$p(x) = 1 + 3x + \frac{6x^2}{2!} + \frac{6x^3}{3!} \quad or \quad (1 + x)^3 \approx 1 + 3x + 3x^2 + x^3$$

*This is the well-known **binomial theorem**!*

Example 9.2.2 *Approximation for e*

Approximate e by a fourth-degree Maclaurin Polynomial, $p_4(x)$.

Solution:
Set $f(x) = e^x$ and seek $f(1)$.
All derivatives of e^x have the unique property of leaving it unchanged.
That is, $f^{(n)}(x) = e^x$, for any value of n.
Four derivatives are required to determine the desired fourth-degree polynomial.

$$f'(0) = f''(0) = f'''(0) = f^{iv}(0) = e^0 = 1$$

⟶

Therefore, the polynomial is

$$p_4(x) = 1 + x + \frac{x^2}{2!} + \frac{x^3}{3!} + \frac{x^4}{4!} = 1 + x + \frac{x^2}{2} + \frac{x^3}{6} + \frac{x^4}{24} \approx e^x$$

To approximate e, set $x = 1$ to yield

$$p_4(1) = 1 + (1) + \frac{(1)^2}{2} + \frac{(1)^3}{6} + \frac{(1)^4}{24} = 2.708333333 \approx e$$

The result is fairly close to the value 2.718281828 obtained with a calculator.

Taylor Polynomials, $p(x)$, are generalizations of Maclaurin Polynomials for an expansion of $f(x)$ about some point, $x = a,\ a \neq 0$. That is,

$$p_n(x) = a_0 + a_1(x-a) + a_2(x-a)^2 + \cdots + a_n(x-a)x^n \approx f(x)$$

The importance of expansion about points of interest is that often fewer power terms are needed to achieve useful approximations to functions in some domain. We illustrate this usage in the following section.

As with the Maclaurin sums, the coefficients are obtained from derivatives. However, instead of evaluating derivatives at $x = 0$, as earlier, they are evaluated at $x = a$.

$$p_n(x) = f(a) + f'(a)(x-a) + \frac{f''(a)}{2!}(x-a)^2 + \cdots + \frac{f^{(n)}(a)}{n!}(x-a)^n$$

The importance of expansion about points of interest is that they can improve approximations to functions. We illustrate this usage in the following section.

Taylor Polynomials

$$\mathbf{p_n(x) = f(a) + f'(a)(x-a) + \frac{f''(a)}{2!}(x-a)^2 + \cdots + \frac{f^{(n)}(a)}{n!}(x-a)^n}$$

$p_n(x)$ is a polynomial approximation for $f(x)$ at $x = a$.

Example 9.2.3 A Taylor Approximation About x = a

Develop a third-degree Taylor Polynomial for $x^{5/2}$ at $x = 1$.

Solution:

To start, the first three derivatives are

$$f'(x) = \frac{5}{2}x^{3/2}, \quad f''(x) = \frac{15}{4}x^{1/2}, \quad f'''(x) = \frac{15}{8}x^{-1/2}$$

⟶

Next, evaluate the function and its derivatives at x = 1.

$$f(1) = 1, \quad f'(1) = \frac{5}{2}, \quad f''(1) = \frac{15}{4}, \quad \text{and} \quad f'''(1) = \frac{15}{8}$$

A third-degree Taylor Polynomial about x = 1 is therefore,

$$p_3(x) = 1 + \frac{5}{2}(x-1) + \frac{15}{4}\frac{(x-1)^2}{2!} + \frac{15}{8}\frac{(x-1)^3}{3!}$$

$$= 1 + \frac{5}{2}(x-1) + \frac{15}{8}(x-1)^2 + \frac{5}{16}(x-1)^3$$

Taylor Polynomials can be used for numerical approximations as in the example.

Example 9.2.4 Approximating a Function

Approximate $\sqrt[4]{17}$ to three decimal places. Hint: use $f(x) = x^{1/4}$.

Solution:
Firstly, as the fourth root of 16 is 2, it is a rough estimate of $\sqrt[4]{17}$.
The first two derivatives evaluated at x = 16 are

$$f'(x) = \frac{1}{4}x^{-3/4}, \quad f'(16) = \frac{1}{4}(16)^{-3/4} = \frac{1}{32}$$
$$f''(x) = \frac{-3}{16}x^{-7/4}, \quad f''(16) = \frac{-3}{16}(16)^{-7/4} = \frac{-3}{2048}$$

A Taylor Polynomial, p(x) to approximate $f(x) = \sqrt[4]{17}$ at x = 16 is

$$p(x) = 2 + \frac{1}{32}(x-16) - \frac{3}{2048}\frac{(x-16)^2}{2}$$

Therefore, p(17) is a three decimal place approximation for f(17)

$$p(17) = 2 + \frac{1}{32}(17-16) - \frac{3}{2048}\frac{(17-16)^2}{2} = 2.031 \approx \sqrt[4]{17}$$

Note that in this case the degree of the polynomial was not known in advance. Rather one adds terms until the desired precision is attained (successive terms uniformly decrease). In this instance, the third-degree term exceeds the required accuracy.

♦ Many so-called “laws” in science and economics are actually empirically determined Taylor Polynomials.

Example 9.2.5 A Law of Thermal Expansion

Early physicists observed that the lengths of many materials tend to increase when heated – the increase being proportional to the temperature change. Imagine a metal rod of length L_0 at an initial temperature T_0. It is heated to a temperature T and its measured length has increased to L. Approximate L by a second-degree Taylor Polynomial.

Solution:

Expand $f(T)$ in a Taylor Polynomial about the initial temperature T_0 so that the polynomial is expressed in terms of the change in temperature, $(T - T_0)$.

That is, $L = f(T) = a_0 + a_1(T - T_0) + a_2(T - T_0)^2$.

Now, when $T = T_0$, $L = L_0$ so $f(T_0) = L_0 = a_0$.

Physics texts denote a_1 by $L_0\alpha$, where α is the "coefficient of linear thermal expansion."

The Law of Thermal Expansion appears in physics texts as $L = L_0(1 + \alpha(T - T_0))$ (in units of length per unit length per degree).

Experiments indicate that a first-order approximation is adequate for many practical purposes. For greater precision, a second-order term can be used.

♦ For the previous example, approximate values for α are 1.9×10^{-5} in/in°C for silver, 2.2×10^{-5} in/in°C for aluminum, and 1.3×10^{-5} in/in°C for nickel. Actually, these values do not hold for all temperatures – especially higher ones. In such instances, the Taylor Polynomial might be expanded to include higher order terms

$$L = L_0 + L_0\alpha(T - T_0) + L_0\alpha(T - T_0)^2 + L_0\alpha(T - T_0)^3, \text{ and so on}$$

Note also that thermal expansion is accounted for in many everyday objects such as the gaps in sidewalk squares, bridge joints, railroad tracks, and so on.

It is interesting that physicists don't know functions as $f(L)$ on theoretical grounds. Their recourse is a Taylor Polynomial. There are many such instances elsewhere in physics, chemistry, economics, psychology, and so on.

Technical information is compiled by the National Institute of Standards and Technology (NIST), the federal agency that aids industry with technology, measurements, and standards.

EXERCISES 9.2

In Exercises 1–10, approximate $f(x)$ by third-degree Maclaurin Polynomials.

1. $f(x) = e^x$
2. $f(x) = e^{-2x}$
3. $f(x) = 2x^3 - 3x^2 + 1$
4. $f(x) = \ln(1 - x)$

5. $f(x) = \ln(x + 1)$

6. $f(x) = (x + 1)^{1/2}$

7. $f(x) = (x + 1)^{3/2}$

8. $f(x) = \sqrt{x + 4}$

9. $f(x) = \dfrac{1}{1 - x}$

10. $f(x) = \dfrac{1}{1 - x^2}$

In Exercises 11–20, approximate f(x) by Taylor Polynomials about the indicated point.

11. $f(x) = \ln x$ about $x = 1$, third degree.

12. $f(x) = \ln x$ about $x = 1$, fifth degree.

13. $f(x) = \dfrac{1}{3 - x}$ about $x = 2$, fourth degree.

14. $f(x) = \dfrac{1}{1 - x^2}$ about $x = 1/2$, third degree.

15. $f(x) = e^x$ about $x = 1$, sixth degree.

16. $f(x) = e^{3x}$ about $x = 2$, fourth degree.

17. $f(x) = e^{2x}$ about $x = 3$, fourth degree.

18. $f(x) = x^{3/2}$ about $x = 1$, third degree.

19. $f(x) = x^{3/2}$ about $x = 4$, third degree.

20. $f(x) = 4x^3 - 3x^2 + x + 5$ about $x = 1$, third degree.

Use Taylor Polynomials to obtain third decimal place approximations.

21. $\sqrt{1.04}$

22. $\sqrt{8.97}$

23. $\sqrt[3]{26.98}$

24. $\sqrt[3]{0.99}$

25. $e^{0.1}$

26. $\ln(3/2)$

9.3 TAYLOR AND MACLAURIN SERIES

The **Taylor Series**, also a power series, is of central importance in applications of mathematics where functional relations are not known on theoretical grounds. As examples, approximations of economic demand and supply functions, the growth rates of animals, elastic behavior of materials, and so on. The coefficients of $a_0, a_1, \ldots$ of

$$f(x) = a_0 + a_1(x - a) + a_2(x - a)^2 + \cdots + a_n(x - a)^n + \cdots$$

are determined by successive differentiation evaluated at the point of expansion, $x = a$. It is assumed that all derivatives of $f(x)$ exist.

$$f'(x) = a_1 + 2a_2(x - a) + \cdots + na_n(x - a)^{n-1} + \cdots$$

$$f''(x) = 2a_2 + 3(2)a_3(x - a) + \cdots + n(n - 1)a_n(x - a)^{n-2} + \cdots$$

$$\vdots$$

$$f^n(x) = n(n - 1)\cdots(3)(2)(1)a_n + (n + 1)(n)\cdots(3)(2)(1)a^{n+1} + \cdots$$

$$\vdots$$

Evaluating these derivatives at the point of expansion, $x = a$, and substituting

$$f(x) = f(a) + f'(a)(x - a) + \frac{f''(a)}{2!}(x - a)^2 + \cdots + \frac{f^{(n)}(a)}{n!}(x - a)^n + \cdots$$

♦ The connection between derivatives of $f(x)$ and the coefficients $a_0, a_1, \ldots$ was discovered by Brook Taylor, a contemporary of Isaac Newton (Historical Notes). The Taylor series is named in his honor.

When $a = 0$, the series becomes

$$f(x) = f(0) + f'(0)x + \frac{f''(x)}{2!}x^2 + \cdots + \frac{f^{(n)}(0)}{n!}x^n + \cdots;$$

the **Maclaurin Series**.

Example 9.3.1 ln *x*

Expand ln x in a Taylor Series about x = 1 to at least fourth-order terms.

Solution:
Firstly, $f(x) = \ln x$ *and* $f(1) = 0$. *The first four derivatives of* $\ln x$ *at* $x = 1$ *are*

$$f'(x) = \frac{1}{x} \quad \text{so} \quad f'(1) = 1 \quad f''(x) = \frac{-1}{x^2} \quad \text{so} \quad f''(1) = -1$$

$$f'''(x) = \frac{2}{x^3} \quad \text{so} \quad f'''(1) = 2 \quad f^{iv}(x) = \frac{-6}{x^4} \quad \text{so} \quad f^{iv}(1) = -6$$

Substituting,

$$f(x) = (x - 1) - \frac{(x - 1)^2}{2} + \frac{(x - 1)^3}{3} - \frac{(x - 1)^4}{4} + \cdots - \frac{(x - 1)^n}{n!} + \cdots$$

Example 9.3.2 e^x

Obtain the Maclaurin Series for $f(x) = e^x$.

Solution:
The successive derivatives of $f(x) = e^x$ *are* $f^n(x) = e^x$ *for* $n = 1, 2, \ldots$.
Therefore, the Maclaurin series is

$$f(x) = 1 + x + \frac{x^2}{2!} + \frac{x^3}{3!} + \cdots + \frac{x^n}{n!} + \cdots = e^x$$

An important infinite series!

Example 9.3.3 $\dfrac{1}{1-x}$

Obtain the Maclaurin Series for $f(x) = \dfrac{1}{1-x}$.

Solution:
The first few derivatives are: $f'(x) = (1-x)^{-2}, f''(x) = 2(1-x)^{-3}, f'''(x) = 6(1-x)^{-4}$, *and* $f^{iv}(x) = 24(1-x)^{-5}$. *The function and these derivatives evaluated at* $x = 0$ *are respectively, 1, 1, 2, 6, and 24. Therefore, the expansion yields*

$$1 + 1(x) + 2\left(\frac{x^2}{2!}\right) + 6\left(\frac{x^3}{3!}\right) + 24\left(\frac{x^4}{4!}\right) + \cdots n!\left(\frac{x^n}{n!}\right) + \cdots = \frac{1}{1-x}$$

Simplification yields:

$$\frac{1}{1-x} = 1 + x + x^2 + x^3 + x^4 + \cdots + x^n + \cdots = \sum_{x=0}^{\infty} x^n \qquad \text{for} \quad -1 < x < 1.$$

This is another infinite series of basic importance.

In applications, Taylor Series are often truncated for ease of usage. For example, it may be written as

$$f(x) = a_0 + a_1(x-a) + a_2(x-a)^2 + \cdots + a_n(x-a)^n + R_n$$

where R_n represents the **Remainder** – the sum of the remaining terms to the series.

Truncating a Taylor Series raises an immediate question: can the remainder terms of a convergent Taylor Series be determined? The answer is a qualified "yes"! Mathematicians have shown that the remainder after n terms in a Taylor Series is

$$R_n = f^{(n+1)}(c)\frac{(x-a)^{n+1}}{(n+1)!} \qquad a < c < x$$

This is important and useful in establishing an upper bound for the remaining terms of a Taylor series. By choosing, c, a value between x and a, where the $n + 1^{\text{st}}$ derivative of $f(x)$ is largest, one obtains a bound on the error by omitting the remaining terms.

Furthermore, if $R_n \to 0$ as $n \to \infty$, the Taylor Series converges.

Example 9.3.4 e^x Revisited

Evaluate the remainder after n terms of the Maclaurin Series for e^x (Example 9.3.2).

Solution:

Using the result of Example 9.3.2

$$f(x) = e^x = 1 + x + \frac{x^2}{2!} + \frac{x^3}{3!} + \cdots + \frac{x^n}{n!} + R_n$$

Now,

$$R_n = \frac{e^c x^{n+1}}{(n+1)!} \qquad 0 < c < x$$

The largest value for $f^{(n+1)}(c)$ occurs at $c = |x|$. Therefore,

$$|R_n| \le \frac{e^{|x|}|x|^{n+1}}{(n+1)!}$$

(The absolute value sign anticipates the possibility of negative values of x).
For any fixed value of x, $|R_n|$ tends to zero as $n \to \infty$.
Therefore, the Maclaurin (Taylor) Series for e^x converges for all values of x.

Example 9.3.5 Approximation Error

What is the largest error in using $x - (x^2/2)$ to approximate $ln(1 + x)$ for $|x| < 0.2$?

Solution:

To obtain the remainder R_2 we seek $f'''(c)$.

$$f(x) = ln(1+x), \quad f'(x) = \frac{1}{1+x}, \quad f''(x) = \frac{-1}{(1+x)^2}, \quad f'''(x) = \frac{2}{(1+x)^3}$$

$\longrightarrow$

Therefore,

$$R_3 = \frac{f'''(c)x^3}{3!} = \frac{2}{(1+c)^3}\frac{x^3}{6} = \frac{x^3}{3(1+c)^3}$$

The largest absolute value is obtained when $c = -0.2$.

Then $|R_3| = \dfrac{(0.2)^3}{3(1-0.2)^3} = \dfrac{0.008}{1.56} = 0.00521.$

This compares with the actual value (estimation error)

$$ln(1-0.2) - \left[(-0.2) + \frac{(-0.2)^2}{2}\right] = 0.00314$$

Although the choice of a point of expansion, such as $x = a$, can be arbitrary, sometimes there are useful choices. We assume that $f(a), f'(a), f''(a), \ldots$ exist and are known. Choosing the point of expansion closer to desired values of $f(x)$ tends to better an approximation with fewer series terms. It is in the neighborhood of $x = a$ that Taylor Series converge more rapidly.

↓ In the use of Taylor series to approximate a function, the fewer terms needed to achieve a desired level of approximation the better. A judicious choice of a point of expansion, $x = a$, can markedly reduce the number of terms needed to achieve a desired precision.

Example 9.3.6 Approximating a Function

Approximate $\sqrt{1.03}$ using a third-degree Taylor series about $x = 1$ and $x = 4$. Which is a better approximation to the actual value?

Solution:

Using $f(x) = x^{1/2}$, the first three derivatives are

$$f'(x) = \frac{1}{2}x^{-\frac{1}{2}}, f''(x) = \frac{-1}{4}x^{\frac{-3}{2}}, \text{ and } f'''(x) = \frac{3}{8}x^{\frac{-5}{2}}$$

The first three derivatives evaluated at 1 are 1/2, $-1/4$, and 3/8, respectively. So $f(x)$ is approximated by

$$f(1) + f'(1)(x-1) + f''(1)\frac{(x-1)^2}{2} + f'''(1)\frac{(x-1)^3}{6}$$

$$= 1 + \left(\frac{1}{2}\right)(1.03-1) + \left(\frac{-1}{4}\right)\left(\frac{(1.03-1)^2}{2}\right) + \frac{3}{8}\left(\frac{(1.03-1)^3}{6}\right)$$

$$= 1 + 0.015 - 0.0001125 + 0.000001688 = 1.014889$$

⟶

Evaluated at $x = 4$ the three derivatives are 1/4, $-1/32$, and 3/256, respectively, so $f(x)$ is approximated by

$$f(4) + f'(4)(x-4) + f''(4)\frac{(x-4)^2}{2} + f'''(4)\frac{(x-4)^3}{6}$$

$$= 2 + \left(\frac{1}{4}\right)(1.03-4) + \left(\frac{-1}{32}\right)\left(\frac{(1.03-4)^2}{2}\right) + \frac{3}{256}\left(\frac{(1.03-4)^3}{6}\right)$$

$$= 2 - 0.7425 - 0.1378 - 0.0512 = 1.0685$$

Both approximations are close to the exact value of 1.0148⋯. The first, expanded about $x = 1$, is closer than expansion about $x = 4$ as a third-degree approximation.

EXERCISES 9.3

In Exercises 1–10, expand functions in Maclaurin Series.

1. $f(x) = \dfrac{1}{1+x}$
2. $f(x) = \dfrac{1}{1-5x}$
3. $f(x) = \dfrac{1}{1+2x}$
4. $f(x) = \dfrac{x}{1-x}$
5. $f(x) = \dfrac{1}{1+x^2}$
6. $f(x) = \dfrac{1}{(1-x)^2}$
7. $f(x) = \ln(1+x)$
8. $f(x) = \ln(1-x)$
9. $f(x) = 4e^{x/2}$
10. $f(x) = 3e^{x/3}$

11. Expand $f(x) = x(e^x - 1)$ in a Maclaurin Series.
12. Expand $f(x) = e^{-x^2}$ in a Maclaurin Series. See Chapter 10 for an important role of e^{-x^2} in probability and statistics.
13. Obtain the Maclaurin series expansion for $\frac{1}{2}(e^x + e^{-x})$.
    Hint: use the known expansion for e^x.

♦ The function 1/2 $(e^x + e^{-x})$ defines the **hyberbolic cosine**. Its graph is a **catenary** (which resembles a parabola). Physically, it is realized by fastening the ends of a flexible string (under gravity) to a horizontal support.

14. Use an expansion of $f(x) = 4e^{x/2}$ (Exercise 9) to show that its derivative is $2e^{x/2}$.
15. Expand $\ln x$ in a Taylor Series about $x = 1/2$.

16. Expand $\dfrac{1}{1+x}$ in a Taylor Series about $x = -2$.
17. Multiply both sides of the Maclaurin expansion for $(1+x)^{-1}$ (Exercise 1) by the differential dx. Then integrate on the left to obtain $\ln(1+x)$ and a term by term integration on the right. Evaluate the integration constant.

9.4 CONVERGENCE AND DIVERGENCE OF SERIES

An infinite series can converge or diverge. A convergent series has a finite sum. The sum of a divergent series increases beyond bounds. Mathematicians have devised a number of tests for a series' behavior.

One immediate observation is whether successive terms eventually decrease. However, successively decreasing terms are not sufficient for convergence. Clearly, successive terms of **the harmonic series** decrease,

$$1 + \frac{1}{2} + \frac{1}{3} + \frac{1}{4} + \cdots + \frac{1}{n} + \cdots \qquad \textit{harmonic series}$$

However, the series diverges so the sum of successive terms increases without bound.

On the other hand, successive terms of the geometric series

$$a + ar + ar^2 + \cdots + ar^n + \cdots \qquad \textit{geometric series}$$

decrease for $|r| < 1$ (see Section 9.1). It is a convergent series. However, if $|r| \geq 1$, it is a divergent series as successive terms clearly increase.

Integral Test

An **integral test** of a series $a_n(x)$, $n = 1, 2, \ldots$ uses the integral

$$\int_1^\infty a_n(x)dx$$

If $a_n(x)$ is positive and decreases as x increases, the series and the integral either both converge or both diverge. A proof is beyond our scope. An example illustrates.

Example 9.4.1 An Integral Test

Test the series $1 + \dfrac{1}{2^2} + \dfrac{1}{3^2} + \cdots + \dfrac{1}{n^2} + \cdots$ using an integral test.

Solution:

$$\int_1^\infty \frac{dx}{x^2} = \lim_{t\to\infty} \int_1^t x^{-2}\,dx = \lim_{t\to\infty}\left(\frac{-1}{t} + \frac{1}{1}\right) = 1$$

Therefore, $\displaystyle\sum_{n=1}^{\infty} \frac{1}{n^2}$ is a convergent series.

Example 9.4.2 Harmonic Series

Test the harmonic series using the integral test.

Solution:

The integral test associated with $1 + \frac{1}{2} + \frac{1}{3} + \frac{1}{4} + \cdots + \frac{1}{n} + \cdots$ *is*

$$\int_1^\infty \frac{1}{x}\,dx = \lim_{t\to\infty} \int_1^t \frac{1}{x}\,dx = \lim_{t\to\infty}(\ln t - \ln 1) = \lim_{t\to\infty} \ln t \to \infty$$

As the integral diverges, so the harmonic series diverges!

↓ ***Recall*** $\ln 1 = 0$

◆ As a rule, the terms of a convergent series tend to zero as $n \to \infty$. However, the converse is not always true. The harmonic series is an oft cited example as $1/n \to 0$ as n increases.

Example 9.4.3 Geometric Series

Use the integral test to verify that the geometric series $a + ar + ar^2 + \cdots + ar^n + \cdots$ *converges for* $|r| < 1$.

Solution:

$$\int ar^x\,dx = \frac{ar^x}{\ln r} + C$$

We seek $\int_1^\infty ar^x\,dx = \lim_{t\to\infty} \int_1^t ar^x\,dx = \lim_{t\to\infty}\left(\frac{ar^t}{\ln r} - \frac{ar}{\ln r}\right) = -\frac{ar}{\ln r}.$

Note: $\lim_{t\to\infty} \frac{ar^t}{\ln r} = 0$ *for* $|r| < 1$.

Comparison Tests

Sometimes, the unknown behavior of a positive term series can be inferred by a **direct comparison test** with another positive term series of known behavior.

Clearly, if a positive term series of unknown behavior is term for term less than another series known to be convergent, one infers that the series in question is also convergent. Similarly, if a positive term series known to be divergent is term for term less than a positive term series of unknown behavior, it is safe to conclude that the unknown series is also divergent.

To make these observations precise, let $\sum a_n = a_0 + a_1 + a_2 + \cdots + a_n + \cdots$ represent a positive term series of unknown behavior and $\sum b_n = b_0 + b_1 + b_2 + \cdots + b_n + \cdots$ a positive term series known to be convergent.

If

$$a_n \leq b_n \quad \text{for all } n = 1, 2, \ldots$$

then $\sum a_n$ is also convergent. Similarly, if $\sum b_n$ is known to be divergent and if

$$a_n \geq b_n \quad \text{for all } n = 1, 2, \ldots$$

then $\sum a_n$ is also divergent.

Example 9.4.4 A Convergent Comparison

Show that the series

$$\sum_{n=1}^{\infty} \frac{1}{1+2^n} = \frac{1}{3} + \frac{1}{5} + \frac{1}{9} + \cdots + \frac{1}{1+2^n} + \cdots$$

is convergent.

Solution:

In this case, $a_n = \frac{1}{1+2^n}$, $n = 1, 2, \ldots$ *is a series of unknown behavior. Consider a similar series* $\sum b_n = \sum \frac{1}{2^n}$, $n = 1, 2, \ldots$

This is a geometric series known to be convergent (as the term ratio, 1/2, is less than unity). In a term-by-term comparison

$$a_n = \frac{1}{1+2^n} \leq \frac{1}{2^n} = b_n \qquad \text{for all } n = 1, 2, \ldots .$$

As the unknown series is less, term for term, than a series known to be convergent, it follows that $\sum_{n=1}^{\infty} \frac{1}{1+2^n}$ *is also a convergent series.*

Example 9.4.5 Another Convergent Comparison

Show that this series is convergent

$$\sum_{n=1}^{\infty} \frac{4}{3n^2+5n+6} = \frac{4}{14} + \frac{4}{28} + \frac{4}{48} + \cdots + \frac{4}{3n^2+5n+6} + \cdots$$

Solution:

A rearrangement as $\dfrac{1}{\frac{3}{4}n^2 + \frac{5}{4}n + \frac{3}{2}}$ *is useful.*

→

Next, note that

$$\frac{1}{\frac{3}{4}n^2 + \frac{5}{4}n + \frac{3}{2}} \leq \frac{1}{\frac{3}{4}n^2} = \frac{4}{3n^2}$$

as each term is positive valued. A comparison with the convergent series $\sum \frac{1}{n^2}$ *is suggested. The factor 4/3 is immaterial. Recall that the series* $\sum \frac{1}{n^2}$, $n = 1, 2, \ldots$ *was shown to be convergent.*

Sometimes, a series may converge for some values of a variable and diverge for others. The values of the variable for which the series converges is the *convergence interval*. The geometric series quickly comes to mind as its behavior is determined by the term ratio, r, with convergence interval $|r| < 1$.

The last two examples illustrate the importance of two series in using comparison tests: $\sum ar^n$, the geometric series, and $\sum \frac{1}{n^p}$, called the p series, $n = 1, 2, \ldots$ Their convergence intervals are $|r| < 1$ (and divergent for $|r| \geq 1$) for the geometric series and $p > 1$ for $\sum \frac{1}{n^p}$ (and divergent for $p \leq 1$).

Example 9.4.6 A Failed Comparison

Can a direct comparison test be used to determine the behavior of the series $\sum_{n=1}^{\infty} \frac{1}{2^n - 1}$?

Solution:

In this case, $a_n = \frac{1}{2^n - 1}$, $n = 1, 2, \ldots$ *is the general term of the series of unknown behavior. A similar and convergent series* $b_n = \frac{1}{2^n}$ *is a possible candidate for comparison. Here,*

$$a_n = \frac{1}{2^n - 1} \geq \frac{1}{2^n} = b_n \quad \text{for } all \ n = 1, 2, \ldots .$$

As the terms of the unknown series are larger than corresponding terms of the convergent series, no conclusion can be drawn. Therefore, some other series is needed for a useful comparison.

Another test, a **limit comparison test** for positive term series uses the limit

$$\lim_{n \to \infty} \frac{a_n}{b_n} = c > 0$$

where a_n and b_n are general terms of two series, $\sum a_n$ and $\sum b_n$, and c is positive. If the limit exists and one of the series is known to be convergent, then the other also converges. Similarly, if one series is known to diverge, the other also diverges.

Example 9.4.7 A Successful Comparison

Determine the behavior of the series $\sum_{n=1}^{\infty} \frac{1}{2^n - 1}$ *(from the previous example) using a limit comparison test.*

Solution:
Again, let $\sum b_n = \sum \frac{1}{2^n}$*, a convergent series. Forming the ratio and taking the limit*

$$\lim_{n\to\infty} \frac{a_n}{b_n} = \lim_{n\to\infty} \frac{\left(\frac{1}{2^n - 1}\right)}{\left(\frac{1}{2^n}\right)} = \lim_{n\to\infty} \frac{2^n}{2^n - 1} = \lim_{n\to\infty} \frac{1}{1 - (1/2)^n} = 1 > 0$$

Therefore, as the limit exists and $\sum b_n$ *is known to be convergent, the limit comparison test indicates that* $\sum_{n=1}^{\infty} a_n = \sum_{n=1}^{\infty} \frac{1}{2^n - 1}$ *is also a convergent series.*

Example 9.4.8 Another Comparison

Is this series convergent?

$$\frac{2}{1} + \frac{2}{2 \cdot 3} + \frac{2}{3 \cdot 5} + \cdots + \frac{2}{n(2n-1)} + \cdots$$

Solution:
The dominant term in the denominator, $\frac{1}{n^2}$*, suggests a comparison to the convergent p-series for* $p = 2$*. Let* a_n *be the general term of the series in question and* b_n *the general term for the p-series with* $p = 2$*,*

$$\frac{1}{1^2} + \frac{1}{2^2} + \frac{1}{3^2} + \cdots + \frac{1}{n^2} + \cdots$$

Therefore, $\sum b_n = \sum \frac{1}{n^2}$ *and* $\sum_{n=1}^{\infty} a_n = \sum_{n=1}^{\infty} \frac{2}{n(2n-1)}$.
Next, consider the ratio and its limit

$$\lim_{n\to\infty} \frac{a_n}{b_n} = \lim_{n\to\infty} \frac{\left(\frac{2}{n(2n-1)}\right)}{\left(\frac{1}{n^2}\right)} = \lim_{n\to\infty} \frac{2n^2}{n(2n-1)} = \lim_{n\to\infty} \frac{2n^2}{2n^2 - n} = 1 > 0$$

As $\sum b_n = \sum \frac{1}{n^2}$ *is known to be convergent, it follows that* $\sum_{n=1}^{\infty} a_n = \sum_{n=1}^{\infty} \frac{2}{n(2n-1)}$ *is also convergent.*

Example 9.4.9 Still Another Comparison

Establish the divergence of the series $\sum_{n=1}^{\infty} a_n = \sum_{n=1}^{\infty} \frac{3n^2 + 5n}{\sqrt{n^6 + 2}}$.

Solution:

Choosing the dominant terms of the numerator and denominator yields

$$b_n = \frac{3n^2}{\sqrt{n^6}} = \frac{3n^2}{n^3} = \frac{3}{n}$$

This suggests a comparison using the divergent harmonic series. The ratio and limit is

$$\lim_{n\to\infty} \frac{a_n}{b_n} = \lim_{n\to\infty} \frac{\left(\dfrac{3n^2 + 5n}{\sqrt{n^6 + 2}}\right)}{\left(\dfrac{3}{n}\right)} = \lim_{n\to\infty} \frac{n(3n^2 + 5n)}{3\sqrt{n^6 + 2}} = \lim_{n\to\infty} \frac{3 + \dfrac{5}{n}}{3\sqrt{1 + \dfrac{2}{n^6}}} = 1 > 0$$

(after dividing numerator and denominator by n^3*)*

As the limit exists and $\sum b_n$ *is known to be the divergent harmonic series, it follows that* $\sum_{n=1}^{\infty} a_n = \sum_{n=1}^{\infty} \frac{3n^2 + 5n}{\sqrt{n^6 + 2}}$ *is also divergent.*

Ratio Test

The **ratio test** is still another series test. The ratio test, as its name implies, relies on an analysis of the ratio of successive terms. If $\lim_{n\to\infty} \left| \frac{a_{n+1}}{a_n} \right| < 1$ the series converges. If the ratio exceeds unity, it diverges. The ratio test fails if the ratio is unity.

The following two examples illustrate the ratio test.

Example 9.4.10 A Ratio Test

Use the ratio test to determine the behavior of the series $\sum_{n=0}^{\infty} a_n = \sum_{n=0}^{\infty} \frac{5^n}{n!}$.

Solution:

In this case, $a_n = \frac{5^n}{n!}$ *and* $a_{n+1} = \frac{5^{n+1}}{(n+1)!}$. *This ratio test relies on the limit:*

$$\lim_{n\to\infty} \left| \frac{\dfrac{5^{n+1}}{(n+1)!}}{\dfrac{5^n}{n!}} \right| = \lim_{n\to\infty} \left| \frac{5^{n+1}}{(n+1)!} \cdot \frac{n!}{5^n} \right| = \lim_{n\to\infty} \left| \frac{5}{n+1} \right| = 0$$

Therefore, as the limit is less than unity, the series is convergent.

Example 9.4.11 Another Ratio Test

Use the ratio test to determine the behavior of the series $\sum_{n=1}^{\infty} a_n = \sum_{n=1}^{\infty} \frac{3^n}{n}$.

Solution:

In this case, $a_n = \frac{3^n}{n}$ *and, therefore,* $a_{n+1} = \frac{3^{n+1}}{n+1}$. *This ratio test relies on the limit:*

$$\lim_{n\to\infty}\left|\frac{\frac{3^{n+1}}{(n+1)}}{\frac{3^n}{n}}\right| = \lim_{n\to\infty}\left|\frac{3^{n+1}}{n+1}\cdot\frac{n}{3^n}\right| = \lim_{n\to\infty}\left|3\cdot\frac{n}{n+1}\right| = 3$$

Therefore, as the limit exceeds unity, the series is divergent.

Ratio Test

Let $\sum_{n=1}^{\infty} a_n$ be a series with nonzero terms.

1. **The series converges if $\lim_{n\to\infty}\left|\frac{a_{n+1}}{a_n}\right| < 1$.**
2. **The series diverges if $\lim_{n\to\infty}\left|\frac{a_{n+1}}{a_n}\right| > 1$.**
3. **The test is inconclusive if $\lim_{n\to\infty}\left|\frac{a_{n+1}}{a_n}\right| = 1$.**

EXERCISE 9.4

In Exercises 1–8, use an integral test to determine the behavior of the series.

1. $\sum_{n=3}^{\infty} \frac{1}{n}$
2. $\sum_{n=1}^{\infty} \frac{n}{n^2+1}$
3. $\sum_{n=1}^{\infty} \frac{\ln n}{n}$
4. $\sum_{n=2}^{\infty} \frac{1}{e^{3n}}$
5. $\sum_{n=2}^{\infty} \frac{1}{n(\ln n)^3}$
6. $\sum_{n=1}^{\infty} \frac{3n^2}{n^3+1}$
7. $\sum_{n=2}^{\infty} 2ne^{-n^2}$
8. $\sum_{n=1}^{\infty} 2e^{3-n}$

In Exercises 9–16, apply an appropriate comparison test for the behavior of the series.

9. $\sum_{n=1}^{\infty} \frac{1}{\sqrt{n^3+1}}$

10. $\sum_{n=1}^{\infty} \sqrt{\frac{n^2+9}{n^3}}$

11. $\sum_{n=2}^{\infty} \frac{1}{\sqrt{n^2-1}}$

12. $\frac{1}{1\cdot 2} + \frac{1}{2\cdot 4} + \frac{1}{3\cdot 8} + \cdots + \frac{1}{n\cdot 2^n} + \cdots$

13. $\frac{1}{2} + \frac{2}{9} + \frac{3}{28} + \frac{4}{65} + \cdots + \frac{n}{n^3+1} + \cdots$

14. $\frac{1}{2} + \frac{1}{1+2^{3/2}} + \frac{1}{1\cdot 3^{3/2}} + \cdots + \frac{1}{1+n^{3/2}} + \cdots$

15. $\frac{\ln 2}{4} + \frac{\ln 3}{9} + \frac{\ln 4}{16} + \cdots + \frac{\ln n}{n^2} + \cdots$ (Hint: $\ln n < n^{1/2}$)

16. $\frac{1}{1} + \frac{1}{\sqrt{2}} + \frac{1}{\sqrt{3}} + \cdots + \frac{1}{\sqrt{n}} + \cdots$

17. For what values of p does the series $\sum_{n=2}^{\infty} \frac{1}{n(\ln n)^p}$ converge? Diverge?

In Exercises 18–23, use a ratio test to determine the behavior of the series.

18. $\sum_{n=1}^{\infty} \frac{n}{3^n} = \frac{1}{3} + \frac{2}{9} + \frac{3}{27} + \cdots$

19. $\sum_{n=0}^{\infty} \frac{3^n}{n!} = \frac{3}{1} + \frac{9}{2} + \frac{27}{6} + \cdots$

20. $\sum_{n=0}^{\infty} \frac{(-1)^{n+1}5^n}{(n+1)3^{2n}}$

21. $\sum_{n=1}^{\infty} \frac{4^n}{n^3} = \frac{4}{1} + \frac{16}{8} + \frac{64}{27} + \cdots$

22. $\sum_{n=1}^{\infty} \frac{n^3}{5^n} = \frac{1}{5} + \frac{8}{25} + \frac{27}{125} + \cdots$

23. $\sum_{n=0}^{\infty} \frac{(-1)^n 6^n}{n!} = \frac{1}{1} - \frac{6}{1} + \frac{36}{2} + \cdots$

9.5 ARITHMETIC AND GEOMETRIC SUMS

Consider the sequence $a, a+d, a+2d, \ldots, a+nd$. Starting with a number a, successively add a number d to generate successive terms.

The sum, S_n, of the first n terms is

$$S_n = \sum_{k=0}^{n-1}(a+kd) = a + (a+d) + (a+2d) + \cdots + (a+(n-1)d)$$

$$= na + n(n-1)d = n\left(a + \frac{n-1}{2}d\right)$$

and is called an **arithmetic sum**.
(Recall that the sum of the first k integers $1+2+\cdots+k=\frac{k(k+1)}{2}$)

Example 9.5.1 Savings Account

On October 1 a child received \$1 for completing a daily chore. Each day the pay increased by 10 cents. How much accrued at Halloween?

Solution:
In this case, a = \$1 and d = \$0.10. Therefore, $S_{31}=31\left(1+\frac{30}{2}(0.1)\right)=\$77.50.$

Example 9.5.2 "The Twelve Days of Christmas"

Express the daily numbers of different gifts in the traditional favorite "The Twelve Days of Christmas" as an arithmetic sum and calculate its total.

Solution:

$$1+2+3+4+5+6+7+8+9+10+11+12=\frac{(12)(13)}{2}=78 \text{ gifts.}$$

♦ The PNC Bank Corporation has estimated the cost of the gifts cited in "The Twelve Days of Christmas" for over 30 years. Christmas 2013 marked the 30th anniversary of the index. In 2013, the estimate totaled \$27,393.18, about a 7.7% increase from 2012.

Consider the sequence $a,\ ar,\ ar^2,\ldots,\ ar^n$. Beginning with a real number, a, successive terms are formed by multiplying by a fixed real number, r.

The sum of the first n terms is

$$S_n=\sum_{k=0}^{n-1} ar^k=a+ar+ar^2+\cdots+ar^{n-1}$$
$$=a\frac{1-r^n}{1-r} \qquad r\neq 1$$

is known as a **geometric sum**. Here, as contrast to the geometric series, there is no restriction on the term ratio, r, other than $r\neq 1$.

Geometric Sum

The sum of the first n terms of a geometric sequence

$$S_n = a, ar, \ldots, ar^{n-1}$$

with first term a and common ratio r is

$$S_n = a\frac{1 - r^n}{1 - r}$$

An example is illustrative:

Example 9.5.3 Sum of a Geometric Sequence

What is the sum of the first seven terms of the geometric sequence $2, \frac{2}{3}, \frac{2}{3}, \ldots, \frac{2}{729}$?

Solution:
In this case, $a = 2$ *and* $r = \frac{1}{3}$. *Therefore, the sum of the first seven terms is*

$$S = 2\left(\frac{1 - \left(\frac{1}{3}\right)^7}{1 - \frac{1}{3}}\right) = 3\left(1 - \frac{1}{2187}\right) = \frac{6558}{2187} \approx 2.9986$$

Note: while S_n is finite for $r \neq 1$, it is undefined for $r = 1$. When $r < 0$, the signs of terms alternate.

Example 9.5.4 Bouncing Ball

A ball is dropped from a height of 6 feet. It recovers 90% of its initial height after each bounce. How high was its fourth bounce?

Solution:
At the first bounce, the ball reaches $6(0.9) = 5.4$ *feet. On the second bounce, the ball reaches 4.86 feet. On the third,* $6(0.9)^3 = 4.37$ *feet, and on the fourth bounce, it reaches* $6(0.9)^4 = 3.94$ *feet.*

Example 9.5.5 Passbook Savings

A 5-year-old child received \$100, placed, in a 3% savings account.

a) The \$100 compounds annually. What is the passbook sum when the child is 21 years?

b) The grandparents match the passbook balance at the end of each year. How much have they contributed?

Solution:

a) Annual balances to age 21 are: 1st year $100(1.03)$; 2nd year, $100(1.03)^2, \ldots,$ 16th year, $100(1.03)^{16}$. The 16th year the passbook shows \$160.47.

b) The "out of pocket" grandparents' expense is the sum of the annual balances, $100(1.03) + 100(1.03)^2 + \cdots + 100(1.03)^{16}$.

This geometric sum has $a = 100(1.03) = 103$ and $r = 1.03$, so the sum is

$$103\left(\frac{1-(1.03)^{16}}{1-1.03}\right) = \$2076.16$$

EXERCISES 9.5

In Exercises 1–12, calculate the finite sums.

1. $1 + 4 + 7 + \cdots + 40$
2. $3 + 8 + 13 + \cdots + 58$
3. $11 + 15 + 19 + \cdots + 35$
4. $6 + 15 + 24 + \cdots + 60$
5. $\frac{3}{2} + \left(\frac{3}{2}\right)^2 + \left(\frac{3}{2}\right)^3 + \left(\frac{3}{2}\right)^4 + \left(\frac{3}{2}\right)^5$
6. $\frac{1}{2} + \frac{1}{2^2} + \frac{1}{2^3} + \frac{1}{2^4} + \cdots + \frac{1}{2^9}$
7. $\frac{1}{5} + \frac{1}{5^2} + \frac{1}{5^3} + \frac{1}{5^4} + \cdots + \frac{1}{5^{12}}$
8. $\frac{3}{2^2} - \frac{3^2}{2^4} + \frac{3^3}{2^6} - \frac{3^4}{2^8} + \cdots - \frac{3^8}{2^{16}}$
9. $\frac{5}{2} - \frac{5^2}{2^4} + \frac{5^3}{2^7} - \frac{5^4}{2^{10}} + \cdots - \frac{5^{12}}{2^{34}}$
10. $12 - 2.4 + 0.48 - 0.096$

11. $10 + 12 + 14.4 + \cdots + 24.8832$

12. $7 + \dfrac{21}{4} + \dfrac{63}{16} + \dfrac{189}{64}$

13. What is the sum of the first six terms of the geometric sequence whose first term is 3 and common ratio is 5?

14. Determine the sum of the first eight terms of the geometric sequence 8, 24, 72, ...

15. What is the sum of the first 10 terms of the geometric sequence whose first term is 3 and common ratio is −2?

16. A ball is dropped from a height of 6 feet on a hard surface and bounces until caught. How high is the ball at the third bounce? What is the total distance traveled by the ball? Assume that the ball rebounds to 85% of its previous height at each bounce. (Physicists refer to 0.85 as the "coefficient of restitution.")

17. A patient is administered 1 mg of a drug daily. If 51% of the drug remains at the time of the next dose, when should a dose be withheld to avoid "double" dosing?

18. A contest advertises a $1000 prize "Free and Clear." The winner, an astute business student, on receiving a $1000 check notes that he must pay a 25% deduction for federal and state taxes. Not exactly "Free and Clear" he thought. The sponsors offer to pay the taxes. The student points out that taxes are also due on the tax supplement. Again, the sponsors generously agree. Help the sponsors calculate the award to fulfill the "Free and Clear" promise.

HISTORICAL NOTES

Augustin-Louis Cauchy (1789–1857) – French mathematician, home schooled by his father, a classical scholar, until the age of 13. Cauchy advanced rapidly in his studies and served as a military engineer for Napolean. At the age of 26, he proved a conjecture by Fermat – a proof that had eluded Euler and Gauss.

Cauchy is credited with establishing the rigor that underlies contemporary mathematics. His contributions include many principles that bear his name; among them the limit concept at the "heart of calculus." The ratio test of this chapter often bears his name.

Colin Maclaurin (1698–1746) – Scottish mathematician, graduated from University of Glasgow at the age 14 and remained to study divinity. He became professor of mathematics at Marischal College in the University of Aberdeen at the age of 19.

As a disciple of Isaac Newton, in 1742 he published *A Treatise of Fluxions* (i.e., derivatives), the first systematic formulation of his mentor's methods. In that work, he developed the expansion of functions about the origin, known as Maclaurin series. Maclaurin is also known for several inventions and for improving maps of the Scottish isles.

Brook Taylor (1685–1731) – An English mathematician and contemporary of Newton, Taylor is credited with expanding functions as polynomials about an arbitrary point. Now known as Taylor series, they appeared in his *Methodus in Crementorum Directa et Inversa* (1715).

He entered St. John's College, Cambridge in 1701 and studied mathematics there. Taylor is also credited with the early recognition of the existence of singular solutions to differential equations.

CHAPTER 9 SUPPLEMENTARY EXERCISES

1. Sum the geometric series: $\frac{2}{3^2} + \frac{2}{3^5} + \frac{2}{3^8} + \cdots$.
2. Sum the geometric series: $1 - \frac{1}{5} + \frac{1}{5^2} - \frac{1}{5^3} + \cdots$.
3. Express the decimal 0.363636 … as a geometric series and a rational fraction.
4. Express the decimal 1.234234 … as a geometric series and a rational fraction.
5. Express $3.0\overline{45}$ as a rational number.
6. Sum $\sum_{x=2}^{\infty} \left(\frac{1}{4}\right)^x$.
7. Sum $\sum_{x=1}^{\infty} \left(\frac{3}{5}\right)^x$.
8. Obtain a fourth-degree Maclaurin Polynomial for $f(x) = e^{4x}$.
9. Obtain a third-degree Taylor Polynomial about $x = 3$ for $f(x) = \frac{2}{4 - x}$.
10. Obtain a fourth-degree Taylor Polynomial for $x^4 + 5$ about $x = 2$.
11. Expand $f(x) = 2e^{x/5}$ in a Maclaurin Series.
12. Determine the power series for $f(x) = \mathrm{e}^{2x}$ about $x = 0$.
    Hint: a power series has the form $a_0 + a_1x + a_2x^2 + a_3x^3 + \cdots + a_nx^n + \cdots$.
13. Use an integral test for the behavior of $\sum_{n=1}^{\infty} \frac{n^2}{n^3 + 2}$.
14. Use a comparison test for the behavior of $\sum_{n=1}^{\infty} \frac{4}{7^n + 3}$.
15. Use a ratio test for the behavior of $\sum_{n=1}^{\infty} \frac{(n + 1)(-1)^n}{n!}$.
16. Use a ratio test for the behavior of $\sum_{n=1}^{\infty} \frac{(-1)^n}{n^3}$.
17. Calculate the finite sum $5 + 9 + 13 + \cdots 45$.
18. Calculate the finite sum $\frac{2}{3} + \frac{2}{3^2} + \cdots \frac{2}{3^{15}}$.
19. Find the sum of the first five terms of a geometric sequence whose first term is 5 and common ratio 3.
20. Find the sum of the first six terms of a geometric sequence whose first term is 2 and common ratio −4.
21. Find the sum of the first 10 terms a geometric sequence 9, 36, 144, …, 2359296.

10 Applications to Probability

The theory of probability is the mathematics of chance phenomena. The subject often divides between **discrete phenomena** such as the number of passing cars, numbers of phone calls, and numbers of kittens in a litter and **continuous phenomena** such as fuel consumption in a winter season, daily temperatures, duration of phone conversations, and

Fundamentals of Calculus, First Edition. Carla C. Morris and Robert M. Stark.

Companion Website: http://www.wiley.com/go/morris/calculus

small variations in the contents of a 12 oz. soda can. The study of continuous phenomena, being calculus based, is the subject of this chapter.

10.1 DISCRETE AND CONTINUOUS RANDOM VARIABLES

A principal characteristic of the mathematics of probability is in the nature of variables. Elsewhere in this text, a variable, say, x, was assigned arbitrary values. Mathematicians know them as Riemann Variables. In probability theory, a Random Variable, X, takes on various values, x. Often, it is simply easier to use the value, x, dropping the subtle distinction and regarding it as both value and random variable. Random variables take their values by chance. One does not know in advance, for example, a child's adult height, the height of a ball about to be thrown, or tomorrow's temperature.

When possible variable values, called **sample points**, are discrete (or countable), as the numbers of heads in coin tosses, number of auto accidents, and so on, the random variable is **discrete valued**. When sample points are on a continuum, as the amount of rainfall, fuel consumption, and so on, the random variable is **continuous valued**.

Example 10.1.1 Classifying Random Variables

Classify these random variables as discrete or continuous.

a) Daily number of accidents at a particular intersection.
b) Ounces of soda in a 2 liter bottle.
c) Time to walk a mile.
d) Months a salesman exceeds quota.

Solution:
In a) and d), the random values are discrete valued.
In b) and c), the random values are continuous valued.

A **probability distribution, $P(x)$,** for a discrete random variable, $X = x$, associates a probability with each value of x. The totality of possible values of X is called the **sample space**. It can be a graph, a table, or a mathematical formula that specifies the probability associated with each possible value of the random variable, that is, on the sample space. Probability distributions for discrete random variables have the following properties:

a. All probabilities are non-negative and bounded by unity.
   That is: $0 \leq P(x) \leq 1$.
b. The sum of probabilities on a sample space is exactly 1.
   $P(x_1) + P(x_2) + \cdots + P(x_n) = 1$, where $x_1, x_2, \ldots, x_n$ are the only possible values of x.

Notation in probability theory can be tricky and often inconsistent. We generally use an upper case to denote a random variable, say, X, and specific values of the random variable with lower case, say x.

Example 10.1.2 Probability Distributions

Identify whether any of the following are probability distributions.

a)

x	0	2	5
$P(x)$	1/3	2/3	1/3

b)

x	−1	1	5
$P(x)$	0.2	0.3	0.5

c)

x	1	2	6	8	10	20
$P(x)$	0.10	0.25	0.05	0.15	0.30	0.10

Solution:

All entries are non-negative and bounded by unity. However, only in b) do the entries properly sum to one. Therefore, a) and c) cannot be probability distributions.

Example 10.1.3 Graphing a Probability Distribution

Display this probability distribution graphically.

x	1	2	4	5
$P(x)$	0.10	0.20	0.40	0.30

Solution:

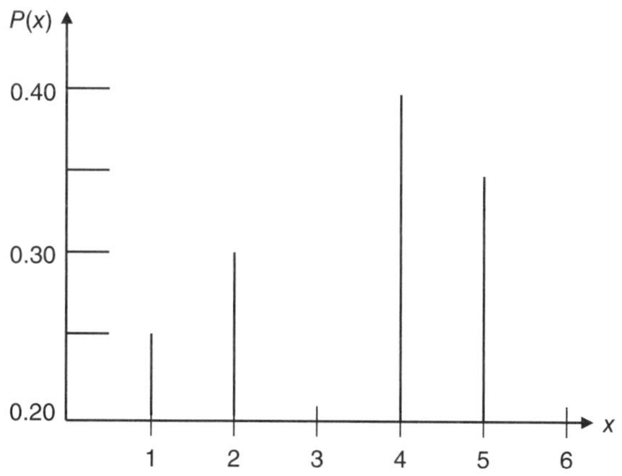

The vertical heights are proportional to the probability.

Example 10.1.4 Coin Tossing

A fair coin is tossed three times, and the number of tails, x, is observed. Display the probability distribution for x in tabular format.

Solution:

The eight possible outcomes of three tosses are:

Outcome	x	Outcome	x
HHH	0	THH	1
HHT	1	THT	2
HTH	1	TTH	2
HTT	2	TTT	3

The random variable, x, takes on values 0, 1, 2, and 3 corresponding to the number of tails. The probability of "no tails," denoted P(0), is 1/8, as there is only one event corresponding to no tails among the eight equally likely outcomes. The probability distribution in tabular format is:

x	0	1	2	3
$P(x)$	1/8	3/8	3/8	1/8

Example 10.1.5 A Discrete Probability Distribution

x	1	2	5	6	8	10
$P(x)$	0.20	0.15	0.05	0.30	0.10	0.20

For the given probability distribution calculate the following:

a) $P(x \geq 3)$

b) $P(x > 5)$

c) $P(x \leq 8)$

d) $P(2 \leq x < 10)$

e) $P(x \leq 7)$

f) $P(x = 9)$

Solution:

a) *The probabilities for* $x = 5$, $x = 6$, $x = 8$, *and* $x = 10$ *are added to yield* $P(x \geq 3) = 0.05 + 0.30 + 0.10 + 0.20 = 0.65$.

b) *The probabilities for* $x = 6$, $x = 8$, *and* $x = 10$ *are added to yield* $P(x > 5) = 0.30 + 0.10 + 0.20 = 0.60$.

⟶

c) It is easier to consider the probability not included ($x = 10$) to yield $P(x \leq 8) = 1 - P(x > 8) = 1 - 0.20 = 0.80$.

d) The probabilities for $x = 2$, $x = 5$, $x = 6$, and $x = 8$ are added to yield $P(2 \leq x < 10) = 0.15 + 0.05 + 0.30 + 0.10 = 0.60$.

e) The probabilities for $x = 1$, $x = 2$, $x = 5$, and $x = 6$ are added to yield $P(x \leq 7) = 0.20 + 0.15 + 0.05 + 0.30 = 0.70$.

f) No probability is given for $x = 9$, so $P(x = 9) = 0$.

For continuous probabilities, tabular formats are not useful. Instead, for a continuous random variable, say, x, a **probability density function**, $f(x)$, is defined. Every probability density function (**pdf**) has the following two properties:

Properties of Probability Density Functions

1. $f(x) \geq 0 \quad a \leq x \leq b$

2. $\int_a^b f(x) = 1$

↓ It is easy to confuse the use of the notation $f(x)$ in this case as a pdf and its "every-day" use in mathematical contexts as a function in prior chapters. Their meanings are different, although operations on them may be similar. Confusion is not likely from the context.

The relation between a probability density function and a **probability** is an integration. Areas under the pdf graph correspond to probabilities. The following figure illustrates a hypothetical pdf and identifies three arbitrary values of x: a, b, and c.

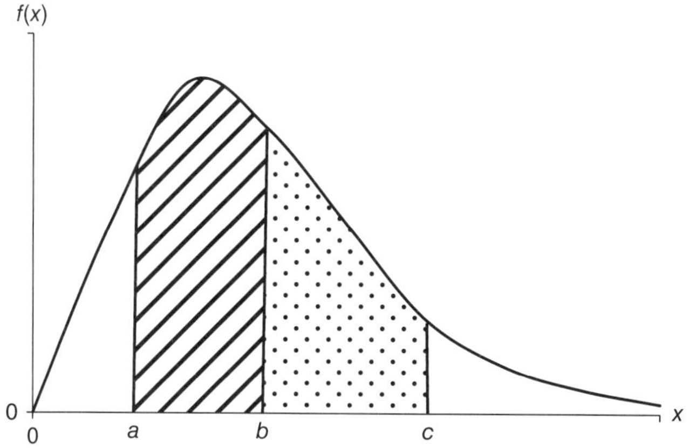

For example, the probability $P(a < x < b)$ is the shaded area above $a < x < b$. Similarly, the probability $P(x < b)$ is the area under the pdf curve for $0 < x < b$. The shaded area on the right corresponds to the probability that x takes a value between $x = b$ and $x = c$, that is, $P(b < x < c)$. It follows that $P(0 < x < \infty) = 1$, as it represents the entire area under the pdf. Note that the probability that x takes on any single value of x is zero for continuous random variables as a line cannot encompass an area.

↓ For this reason, it makes no difference whether the equality sign is, or is not, included in an area. That is, $P(x < b) = P(x \leq b)$ in this text. (No discontinuity is assumed).

Example 10.1.6 A Probability Density Function

Is $f(x) = x^2, 0 \leq x \leq \sqrt[3]{3}$ a probability density function?

Solution:

There are two properties for a pdf. The first is that $f(x)$ be non-negative. Here, $f(x) = x^2$ is non-negative.

An integration over the sample space $\int_0^{\sqrt[3]{3}} x^2\,dx = \left.\frac{x^3}{3}\right|_{x=0}^{x=\sqrt[3]{3}} = \frac{\left(\sqrt[3]{3}\right)^3}{3} - \frac{0^3}{3} = 1$. *This satisfies the second property.*

Therefore, $f(x) = x^2, 0 \leq x \leq \sqrt[3]{3}$ qualifies as a pdf.

Note: formally, we add that $f(x) = 0$ for values other than $0 \leq x \leq \sqrt[3]{3}$. The integral of $f(x) = 0$ being zero, it is common to omit $f(x) = 0$ for other values $\left(x < 0 \text{ or } x > \sqrt[3]{3}\right)$ of the random variable.

It is important to realize that the use of $f(x)$ as a probability density function is different conceptually from its use in earlier chapters. It has the same appearance, as in the previous example, $f(x) = x^2$; however, its use in probability is to calculate probabilities. It usually is not used to evaluate $f(x)$ for various values of x. Any function having the two required properties is a probability density function. Customarily, probability density functions are also known as *probability distributions*.

Example 10.1.7 Obtaining a Probability

What is the probability that $x < 1$ if its pdf is $f(x) = x^2, 0 \leq x \leq \sqrt[3]{3}$?

Solution:

$$P(x \leq 1) = \int_0^1 x^2 dx = \left.\frac{x^3}{3}\right|_0^1 = \frac{1^3}{3} - \frac{0^3}{3} = \frac{1}{3}.$$

The required probability is $\frac{1}{3}$. The notation $P(x \leq 1)$ is widely used to denote probability of the event in parenthesis, $x \leq 1$, in this case. It is the shaded area in the following sketch.

⟶

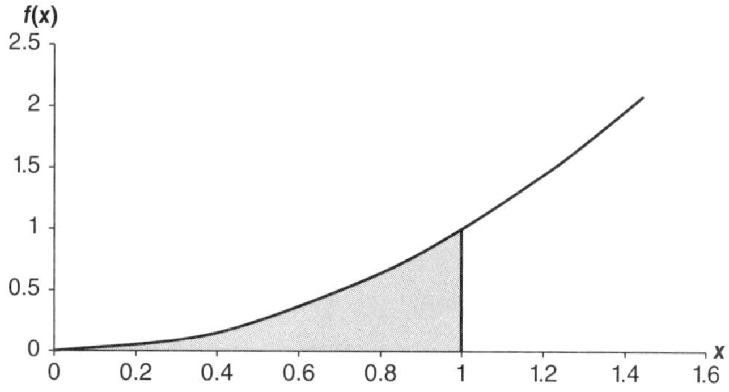

Example 10.1.8 Phone Calls

The random duration of public telephone conversations, t, is often described by an exponential probability density function, $f(t) = ae^{-at}, t > 0$ *where a is a constant. Calculate the probabilities, using* $a = 2$*, that a conversation will last*

a) less than 1 minute. *b) more than 3 minutes.*

Solution:

a) In this case, we seek $P(t \leq 1) = \int_0^1 2e^{-2t}dt = -e^{-2t}\Big|_0^1 = -e^{-2} + 1 = 0.865.$

b) In this case, we seek $P(t \geq 3) = \int_3^{\infty} 2e^{-2t}dt = \lim_{t\to\infty} \int_3^t 2e^{-2t}dt = e^{-6} = 0.0025.$

Example 10.1.9 Soft Drink Bottling

An automated machine is set to fill bottles with 2 liters of soft drink. For various reasons (timing mechanism, gear wear, water pressure, etc.) the actual amount dispensed varies uniformly between 1.95 and 2.05 liters. What proportion of production exceeds 2.01 liters?

Solution:

That the variation is uniform between 1.95 and 2.05 liters implies a probability density function $f(x) = c, \quad 1.95 \leq x \leq 2.05$

To determine c, recall that a pdf must satisfy a normality condition that $\int_{all\ x} f(x)dx = \int_{1.95}^{2.05} cdx = 1$*. Integration yields* $cx|_{1.95}^{2.05}$*, and one concludes that* $c = 10$*.*

$$P(2.01 \leq x \leq 2.05) = \int_{2.01}^{2.05} 10dx = 10x|_{2.01}^{2.05} = 0.40.$$

There is a 40% chance for production to exceed 2.01 liters.

Example 10.1.10 Stock Price Fluctuations

Stock prices fluctuate. A stock trader tracking daily variation of a stock concluded better buying chances occurred when volume was lower. Chance variation of the stock's daily volume, x, (in millions of shares) is estimated by an empiric pdf, say

$$f(x) = \frac{3}{19}x^2 \qquad 2 \leq x \leq 3$$

What is the probability of a day's volume will be less than 2.2 million shares?

Solution:

In this case, $P(2 \leq x \leq 2.2) = \int_2^{2.2} \frac{3}{19}x^2 dx = \left.\frac{x^3}{19}\right|_2^{2.2} = \frac{(2.2)^3 - (2)^3}{19} = 0.14$, *or about a 14% chance.*

EXERCISES 10.1

1. Classify these random variables as discrete or continuous:
   a) Time to type a report.
   b) Number of typing errors in a report.
   c) Number of employees calling in sick on a particular day.
   d) Amount of milk in a quart container.
   e) Weight of a watermelon.
2. The approximate distribution of blood types in the U.S. population follows:

Blood type	Percentage
O Rh – positive	
O Rh – negative	7
A Rh – positive	34
A Rh – negative	6
B Rh – positive	9
B Rh – negative	2
AB Rh – positive	3
AB Rh – negative	1

   What percentage of the U.S. population has blood type O and Rh-positive?
3. Which of the following are proper probability distributions?

a)

x	10	20	30	40
$P(x)$	0.14	0.16	0.30	0.50

→

b)

x	-3	1	4	9
$P(x)$	0.4	0.3	0.2	0.1

c)

x	14	19	21	23
$P(x)$	0.12	0.18	0.14	0.16

4. Complete these to depict proper probability distributions.

a)

x	100	200	300	400
$P(x)$	0.20	0.10	–	0.40

b)

x	21	25	29	35	40
$P(x)$	0.05	0.35	0.25	0.20	–

c)

x	-2	6	8	9	11	24
$P(x)$	1/16	1/8	3/16	5/16	–	1/8

5. For this probability distribution determine

x	51	52	55	56	58	60
$P(x)$	0.20	0.15	0.05	0.30	0.10	0.20

a) $P(x \geq 53)$

b) $P(x > 55)$

c) $P(x \leq 58)$

d) $P(52 \leq x < 60)$

e) $P(x \leq 57)$

f) $P(x = 59)$

6. Four coins are tossed. Let x be the number of heads observed. Display the probability distribution for x graphically.

Verify that the functions in Exercises 7–16 are probability density functions.

7. $f(x) = \dfrac{1}{7} \qquad 1 \leq x \leq 8$

8. $f(x) = \dfrac{1}{10} \qquad 1 \leq x \leq 11$

9. $f(x) = \dfrac{1}{50}x \qquad 0 \leq x \leq 10$

10. $f(x) = \dfrac{1}{8}x \qquad 0 \leq x \leq 4$

11. $f(x) = 3x^2 \qquad 0 \leq x \leq 1$

12. $f(x) = \dfrac{x^2}{3} \qquad -1 \leq x \leq 2$

13. $f(x) = 4x^3 \qquad 0 \leq x \leq 1$

14. $f(x) = \dfrac{1}{20}x^3 \qquad 1 \leq x \leq 3$

15. $f(x) = 3e^{-3x} \qquad x \geq 0$

16. $f(x) = \dfrac{5000}{x^3} \qquad x \geq 50$

In Exercises 17–26, calculate probabilities for the aforementioned exercises cited.

17. $P(2 \leq x \leq 7)$ for Exercise 7.
18. $P(2.5 \leq x \leq 8.5)$ for Exercise 8.
19. $P(3 \leq x \leq 7)$ for Exercise 9.
20. $P(1 \leq x \leq 3)$ for Exercise 10.
21. $P(0.5 \leq x \leq 1)$ for Exercise 11.
22. $P(0 \leq x \leq 1)$ for Exercise 12.
23. $P(0.1 \leq x \leq 0.8)$ for Exercise 13.
24. $P(2 \leq x \leq 2.5)$ for Exercise 14.
25. $P(1/3 \leq x \leq 5)$ for Exercise 15.
26. $P(60 \leq x \leq 80)$ for Exercise 16.

10.2 MEAN AND VARIANCE; EXPECTED VALUE

Mean and Variance – Discrete Variables

The long-run **average**, or **mean** value, of a random variable, X, is its **expected value**. This is unfortunate terminology, as an average need not be "expected" nor even a possible outcome. For example, the average number of tails when a fair coin is tossed repeatedly in groups of three tosses, 3/2 tails is the average or expected value. Of course, a "fraction of a tail" is impossible as an actual outcome.

The expected value (mean) of a random variable, X, is universally denoted as $\boldsymbol{E(X)}$. It is obtained by multiplying each value of a random variable by its associated probability and summing over every value of the random variable. For example, for the toss of three coins $E(X) = 0(1/8) + 1(3/8) + 2(3/8) + 3(1/8) = 12/8 = 3/2$.

To Calculate a Mean

For a random variable, X, with probability distribution, $P(x)$, multiply each value of x by $P(x)$ and sum over all values of x. That is,

$$E(X) = \mu = x_1P(x_1) + x_2P(x_2) + \cdots + x_nP(x_n)$$

is the mean or expected value of X.

Example 10.2.1 Gambling and Expected Value

Three fair coins are tossed. A gambler pays \$3 to play a game that pays \$2 for each tail. So, the gambler wins or loses \$1 or \$3 at each toss. Is the game fair?

Solution:

*A **fair** game has an expected gain/loss of zero in long-term play.*

If no tails are tossed, the gambler loses \$3; if one tail, the loss is \$1; if two tails, a win of \$1; and if three tails are tossed, \$3 is won.

⟶

x	-3	-1	1	3
$P(x)$	$1/8$	$3/8$	$3/8$	$1/8$

These values of the random variable are multiplied by the corresponding probability to yield

$$E(X) = (-3)(1/8) + (-1)(3/8) + (1)(3/8) + (3)(3/8) = \$0.$$

The expected value is zero, so the game is fair!

Averages or expected values are a useful "shorthand" to describe a random variable. Average temperature, for example, or average cost, average mileage, and so on are more convenient for everyday use than a table of data.

Besides the average value, the **variability** in data is of importance. It is a measure of its **dispersion** or **spread**. The simplest measure of variability is the **range** – the difference of the largest and smallest values.

Although the range is easy to calculate and interpret, it is not particularly useful for statistical theory. It is easy to understand why! The range, depending on only two observations, does not utilize the information in the other sample values. An unusual value for either the largest or the smallest observation dramatically alters the range.

Another more useful measure for variability is the **variance**. A **population variance**, σ^2, is defined by

$$\sigma^2 = \sum_{\text{all } x} (x - \mu)^2 P(x)$$

where $P(x)$ is the probability distribution for x, and μ is the population mean (expected value). The variance is the expected value of the squared fluctuations about the mean. When all n elements of a population are equally likely, $P(x)$ is simply $1/n$. The square root of the variance is the **standard deviation,** σ. Sometimes, σ is called the **standard error** and $\text{Var}(x)$ is used for σ^2.

Example 10.2.2 Variance and Standard Deviation

Given the probability distribution for a discrete random variable x, calculate the variance and standard deviation.

x	35	50	90	120	160
$P(x)$	0.30	0.25	0.10	0.20	0.15

Solution:
Firstly, calculate the mean as

$$E(X) = \mu = 35(0.30) + 50(0.25) + 90(0.10) + 120(0.20) + 160(0.15) = 80$$

⟶

Next, the variance is

$$\sigma^2 = (35-80)^2(0.30) + (50-80)^2(0.25) + (90-80)^2(0.10) + (120-80)^2(0.20)$$
$$+ (160-80)^2(0.15) = 2,122.5.$$

Finally, the standard deviation is $\sigma = \sqrt{2,122.5} = 46.07$.

Mean and Variance – Continuous Case

It is common in everyday usage to express chance phenomena in terms of an **average** and, perhaps, a **variability** as a kind of simplification of a governing probability distribution.

Outdoor temperatures or the duration of phone conversations are continuous valued random variables and observations over an extended period enable one to empirically fit a pdf to data. However, it is more convenient to answer a query about temperature, say, by citing an average temperature and perhaps by adding some measure of variation such as ±10°.

♦ Travel brochures usually cite averages of day and night temperatures and perhaps some variation for the time of year.

For a continuous random variable, x, governed by a probability density function, $f(x)$, the population mean and variance are defined as:

Population Mean and Variance

Definitions:

$$\mu = \int_a^b xf(x)dx \qquad \textbf{mean}$$

$$\sigma^2 = \int_a^b (x-\mu)^2 f(x)dx \text{ or } \int_a^b x^2 f(x)dx - \mu^2 \qquad \textbf{variance}$$

where x is a continuous random variable and $f(x), a \le x \le b$, its probability density function.

Example 10.2.3 Calculating a Mean and Variance

For $f(x) = cx, 0 \le x \le 4$, *determine* c *so* $f(x)$ *is a pdf. Calculate the mean and variance of* x.

Solution:

Firstly, set $\int_0^4 cxdx = \left.\frac{cx^2}{2}\right|_0^4 = \frac{16c}{2} - 0 = 8c = 1$ *to conclude that* $c = \frac{1}{8}$.

⟶

Next,

$$\mu = \int_0^4 x\left(\frac{1}{8}x\right)dx = \frac{x^3}{24}\bigg|_0^4 = \frac{64}{24} - 0 = \frac{8}{3} \quad \text{and}$$

$$\sigma^2 = \int_0^4 x^2\left(\frac{1}{8}x\right)dx - \left(\frac{8}{3}\right)^2 = \frac{x^4}{32}\bigg|_0^4 - \frac{64}{9} = \left(\frac{256}{32} - 0\right) - \frac{64}{9} = \frac{8}{9}.$$

Example 10.2.4 Phone Call Durations

The duration of phone conversations is approximated by $f(x) = ae^{-ax}$, $x > 0$, a is a constant (Example 10.1.7).

Determine the mean and variance for an exponential random variable.

Solution:

$$\textit{Mean}:\ \mu = \int_0^\infty x(ae^{-ax})dx = \lim_{t\to\infty}\int_0^t axe^{-ax}dx = \lim_{t\to\infty}\int_0^t axe^{-ax}dx.$$

Using integration by parts with $f(x) = x$ and $g'(x) = e^{-ax}$ yields

$$\lim_{t\to\infty}\left(-te^{-at} - \frac{e^{-at}}{a}\right) - \left(0 - \frac{1}{a}\right) = \frac{1}{a}.$$

$$\textit{Variance}:\ \sigma^2 = \int_0^\infty x^2(ae^{-ax})dx - \left[\frac{1}{a}\right]^2.$$

In this case, successive integrations by parts are required to yield

$$\lim_{t\to\infty}\left[\left(-t^2e^{-at} - \frac{2te^{-at}}{a} - \frac{2e^{-at}}{a^2}\right) - \left(-\frac{2}{a^2}\right)\right] = \frac{2}{a^2}.$$

The variance is $\dfrac{2}{a^2} - \dfrac{1}{a^2} = \dfrac{1}{a^2}.$

◆ The exponential distribution is widely used for the life of electrical equipment and the time gap between consecutive vehicles in ordinary traffic.

Example 10.2.5 Uniform PDF

Calculate the mean and variance for the automated soft drink machine in Example 10.1.9.

Solution:

$$\mu = \int_a^b \frac{x}{b-a}dx = \frac{x^2}{2\,(b-a)}\bigg|_a^b = \frac{b^2}{2(b-a)} - \frac{a^2}{2(b-a)} = \frac{b^2-a^2}{2(b-a)} = \frac{b+a}{2}.$$

⟶

$$\sigma^2 = \int_a^b \frac{x^2}{(b-a)} dx - \left[\frac{b+a}{2}\right]^2 = \frac{x^3}{3\,(b-a)}\bigg|_a^b - \frac{b^2+2ab+a^2}{4}$$
$$= \frac{b^3-a^3}{3(b-a)} - \frac{b^2+2ab+a^2}{4} = \frac{b^2+2ab+a^2}{3} - \frac{b^2+2ab+a^2}{4}$$
$$= \frac{b^2-2ab+a^2}{12} = \frac{(b-a)^2}{12}.$$

Graphically, the pdf is a rectangle with the mean $\mu = a + \frac{b-a}{2} = \frac{a+b}{2}$ *at its center. This agrees with the formal calculation shown previously. Unfortunately, the variance does not lend itself to an easy graphical description.*

Example 10.2.6 Stock Price Fluctuations – revisited

Determine the mean and variance for the stock price fluctuations in Example 10.1.10.

Solution:

The mean, $\mu = \int_2^3 x\left(\frac{3}{19}x^2\right) dx = \int_2^3 \left(\frac{3x^3}{19}\right) dx = \frac{3x^4}{76}\bigg|_2^3 = \frac{243}{76} - \frac{48}{76} = \frac{195}{76}.$

The variance,

$$\sigma^2 = \int_2^3 x^2\left(\frac{3x^2}{19}\right) dx - \left[\frac{195}{76}\right]^2$$
$$= \int_2^3 \left(\frac{3x^4}{19}\right) dx - \left[\frac{195}{76}\right]^2$$
$$= \frac{3x^5}{95}\bigg|_2^3 - \left(\frac{195}{76}\right)^2 = \frac{729}{95} - \frac{96}{95} - \frac{38{,}025}{5{,}776} = \frac{2{,}307}{28{,}880} \approx 0.08.$$

EXERCISES 10.2

In Exercises 1–4, calculate means, variances, and standard deviations for these probability distributions.

1.

x	50	100	150	200
$P(x)$	0.20	0.10	0.30	0.40

2.

x	5	8	12	20	30
$P(x)$	0.20	0.10	0.20	0.30	0.20

3.

x	1	4	7	10	12
$P(x)$	0.15	0.15	0.25	0.20	0.25

4.

x	12	24	36	48	60	72
$P(x)$	1/6	1/6	1/6	1/6	1/6	1/6

Determine the means, variances, and standard deviations for these probability distributions (Exercises 10.1.7–10.1.15).

5. $f(x) = \frac{1}{7} \quad 1 \le x \le 8$
6. $f(x) = \frac{1}{10} \quad 1 \le x \le 11$
7. $f(x) = \frac{1}{50}x \quad 0 \le x \le 10$
8. $f(x) = \frac{1}{8}x \quad 0 \le x \le 4$
9. $f(x) = 3x^2 \quad 0 \le x \le 1$
10. $f(x) = \frac{x^2}{3} \quad -1 \le x \le 2$
11. $f(x) = 4x^3 \quad 0 \le x \le 1$
12. $f(x) = \frac{1}{20}x^3 \quad 1 \le x \le 3$
13. $f(x) = 3e^{-3x} \quad x \ge 0$

10.3 NORMAL PROBABILITY DENSITY FUNCTION

The **Normal Distribution,** a pdf, arises repeatedly in nature and in everyday human affairs. Remarkably, the normal distribution closely approximates such diverse everyday phenomena as project completion times, heights, weights, and IQ's of groups of animals and peoples, grade point averages, shots about a target, tire mileage and much more.

The normal pdf is defined by $f(x) = \frac{1}{\sqrt{2\pi}\sigma} e^{-\frac{1}{2}\left(\frac{x-\mu}{\sigma}\right)^2} \qquad -\infty < x < \infty$

where parameters μ and σ^2 are, as earlier, the mean and variance. The normal random variable, X, can assume any value on the entire real line, $-\infty < x < \infty$.

Fortunately, there is no need to remember this important and awkward appearing pdf. Clearly, $f(x) \ge 0$ and it can be shown that $\int_{-\infty}^{\infty} f(x)dx = 1$, although a proof is beyond our scope. Not only is it unnecessary to remember the pdf of the normal distribution, it is actually unnecessary to calculate probabilities corresponding to areas on its graph; they are widely tabulated.

You may recognize the following sketch, commonly called the **bell curve**. This sketch of the normal distribution is the special case when $\mu = 0$ and $\sigma = 1$ and is known as the **Standard Normal Distribution**. Tabulations for it are on an accompanying page. Every normal distribution, regardless of the values of μ and σ, can be placed into the Standard Form by a change of variable. Therefore, the accompanying standard table permits calculation for almost any value of μ and σ.

The variable transformation $z = \frac{x-\mu}{\sigma}$ transforms f(x) into $g(z) = \frac{1}{\sqrt{2\pi}} e^{-(1/2)z^2}$, the standard normal distribution. It is important to note the symmetry of the normal curve about its mean, μ. The symmetry for the Standard Normal Variable is about $\mu = 0$

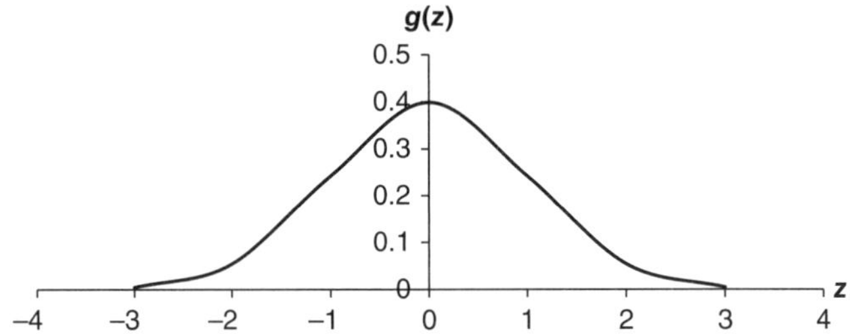

Standard Normal Random Variable, Z

$$g(z) = \frac{1}{\sqrt{2\pi}} e^{-(1/2)z^2} \qquad -\infty < z < \infty$$

Firstly, note in the following table that the values of z appear in the left column, while second decimal places head the columns. So, for example, the table value for $z = 1.60$ is 0.4452 and for $z = 1.63$, move horizontally from $z = 1.60$ to the 0.03 column to read 0.4484. This is the area under the normal curve between $z = 0$ and $z = 1.63$.

Note that values of z span $z = 0$ to 3.09. Note also that the area between $z = 0$ and 3.09 is 0.4990, nearly the total area of 0.5000 (for the right half of the bell curve). For practical purposes, there is little need for values of z greater than 3.09, so they are omitted in most tables. Symmetry permits the Table to be used for negative values of z. The area, 0.4484, for $z = -1.63$ to $z = 0$, matches $z = 0$ to $z = 1.63$. The following example illustrates.

Example 10.3.1 Some Standard Normal Probabilities

Determine probabilities for a Standard Normal random variable with values

a) $-1 \leq z \leq 1$ *b)* $-2 \leq z \leq 2$ *c)* $-3 \leq z \leq 3$

Solution:

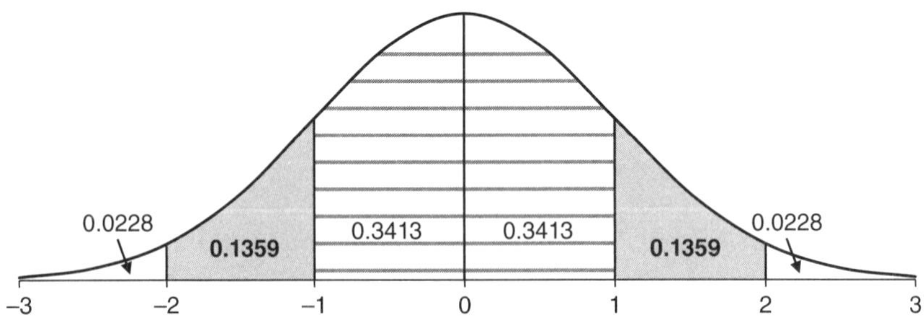

a) The area for $0 \leq z \leq 1$ *is 0.3412 in the Standard Normal Table. Using symmetry, the area encompassed by* $-1 \leq z \leq 1$ *is 0.3413 + 0.3413 = 06826 (figure).*

→

b) Similarly, between $z = 0$ and $z = 2$, the area is 0.4772. So the probability $P(-2 < z < 2) = 0.4772 + 0.4772 = 0.9544$; or about 95% (sum of shaded and lined areas).

c) Finally, $P(-3 < z < 3) = 0.4990 + 0.4990 = 0.9980$; nearly the entire area.

Note these results as they arise often.

The Standard Normal Table ($\mu = 0, \sigma = 1$) has many useful features.

Standard Normal Probabilities

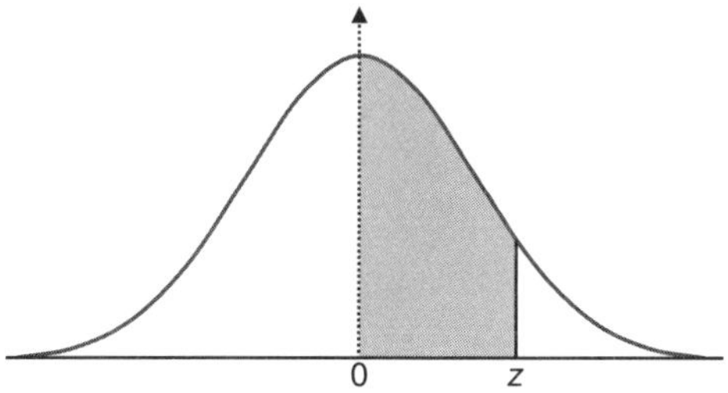

Z	0.00	0.01	0.02	0.03	0.04	0.05	0.06	0.07	0.08	0.09
0.0	0.0000	0.0040	0.0080	0.0120	0.0160	0.0199	0.0239	0.0279	0.0319	0.0359
0.1	0.0398	0.0438	0.0478	0.0517	0.0557	0.0596	0.0636	0.0675	0.0714	0.0753
0.2	0.0793	0.0832	0.0871	0.0910	0.0948	0.0987	0.1026	0.1064	0.1103	0.1141
0.3	0.1179	0.1217	0.1255	0.1293	0.1331	0.1368	0.1406	0.1443	0.1480	0.1517
0.4	0.1554	0.1591	0.1628	0.1664	0.1700	0.1736	0.1772	0.1808	0.1884	0.1879
0.5	0.1915	0.1980	0.1985	0.2019	0.2054	0.2088	0.2123	0.2157	0.2190	0.2224
0.6	0.2257	0.2291	0.2324	0.2357	0.2389	0.2422	0.2454	0.2486	0.2517	0.2549
0.7	0.2580	0.2611	0.2642	0.2673	0.2704	0.2734	0.2764	0.2794	0.2823	0.2852
0.8	0.2881	0.2910	0.2939	0.2967	0.2995	0.3023	0.3051	0.3078	0.3106	0.3133
0.9	0.3159	0.3186	0.3212	0.3238	0.3264	0.3289	0.3315	0.3340	0.3365	0.3389
1.0	0.3413	0.3438	0.3461	0.3485	0.3508	0.3531	0.3554	0.3577	0.3599	0.3621
1.1	0.3643	0.3665	0.3686	0.3708	0.3729	0.3749	0.3770	0.3790	0.3810	0.3830
1.2	0.3849	0.3869	0.3888	0.3907	0.3925	0.3944	0.3962	0.3980	0.3997	0.4015
1.3	0.4032	0.4049	0.4066	0.4082	0.4099	0.4115	0.4131	0.4147	0.4162	0.4177
1.4	0.4192	0.4207	0.4222	0.4236	0.4251	0.4265	0.4279	0.4292	0.4306	0.4319
1.5	0.4332	0.4345	0.4357	0.4370	0.4382	0.4394	0.4406	0.4418	0.4429	0.4441
1.6	0.4452	0.4463	0.4474	0.4484	0.4495	0.4505	0.4515	0.4525	0.4535	0.4545
1.7	0.4554	0.4564	0.4573	0.4582	0.4591	0.4599	0.4608	0.4616	0.4625	0.4633
1.8	0.4641	0.4649	0.4656	0.4664	0.4671	0.4678	0.4686	0.4693	0.4699	0.4706
1.9	0.4713	0.4719	0.4726	0.4732	0.4738	0.4744	0.4750	0.4756	0.4761	0.4767
2.0	0.4772	0.4778	0.4783	0.4788	0.4793	0.4798	0.4803	0.4808	0.4812	0.4817
2.1	0.4821	0.4826	0.4830	0.4834	0.4838	0.4842	0.4846	0.4850	0.4854	0.4857
2.2	0.4861	0.4864	0.4868	0.4871	0.4875	0.4878	0.4881	0.4884	0.4887	0.4890
2.3	0.4893	0.4896	0.4898	0.4901	0.4904	0.4906	0.4909	0.4911	0.4913	0.4916
2.4	0.4918	0.4920	0.4922	0.4925	0.4927	0.4929	0.4931	0.4932	0.4934	0.4936
2.5	0.4938	0.4940	0.4941	0.4943	0.4945	0.4946	0.4948	0.4949	0.4951	0.4952
2.6	0.4953	0.4955	0.4956	0.4957	0.4959	0.4960	0.4961	0.4962	0.4963	0.4964
2.7	0.4965	0.4966	0.4967	0.4968	0.4969	0.4970	0.4971	0.4972	0.4973	0.4974
2.8	0.4974	0.4975	0.4976	0.4977	0.4977	0.4978	0.4979	0.4979	0.4980	0.4981
2.9	0.4981	0.4982	0.4982	0.4983	0.4984	0.4984	0.4985	0.4985	0.4986	0.4986
3.0	0.4987	0.4987	0.4987	0.4988	0.4988	0.4989	0.4989	0.4989	0.4990	0.4990

Example 10.3.2 More Standard Normal Probabilities

Find the indicated probabilities for the Standard Normal random variable Z and illustrate them in a sketch.

a) $P(0 < z < 1.61)$ b) $P(z > -1.35)$ c) $P(-1.32 < z < 2.94)$

Solution:

a) *An area of interest is shaded as follows. The probability read from the Standard Table for* $0 \le z \le 1.61$ *is 0.4463.*

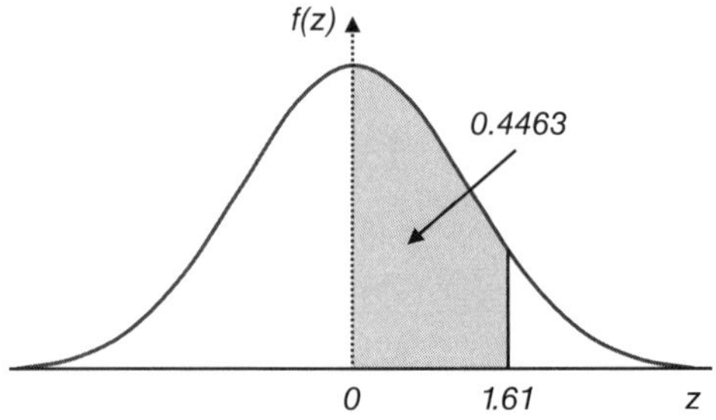

b) *For* $z > -1.35$, *the desired probability is the shaded areas. Using the symmetry of the normal distribution, the area above* $z > 0$ *is 0.5000. The area for* $-1.35 < z < 0$ *is the same for* $0 < z < 1.35$ *(again, due to symmetry). Therefore, the desired probability is* $0.4115 + 0.5000 = 0.9115$.

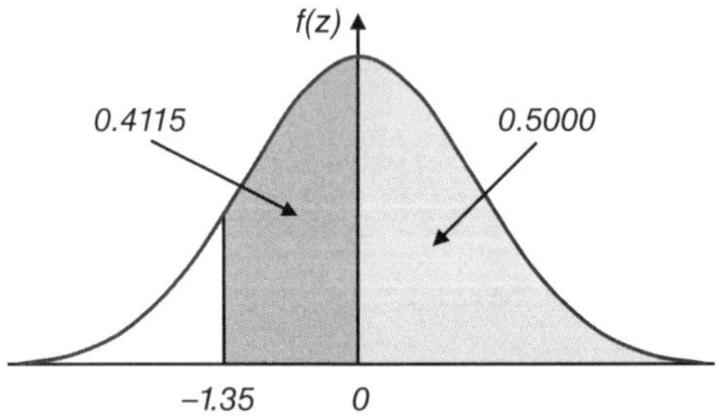

c) *As the z values are opposite in sign, the probabilities associated with* $-1.31 \le z \le 0$ *and* $0 \le z \le 2.93$ *are, respectively, 0.4066 and 0.4984.*
Adding, $0.4066 + 0.4984 = 0.9050$ *(see sketch).*

→

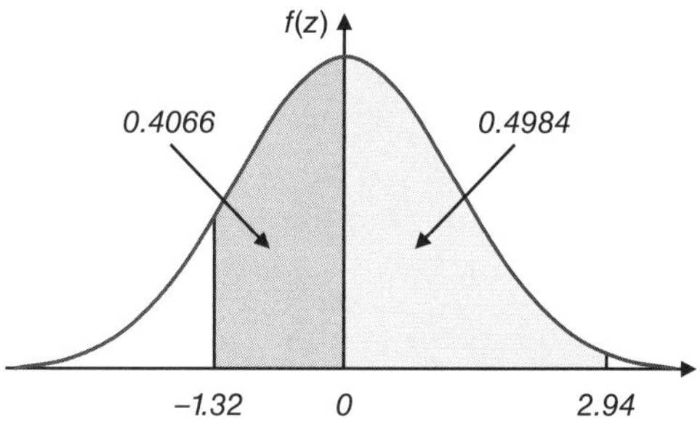

Standard Normal Probabilities *($\mu = 0, \sigma = 1$)*

1. **Sketch a standard normal curve.**
2. **Shade the area of interest.**
3. **Areas under the curve are values (probability) in the Standard Normal Table. For negative values of z, use the symmetry of the Normal Distribution.**
4. **Useful properties:**
   a) $P(z \geq 0) = (z \leq 0) = 0.50$
   b) $P(-a \leq z \leq 0) = P(0 \leq z \leq a) \qquad a > 0$
   c) $P(z = a) = 0$

Transforming to a Standard Normal Distribution

In practice, normal random variables usually have means and variances other than the standard, $\mu = 0$ and $\sigma^2 = 1$. Fortunately, every normal pdf can be transformed into a standard normal pdf using the linear **z-transformation**, or **z-score**,

$$z = \frac{x - \mu}{\sigma}$$

Therefore, one can determine areas, and hence probabilities, for any normal distribution using the same standard normal table as earlier.

To apply the normal probability table to a normal random variable X with mean μ and standard deviation σ, first convert the value X to its corresponding z-score. A negative z-score indicates that X is less than the mean, while a positive z-score that X exceeds the mean.

The following couple of examples illustrate the calculation.

Example 10.3.3 *$\mu = 2$ and $\sigma = 1/2$*

Find the probability that $1.5 \leq x \leq 3$ for the normal random variable x with $\mu = 2$ and $\sigma = 1/2$.

⟶

Solution:
The interval $1.5 \leq x \leq 3.0$ is transformed into an equivalent interval for z. Next, subtract the mean from each term in the inequality to yield

$$1.5 - 2 \leq x - \mu \leq 3 - 2 \text{ or} - 0.5 \leq x - \mu \leq 1.0.$$

Next, divide each term by the standard deviation to yield

$$\frac{-0.5}{1/2} \leq \frac{x - \mu}{\sigma} \leq \frac{1.0}{1/2} \text{ or} - 1 \leq z \leq 2$$

Therefore, $P(1.5 \leq x \leq 3.0) = P(-1 \leq z \leq 2)$. The two z values are on opposite sides of $z = 0$, so the desired probability is obtained by addition.

$$P(-1 \leq z \leq 0) + P(0 \leq z \leq 2) = 0.3413 + 0.4772 = 0.8185. \text{ (a sketch is useful)}$$

Example 10.3.4 $\mu = 8$ **and** $\sigma = 2$

The random variable, x, has a normal distribution with $\mu = 8$ and $\sigma = 2$. What is the probability that $x < 6$?

Solution:
Transforming to a z value $x < 6$ is equivalent to $z < (6 - 8)/2 = -1$; the lower tail of the normal curve. For $P(z < -1) = 0.50 - P(-1 \leq z \leq 0) = 0.500 - 0.3413 = 0.1587$; the required probability.

Note the use made of the complement, $P(z < -1) + P(-1 < z < 0) = 0.5000$.

The normal distribution is the most important of all probability distributions. It is fundamental to probability and statistical theory and arises repeatedly in practical problems. The normal curve provides a good approximation for many other probability density functions. For example, the normal distribution has been used as an approximation for the relative frequency distributions for heights and weights of animals, IQ scores, manufacturing tolerances, measurement errors, and many others.

Example 10.3.5 Vending Machine Operation

A vending machine is set to dispense 12 ounces of liquid. The amount actually dispensed is normally distributed with $\sigma = 0.25$ oz. What is the probability the machine dispenses at least 11.5 ounces of liquid?

⟶

Solution:

Let x be the random amount of dispensed liquid with a mean of 12 ounces. Here, $z = \dfrac{11.5 - 12}{0.25} = -2$*. We seek* $P(z \geq -2)$*.(figure)*

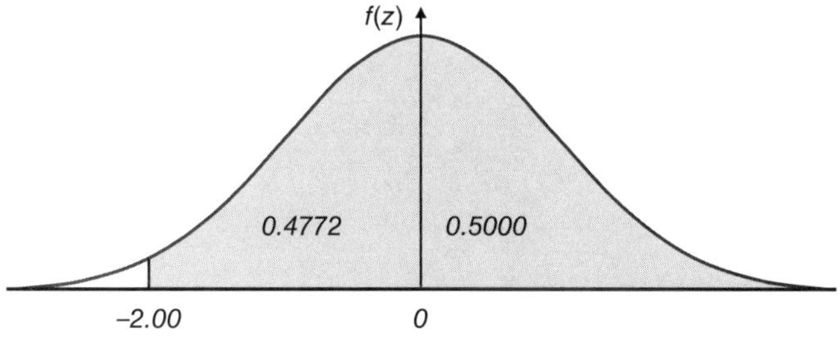

The probability the machine dispenses at least 11.5 ounces of liquid is $0.4772 + 0.5000 = 0.9772$*. There is nearly a 98% chance, on average, that at least 11.5 ounces is dispensed.*

Example 10.3.6 Rescuing a Downed Flier

A downed flier is spotted. A rectangular 250 ft × 200 ft rectangular area is outlined with the flier at the center. A life line is to be thrown from the air and, to be effective, should land within 25 ft of the flier horizontally and 15 ft on the vertical. Unfortunately, its landing place is uncertain and described by normal distributions in the x and y directions with respective means of zero and variances 1600 ft^2 *and 400* ft^2*. What is the probability that a single "throw" will succeed?*

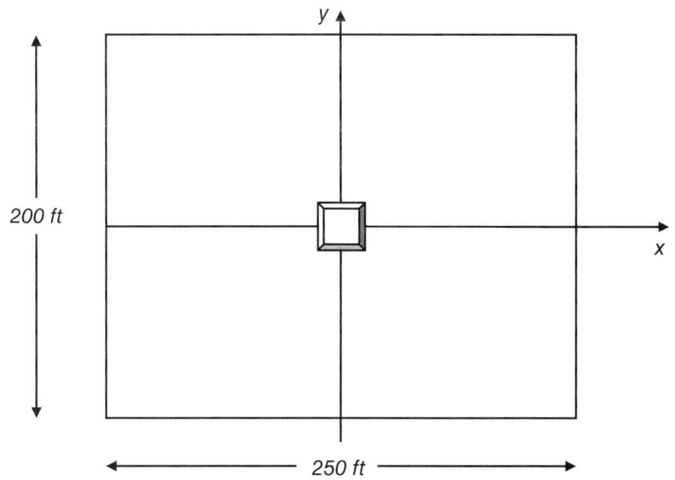

Solution:

Firstly, the square roots of the variances yield standard deviations of 40 ft and 20 ft, respectively.

⟶

Horizontally, transforming to a z-score,

$$P(-25 < x < 25) = P\left(\frac{-25-0}{40} < z < \frac{25-0}{40}\right) = P(-0.625 < z < 0.625) = 0.4680$$

and vertically,

$$P(-15 < x < 15) = P\left(\frac{-15-0}{20} < z < \frac{15-0}{20}\right) = P(-0.75 < z < 0.75) = 0.5468.$$

The probability of both x and y taking values within the target area is their product (0.4680)(0.5468) = 0.2559.

Note: students of probability will recognize that x and y are regarded as independent random variables, so the target probability is the product of the individual probabilities.

An interesting situation arises when there are alternate ways to complete a task such as a construction job or travel between two cities. Suppose that the random times to complete each of two jobs have Normal pdfs. The four possible arrangements of their means and variances are:

$\mu_1 > \mu_2$	$\mu_1 < \mu_2$	$\mu_1 > \mu_2$	$\mu_1 < \mu_2$
$\sigma_1 > \sigma_2$	$\sigma_1 > \sigma_2$	$\sigma_1 < \sigma_2$	$\sigma_1 < \sigma_2$
(a)	(b)	(c)	(d)

If the object is to choose the method that is most likely to complete the job in the shortest time, arrangements (a) and (d) are easy decisions! In those arrangements, one of the choices not only has a smaller mean, but also has a smaller variance, and that is the clear choice. However, arrangements (b) and (c) present a dilemma! While one of the jobs has a mean that is smaller than the other, its variance is larger, so the shorter average time is compromised by the greater variability. That implies greater risk due to the increased variability. The choice is not a trivial decision. The following example illustrates the possibilities.

Example 10.3.7 Which Construction Plan?

A construction superintendent is considering two alternative plans for a project. The time of completion is believed to be approximately normally distributed for each alternative and with the following parameters:

Plan I: $\mu_1 = 33.9$ *months and* $\sigma_1 = 11.8$ *months*
Plan II: $\mu_2 = 29.3$ *months and* $\sigma_2 = 17.5$ *months*

The target completion time is 42 months. Which plan is the better choice?

Solution:
Let x_1 *and* x_2 *be the respective random completion times. The means and variances, of themselves, are conflicting signals. However, as a target completion time is specified, it is possible to calculate the probabilities of meeting the target time. This suggests that the*

→

better choice is the plan with the higher probability of timely completion. We seek the probabilities:

$$P(x_1 < 42) = P\left(\frac{x_1 - \mu_1}{\sigma_1} < \frac{42 - 33.9}{11.8} = 0.69\right) = 0.7549 \text{ and}$$

$$P(x_2 < 42) = P\left(\frac{x_2 - \mu_2}{\sigma_2} < \frac{42 - 29.3}{17.5} = 0.73\right) = 0.7673.$$

The two probabilities are close so that other factors, not considered here, may influence the superintendent's choice. However, other things equal, Plan II has a slightly better chance of meeting the 42-month target.

Example 10.3.8 Change in Target Date

Suppose that the target date of the previous example has been extended to 44 months. Should this affect the superintendent's decision? If so, how?

Solution:

$$P(x_1 < 44) = P\left(\frac{x_1 - \mu_1}{\sigma_1} < \frac{44 - 33.9}{11.8} = 0.86\right) = 0.8051$$

$$P(x_2 < 44) = P\left(\frac{x_2 - \mu_2}{\sigma_2} < \frac{44 - 29.3}{17.5} = 0.84\right) = 0.7995.$$

Again, the difference in the two probabilities is small. However, now Plan I, possibly a much different construction plan, is preferred.

The aforementioned two examples illustrate a powerful use of probability in choosing among competing alternatives.

The examples in this Section have specified that the random variables (liquid dispensed, landing spot, travel times, etc.) are normally distributed. In practice, the governing probability distribution is usually not known – (although a normal distribution is often assumed). What is known is that the Central Limit Theorem – one of the most remarkable in mathematical theory – is the basis for the ubiquitousness of the normal distribution. Roughly said, the theorem, under rather general conditions, assures that the average of repeated observations (the sample average) is approximated by a normal distribution; the approximation improves as the number of observations (the sample size) in the average increases. Note that the Central Limit Theorem makes no assumption as to the distribution of the individual observations of the constituent variables.

EXERCISES 10.3

1. Calculate probabilities for the standard normal random variable, z:
   a) $P(0 \le z \le 1.47)$
   b) $P(0 \le z \le 0.97)$
   c) $P(-2.36 < z < 0)$
   d) $P(-1.24 < z < 0)$
   e) $P(-2.13 \le z \le 0)$
   f) $P(-0.19 \le z \le 0)$

2. Calculate probabilities for the standard normal random variable z:

   a) $P(-1.23 \leq z \leq 2.45)$
   b) $P(-2.59 \leq z \leq 1.04)$
   c) $P(-0.41 \leq z \leq 2.19)$
   d) $P(-2.67 \leq z \leq 0.83)$
   e) $P(-1.23 \leq z \leq 1.23)$
   f) $P(-2.33 \leq z \leq 2.17)$

3. Calculate these probabilities for the standard normal random variable z:

   a) $P(z > -1.55)$ b) $P(z \geq 1.86)$ c) $P(z < -1.47)$ d) $P(z < 1.30)$

4. The random variable X has a normal distribution, mean 80 and standard deviation 5. Determine the z-score that corresponds to these X values.

   a) 85 b) 77 c) 69 d) 71 e) 86 f) 80

5. If SAT math scores are normally distributed with a mean of 550 and standard deviation of 100, determine the probability that a score is
   a) between 460 and 640.
   b) below 730.
   c) at least 410.

6. Grades on a test are normally distributed with a mean of 75 and variance of 16. Determine the grade that
   a) 85% of the class did not achieve.
   b) 70% of the class exceeded.

7. A machine that regulates the volume dispensed for a gallon of paint is set for 128.4 ounces on average. The volume dispensed is normally distributed. If less than 128 ounces is dispensed, it is unacceptable; as the claim is false. Determine the necessary standard deviation so that no more than 1% of paint cans are under filled.

8. A machine used to regulate the amount of sugar dispensed to make a new soft drink can be set so that it dispenses μ grams of sugar per liter on average. The amount of sugar dispensed is normally distributed with a standard deviation 0.36. If more than 8.5 g of sugar is dispensed, the liter of soft drink is unacceptable. What setting of μ will allow no more than 4% of output to be unacceptable?

HISTORICAL NOTES

Karl Friedrich Gauss (1777–1855) — German-born Johann Friedrich Karl Gauss distinguished himself as mathematician, physicist, and astronomer. Called "The Prince of Mathematicians," Gauss made profound contributions to number theory, the theory of functions, probability, statistics, electromagnetic theory, and much more!

His doctoral dissertation (University of Göttingen, 1795–1798) established that every algebraic equation has at least one root (solution). This theorem that had challenged mathematicians for centuries is "the fundamental theorem of algebra." Gauss insisted on a complete proof of any result before publication. Consequently, many discoveries were not credited to him, at least initially, including the method of least squares. However, his published works were sufficient to establish his reputation as perhaps the greatest mathematician of all time.

CHAPTER 10 SUPPLEMENTARY EXERCISES

1. Which of these are probability distributions?

a)

x	4	9	11	13
$P(x)$	0.13	0.17	0.35	0.45

b)

x	3	6	8	9
$P(x)$	0.33	0.27	0.22	0.18

2. Complete the tables to form probability distributions.

a)

x	1	2	3	4
$P(x)$	0.25	0.15	0.40	–

b)

x	1	5	9	15	20
$P(x)$	0.07	0.33	0.20	–	0.09

3. For this probability distribution determine:

x	10	12	15	16	18	20
$P(x)$	0.15	0.20	0.30	0.05	0.20	0.10

a) $P(x \geq 15)$

b) $P(x > 17)$

c) $P(12 \leq x < 16)$

d) $P(x \geq 18)$

4. Verify that $f(x)$ is a probability density function.

$$f(x) = \frac{1}{40}x \qquad 1 \leq x \leq 9$$

5. Verify that $f(x)$ is a probability density function.

$$f(x) = 5x^4 \qquad 0 \leq x \leq 1$$

6. Calculate the Expected value, variance, and standard deviation for Supplementary Exercise 4.

7. Calculate the Expected value, variance, and standard deviation for Supplementary Exercise 5.

8. Calculate these probabilities for the standard normal random variable, z:

a) $P(0 \leq z \leq 1.47)$
b) $P(-1.38 \leq z \leq 0)$
c) $P(-2.12 < z < 1.45)$
d) $P(-2.4 < z < 1.53)$
e) $P(1.13 \leq z \leq 2.27)$
f) $P(z \leq 1.69)$

Answers to Odd Numbered Exercises

For complete solutions to these Exercises see the companion "Student Solutions Manual" by Morris and Stark.

EXERCISES 1.1

1. conditional equation. 3. Contradiction. 5. Contradiction. 7. $x = 4$. 9. $x = 5$. 11. $x = 1$. 13. $x = 50$. 15. $x = 5/4$. 17. No solution. 19. $s = 10$. 21. $t = 3$. 23. $x = 5$. 25. $x = 6$. 27. $x = 2$. 29. $x = \frac{1}{2}y + \frac{3}{2}$. 31. $y = 3 - \frac{3}{5}x$. 33. $V/LH = W$. 35. $x = Z\sigma + \mu$.

37. \$200 monthly installment. 39. $C(x) = 0.75x + 9.5$. 41. a) $d = 9$ miles b) 4 seconds.
43. $T = 0.062x \quad 0 \le x \le 87{,}000$. 45. a) 8187 sq cm b) Wt = 26.2kg

EXERCISES 1.2

1. a) x-intercept 3, y-intercept -5 b) x-intercept 5/4, y-intercept -5
c) x-intercept 12, y-intercept 8 d) x-intercept 2, y-intercept -18
e) x-intercept 4, no y-intercept (vertical line) f) no x-intercept, y-intercept -2

3. a) $m = 1/2$ b) $m = 5$ c) $m = 2/3$ d) m is undefined e) $m = 0$ f) $m = 1/5$

5. a) x-intercept 5/2 and y-intercept -5 b) x-intercept 4 and no y-intercept

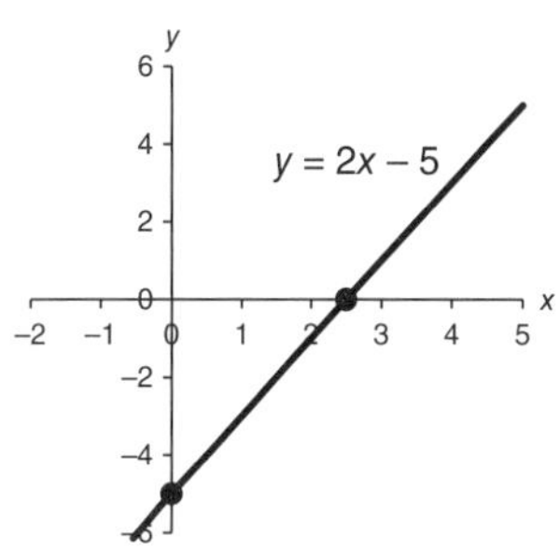

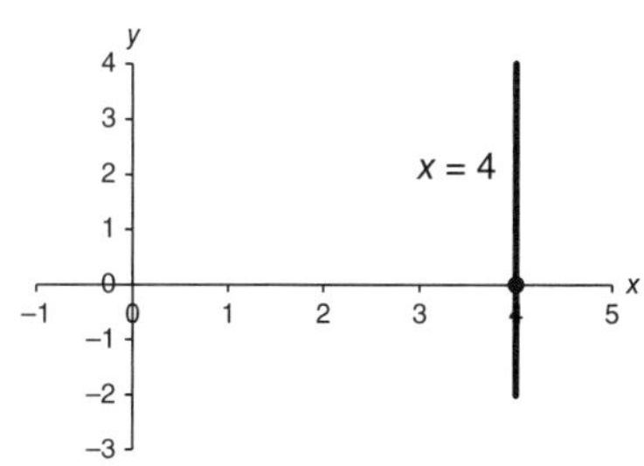

→

Fundamentals of Calculus, First Edition. Carla C. Morris and Robert M. Stark.

Companion Website: http://www.wiley.com/go/morris/calculus

c) x-intercept 5 and y-intercept 3

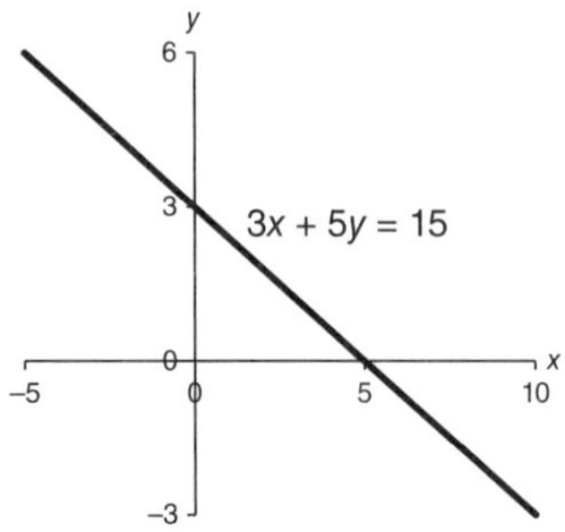

d) x-intercept 7 and y-intercept 2

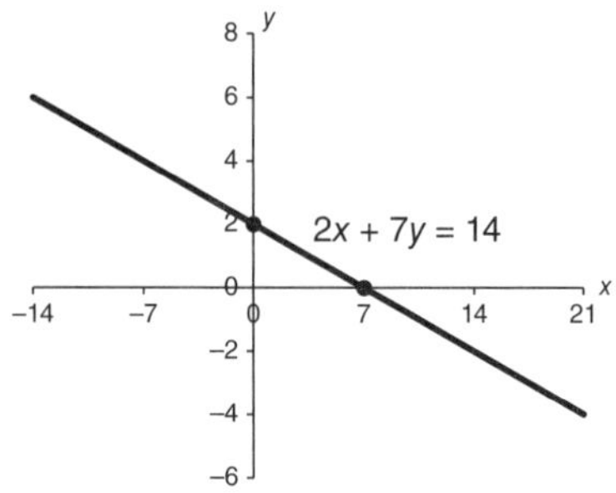

7. a) parallel b) perpendicular c) parallel d) neither e) perpendicular
9. $y = 0$ (the x-axis) it has an infinite number of x-intercepts. Any horizontal line (except $y = 0$) has no x-intercepts. If $x = 0$ (the y-axis) it has an infinite number of y-intercepts. Any vertical line (except $x = 0$) has no y-intercepts.
11. $V(t) = -6000t + 75{,}000$ 13. $y - 245 = 35(x - 7)$ or $y = 35x$
15. $C(x) = 1100 + 5x$ 17. a) $C(x) = 50 + 0.28x$ b) 150 miles
19. $R - 84 = (6/5)(C - 70)$ or $R = (6/5)C$

EXERCISES 1.3

1. $8(x - 3)$ 3. $5x(x^2 - 2x + 3)$ 5. $5a^3bc^3(bc + 2)$ 7. $5x^2y^3z^5(4xy^2z + 3x^2z^2 + 4y)$
9. $(x - 5)(x + 5)$ 11. $3(x^2 + 9)$ 13. $2(x - 2)(x^2 + 2x + 4)$ 15. $7(a + b)(x + 2)(x - 2)$
17. $(x + 4)(x + 1)$ 19. not factorable 21. $(x - 8)(x + 2)$ 23. $2(x + 4)(x + 2)$
25. $(ab + 4)(ab + 5)$ 27. $2(xy + 9)(xy + 5)$ 29. not factorable
31. $(x - 2)(x + 2)(x - 1)(x + 1)$ 33. $(x - a)[(x + a) + 5]$ 35. $2[(b - 2x)(2a + 3y)]$
37. $x = -8$ or $x = -1$ 39. $x = -8$ or $x = -9$ 41. no real solutions 43. $x = 6$ or $x = 3$
45. $x = 1$ or $x = 1/2$

EXERCISES 1.4

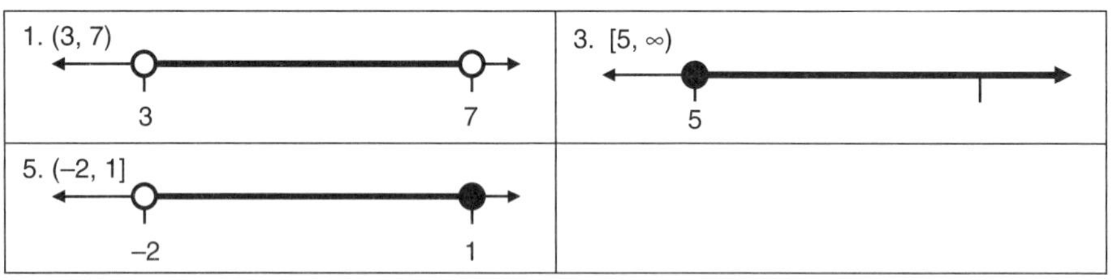

7. $(4, \infty)$ 9. $(-3, 7)$ 11. $[1, 8)$ 13. $[5, 8)$ 15. $(-\infty, \infty)$ 17. $[5/2, \infty)$
19. $(-\infty, -3) \cup (-3, 1) \cup (1, \infty)$
21. a) $f(0) = 3$ b) $f(1) = 15$ c) $f(x + 3) = 7(x + 3)^3 + 5(x + 3) + 3$
23. a) $f(-1) = 10$ b) $f(a^2) = a^{10} + 11$ c) $f(x + h) = (x + h)^5 + 11$.

25. It is not a function. It fails the vertical line test.

27. It is not a function. It fails the vertical line test.

29. $f(x) = x^2 - 4$

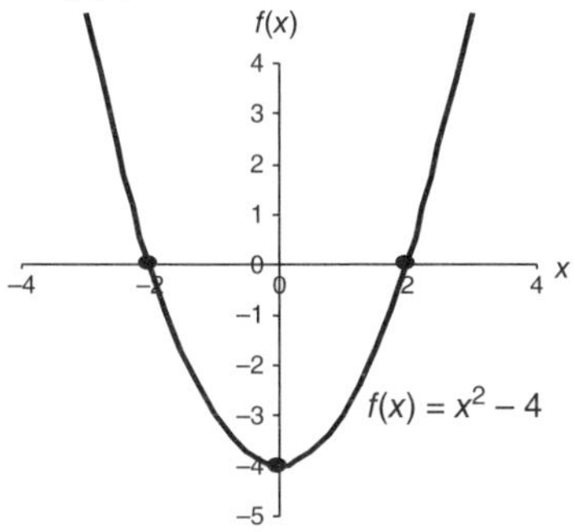

31. $f(x) = x^3 - 8$

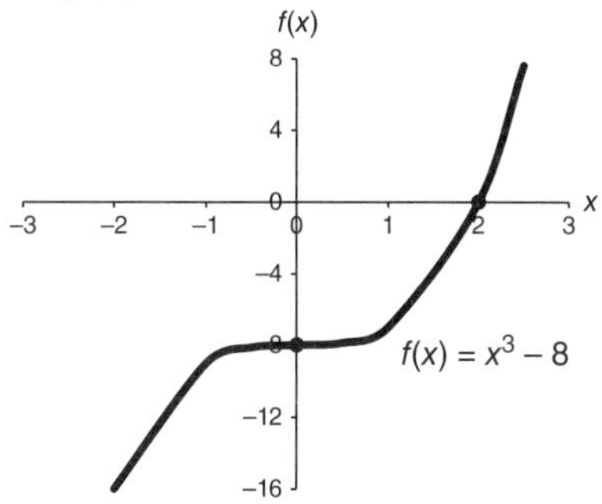

33. It is the piecewise graph shown as follows.

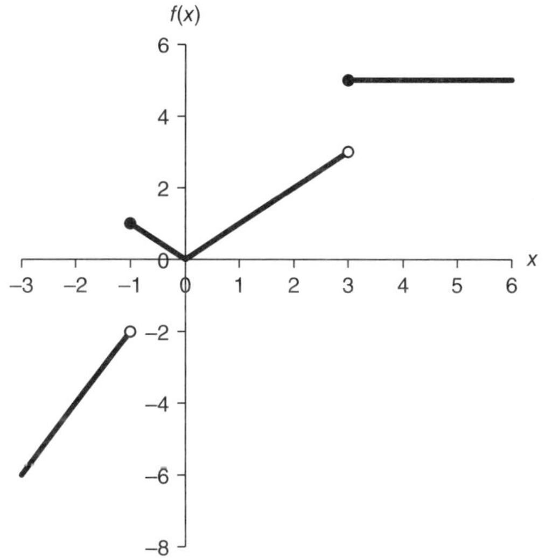

35. a) $-x^5 + 9x^3 - 2x + 8$
    b) $x^5 - 9x^3 + 2x - 8$
    c) $(4x^5 - 2x^3 + 2x)(3x^5 + 7x^3 + 8)$
    d) $3(4x^5 - 2x^3 + 2x)^5 + 7(4x^5 - 2x^3 + 2x)^3 + 8$

37. a) $2x^5 + h$ b) $(x + h)^2 + 4$ c) $(2a^5)(a^2 + 4)$ d) $2(x + 1)^5[(x + 2)^2 + 4]$

EXERCISES 1.5

1. 1 3. 125 5. 32 7. 9/4 9. 0.20 11. 27 13. x^8 15. $8x^3y^3$ 17. x^{12} 19. x^2y^7
21. $\dfrac{4x^6}{y^4}$ 23. x^3 25. $3xy^2$ 27. $64x^6y^7$ 29. $4x^3y^4$

EXERCISES 1.6

1. a) $\frac{6h}{h} = 6$ b) $\frac{6x - 6a}{x - a} = \frac{6(x - a)}{x - a} = 6$

3. a) $\frac{7h}{h} = 7$ b) $\frac{7x - 7a}{x - a} = \frac{7(x - a)}{x - a} = 7$

5. a) $\frac{[(x+h)^2 - 7(x+h) + 4] - [x^2 - 7x + 4]}{h} = \frac{h(2x + h - 7)}{h} = 2x + h - 7$

   b) $\frac{[x^2 - 7x + 4] - [a^2 - 7a + 4]}{x - a} = \frac{(x - a)(x + a) - 7(x - a)}{x - a} = x + a - 7$

7. a) $\frac{[(x+h)^2 + 6(x+h) - 8] - [x^2 + 6x - 8]}{h} = \frac{h(2x + h + 6)}{h} = 2x + h + 6$

   b) $\frac{[x^2 + 6x - 8] - [a^2 + 6a - 8]}{x - a} = \frac{(x - a)(x + a) + 6(x - a)}{x - a} = x + a + 6$

9. a) $\frac{[5(x+h)^2 - 2(x+h) - 3] - [5x^2 - 2x - 3]}{h} = \frac{h(10x + 5h - 2)}{h} = 10x+5h-2$

   b) $\frac{[5x^2 - 2x - 3] - [5a^2 - 2a - 3]}{x - a} = \frac{5(x - a)(x + a) - 2(x - a)}{x - a} = 5(x + a) - 2$

11. a) $\frac{[(x+h)^3 - 4(x+h) + 5] - [x^3 - 4x + 5]}{h} = \frac{h(3x^2 + 3xh + h^2 - 4)}{h}$

    $= 3x^2 + 3xh + h^2 - 4$

    b) $\frac{[x^3 - 4x + 5] - [a^3 - 4a + 5]}{x - a} = \frac{(x - a)(x^2 + ax + a^2) - 4(x - a)}{x - a}$

    $= x^2 + ax + a^2 - 4$

13. a) $\frac{[2(x+h)^3 - 7(x+h) + 3] - [2x^3 - 7x + 3]}{h} = \frac{h(6x^2 + 6xh + 2h^2 - 7)}{h}$

    $= 6x^2 + 6xh + 2h^2 - 7$

    b) $\frac{[2x^3 - 7x + 3] - [2a^3 - 7a + 3]}{x - a} = \frac{2(x - a)(x^2 + ax + a^2) - 7(x - a)}{x - a}$

    $= 2(x^2 + ax + a^2) - 7$

15. a) $\frac{\frac{3}{(x+h)^3} - \frac{3}{x^3}}{h} = \frac{3x^3 - 3(x+h)^3}{x^3h(x+h)^3} = \frac{-3h(3x^2 + 3xh + h^2)}{x^3h(x+h)^3}$

    $= \frac{-3(3x^2 + 3xh + h^2)}{x^3(x+h)^3}$

    b) $\frac{\frac{3}{x^3} - \frac{3}{a^3}}{x - a} = \frac{\frac{3a^3 - 3x^3}{a^3x^3}}{x - a} = \frac{3(a - x)(a^2 + ax + x^2)}{(x - a)a^3x^3} = \frac{-3(a^2 + ax + x^2)}{a^3x^3}$

CHAPTER 1 SUPPLEMENTARY EXERCISES

1. $x = 7/2$
3. $\mu = x - Z\sigma$
5. $3x + 5y = 15$

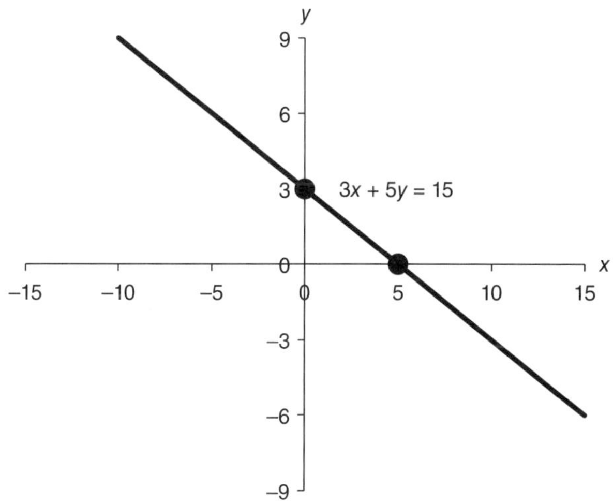

7. $y - 7 = 2(x - 5)$ 9. $y - 5 = \frac{4}{3}(x - 2)$
11. $2x(x - 10)(x + 1)$ 13. $(x - y)(2a + b)$
15. $(4, \infty)$ 17. $x \neq 0,\ x \neq -8$ and $x \neq -1$
19. a) $-x^3 + 4x^2 + 3x + 5$
b) $(x^2 + 3x + 1)(x^3 - 3x^2 - 4)$ c) $(x^3 - 3x^2 - 4)^2 + 3(x^3 - 3x^2 - 4) + 1$
21. $\frac{-1}{2x^4}$
23. $\frac{27z^{12}}{8x^9y^6}$
25. a) $\frac{[(x+h)^3 + 3(x+h) + 1] - [x^3 + 3x + 1]}{h} = \frac{h(3x^2 + 3xh + h^2 + 3)}{h}$
$= 3x^2 + 3xh + h^2 + 3$
b) $\frac{[x^3 + 3x + 1] - [a^3 + 3a + 1]}{x - a} = \frac{(x-a)(x^2 + ax + a^2) + 3(x - a)}{x - a}$
$= x^2 + ax + a^2 + 3$

For complete solutions to these Exercises see the companion “Student Solutions Manual” by Morris and Stark.

EXERCISES 2.1

1. $y'(0) = 0$ 3. $y'(5) = 10$ 5. $y'(1/2) = 1$ 7. $y - 25/4 = 5(x - 5/2)$
9. $y - 49/4 = -7(x + 7/2)$ 11. $(7/8, 49/64)$ 13. $(2/5, 4/25)$
15. $(1/3, 1/9)$ 17. $m = 12$ 19. $m = 3/4$ 21. $y - 64 = 48(x - 4)$

⟶

23. $y - 27/64 = (27/16)(x - 3/4)$ 25. (4, 64) and (−4, −64)

27. $\left(\frac{2}{3}, \frac{8}{27}\right)$ and $\left(-\frac{2}{3}, -\frac{8}{27}\right)$.

EXERCISES 2.2

1. a) 10 b) 4 c) does not exist 3. a) −3 b) −3 c) −3 5. a) 0 b) 0 c) 0 7. 4 9. 17 11. 41 13. 17 15. 3 17. −1 19. does not exist 21. 6 23. 12 25. 0 27. does not exist 29. −1 31. 7 33. 0 35. 10

EXERCISES 2.3

1. a) $\lim\limits_{h\to 0} \dfrac{[4(x+h)+11]-[4x+11]}{h} = \lim\limits_{h\to 0} \dfrac{4h}{h} = 4 = f'(x).$

   b) $\lim\limits_{x\to a} \dfrac{[4x\ +11]-[4a+11]}{x-a} = \lim\limits_{x\to a} \dfrac{4x-4a}{x-a} = \lim\limits_{x\to a} \dfrac{4(x-a)}{x-a} = 4 = f'(a).$

   Therefore, $f'(x) = 4$.

3. a) $\lim\limits_{h\to 0} \dfrac{[(x+h)^2+5(x+h)+1]-[x^2+5x+1]}{h}$

   $= \lim\limits_{h\to 0} \dfrac{2xh+h^2+5h}{h} = \lim\limits_{h\to 0} \dfrac{h(2x+h+5)}{h} = 2x+5 = f'(x)$

   b) $\lim\limits_{x\to a} \dfrac{[x^2+5x+1]-[a^2+5a+1]}{x-a} = \lim\limits_{x\to a} \dfrac{(x^2-a^2)+(5x-5a)}{x-a}$

   $= \lim\limits_{x\to a} \dfrac{(x-a)(x+a)+5(x-a)}{x-a} = 2a+5 = f'(a).$

   Therefore, $f'(x) = 2x + 5$.

5. a) $\lim\limits_{h\to 0} \dfrac{[(x+h)^2-6(x+h)+1]-[x^2-6x+1]}{h}$

   $= \lim\limits_{h\to 0} \dfrac{2xh+h^2-6h}{h} = \lim\limits_{h\to 0} \dfrac{h(2x+h-6)}{h} = 2x-6 = f'(x)$

   b) $\lim\limits_{x\to a} \dfrac{[x^2-6x+1]-[a^2-6a+1]}{x-a} = \lim\limits_{x\to a} \dfrac{(x^2-a^2)-(6x-6a)}{x-a}$

   $= \lim\limits_{x\to a} \dfrac{(x-a)(x+a)-6(x-a)}{x-a} - 2a-6 = f'(a).$

   Therefore, $f'(x) = 2x - 6$.

7. a) $\lim\limits_{h\to 0} \dfrac{[(x+h)^3+5]-[x^3+5]}{h} = \lim\limits_{h\to 0} \dfrac{3x^2h+3xh^2+h^3}{h}$

   $= \lim\limits_{h\to 0} \dfrac{h(3x^2+3xh+h^2)}{h} = 3x^2 = f'(x)$

b) $\lim_{x\to a} \frac{[x^3+5]-[a^3+5]}{x-a} = \lim_{x\to a} \frac{(x^3-a^3)}{x-a} = \lim_{x\to a} \frac{(x-a)(x^2+ax+a^2)}{x-a}$
$= 3a^2 = f'(a)$
Therefore, $f'(x) = 3x^2$

9. $f'(x) = 0$ 11. $f'(x) = 0$ 13. $f'(x) = 7$ 15. $f'(x) = 14$ 17. $f'(x) = 4x + 7$
19. $f'(x) = 30x^2 - 18x + 3$ 21. $35x^4 - 16x^3 + 6x$ 23. $9 - \frac{4}{x^2}$ 25. $\frac{5}{6}x^{-1/6} - \frac{8}{x^5}$
27. $y'(1) = 10$ 29. $f(2) = 14$ and $f'(2) = 8$. 31. $f(-1) = 6$ and $f'(-1) = -7$

33. $y - 24 = 17(x - 2)$ 35. $y - 25 = 8(x - 0)$.

EXERCISES 2.4

1. not differentiable 3. not differentiable 5. differentiable 7. not continuous 9. continuous
11. continuous 13. It is a polynomial so differentiable and continuous at $x = 0$.

15. a) not differentiable b) continuous
17. a) not differentiable b) continuous
19. It is a polynomial so differentiable and continuous at $x = 1$.
21. a) not differentiable b) continuous
23. It is a polynomial so differentiable and continuous at $x = 2$.
25. a) not differentiable b) continuous

27. $f(x) = \begin{cases} \frac{x^2-36}{x+6} & x \neq -6 \\ x - 6 & x = -6 \end{cases}$ 29. $f(x) = \begin{cases} \frac{x^3-25x}{x-5} & x \neq 5 \\ x^2 + 5x & x = 5 \end{cases}$

EXERCISES 2.5

1. $32x(4x^2 + 1)^3$ 3. $70x(5x^2 + 3)^6$ 5. $168x(3x^2 + 1)^3$

7. $5(9x^{10} + 6x^5 - x)^4(90x^9 + 30x^4 - 1)$ 9. $10(12x^7 + 3x^4 - 2x + 5)^9(84x^6 + 12x^3 - 2)$
11. $\frac{-20}{(9x^3+\sqrt{x}+3)^6}\left(27x^2 + \frac{1}{2}x^{-1/2}\right)$ 13. $6(7x^{8/5} + 5x + 1)^5\left(\frac{56}{5}x^{3/5} + 5\right)$

15. $5(\sqrt{x}+1)^4\left(\frac{1}{2}x^{-1/2}\right)$ 17. $6\left(4x^5 + \sqrt[3]{x^2} + 1\right)^5\left(20x^4 + \frac{2}{3}x^{-1/3}\right)$

19. $21x^2 + 16x - \frac{140}{(4x-3)^8}$ 21. $y'(1) = 60$ 23. $y - 27 = 18(x - 2)$
25. $y - 5 = -15(x - 1)$.

EXERCISES 2.6

1. $4z^3 + 6z$ 3. $12r^3 + 3r^2 + 4r$ 5. $24p^7 + 30p^5 + 6p^2 + 4$ 7. $10(5t^2 + 3t + 1)^9(10t + 3)$
9. $12(3p^{10} + \sqrt{p} + 5)^{11}\left(30p^9 + \frac{1}{2}p^{-1/2}\right)$

11. $\frac{21}{5}t^{2/5} - 5$ 13. $20p^3 - 2p^{-1/3}$ 15. $\frac{d}{dt} = 10a^7t^4 - 27bt^2 + 2t$ and $\frac{d}{db} = -9t^3$

17. $f'(x) = 27x^2 + 4$ and $f''(x) = 54x$

19. $y' = 6x^2 + 3 + \frac{3}{4}x^{-1/4}$ and $y'' = 12x - \frac{3}{16}x^{-5/4}$ 21. $v' = 6t^2 + 18$ and $v'' = 12t$

23. -4 25. 30 27. 45 29. $y' = 5x^4 + 9x^2 + 9$, $y'' = 20x^3 + 18x$, $y''' = 60x^2 + 18$, $y^{iv} = 120x$.

EXERCISES 2.7

1. $\Delta f(x) = [6(x + 1) + 4] - [6x + 4] = 6x + 6 + 4 - 6x - 4 = 6$
3. $\Delta f(x) = [4(x + 1)^2] - [4x^2] = 4x^2 + 8x + 4 - 4x^2 = 8x + 4$

5. $\Delta f(x) = [5(x + 1)^2 + 2(x + 1) + 3] - [5x^2 + 2x + 3]$
$= 5x^2 + 10x + 5 + 2x + 2 + 3] - [5x^2 + 2x + 3] = 10x + 7$

7. $\Delta f(x) = [(x + 1)^3 + 3(x + 1) + 1] - [x^3 + 3x + 1]$
$= [x^3 + 3x^2 + 3x + 1 + 3x + 3 + 1] - [x^3 + 3x + 1] = 3x^2 + 3x + 4$

9. $\Delta f(x) = [9(x + 1) - 1] - [9x - 1] = 9x + 9 - 1 - 9x + 1 = 9$
as $\Delta f(x)$ is a constant $\Delta^2 f(x) = 0$. Alternatively,

$$\Delta^2 f(x) = f(x + 2) - 2f(x + 1) + f(x)$$
$$= [9(x + 2) - 1] - 2[9(x + 1) - 1] + [9x - 1] = 0.$$
$$\Delta^2 f(x) = f(x + 2) - 2f(x + 1) + f(x)$$
$$= [8(x + 2) + 3] - 2[8(x + 1) + 3] + [8x + 3] = 0.$$

11. $\Delta^2 f(x) = [2(x + 2)^2 + 5] - 2[2(x + 1)^2 + 5] + [2x^2 + 5]$
$= 2x^2 + 8x + 8 + 5 - 4x^2 - 8x - 4 - 10 + 2x^2 + 5 = 4$

13. $\Delta^2 f(x) = [5(x + 2)^3 + 2] - 2[5(x + 1)^3 + 2] + [5x^3 + 2]$
$= 5x^3 + 30x^2 + 60x + 40 + 2 - 10x^3 - 30x^2 - 30x - 10 - 4 + 5x^3 + 2$
$= 30x + 30$

CHAPTER 2 SUPPLEMENTARY EXERCISES

1. $m = 1/2$ 3. $(3/8, 9/64)$ 5. $\lim_{x\to 2} 3x + \frac{4}{x} = 3(2) + \frac{4}{2} = 8$

7. $\lim_{x\to 4} \frac{x^2 + 2x - 24}{x - 4} = \lim_{x\to 4} \frac{(x - 4)(x + 6)}{x - 4} = 4 + 6 = 10$

9. a) $\lim_{h\to 0} \frac{[3(x+h)^2+5(x+h)+1]-[3x^2+5x+1]}{h}$

$= \lim_{h\to 0} \frac{6xh+3h^2+5h}{h} = \lim_{h\to 0} \frac{h(6x+3h+5)}{h} = 6x+5 = f'(x)$

b) $\lim_{x\to a} \frac{[3x^2+5x+1]-[3a^2+5a+1]}{x-a} = \lim_{x\to a} \frac{(3x^2-3a^2)+(5x-5a)}{x-a}$

$= \lim_{x\to a} \frac{3(x-a)(x+a)+5(x-a)}{x-a} = 6a+5 = f'(a).$

Therefore, $f'(x) = 6x+5$.

11. $f(1) = 10$ and $f'(1) = 19$ 13. $y-2 = 29(x-1)$ 15. continuous

17. $25(9x^3+4x^2+3x+1)^{24}(27x^2+8x+3)$ 19. $4\left(2\sqrt[3]{x^2}+3x+1\right)^3\left(\frac{4}{3}x^{-1/3}+3\right)$

21. $y-3{,}125 = 1{,}250(x-6)$ 23. $20a^3p^3+6ap+2b$ and $15a^2p^4+3p^2$

25. $y''(1) = 162$

For complete solutions to these Exercises see the companion "Student Solutions Manual" by Morris and Stark.

EXERCISES 3.1

1. c and d 3. e 5. b, d, g, and h
7. The graph is increasing on $(2, \infty)$ and decreasing on $(-\infty, 2)$. It has a local and absolute minimum at $(2, -1)$. The graph is concave up on $(-\infty, \infty)$ with no inflection points. The y-intercept is $(0, 3)$ and x-intercepts $(1, 0)$ and $(3, 0)$. There are no undefined points and no asymptotes.
9. The graph is decreasing on $(-\infty, -1) \cup (0, 1)$ and increasing on $(-1, 0) \cup (1, \infty)$. There is a local maximum at $(0, 0)$ and local and absolute minimums at $(\pm 1, -1)$. The graph is concave down on $(-3/4, 3/4)$ and concave up on $(-\infty, -3/4) \cup (3/4, \infty)$. There are inflection points at $(\pm 3/4, -1/2)$. The y-intercept is $(0, 0)$ and x-intercepts at $(0, 0)$, $(\pm 1.4, 0)$. There are no asymptotes and no undefined points.
11. There are endpoint extrema at A and G. There are local extrema at B, D, and F.
13. There are inflection points at C and E.
15. There is an absolute maximum at D and absolute minimum at F.

EXERCISES 3.2

1. c and d 3. e
5. One possibility is depicted in the following graph.

→

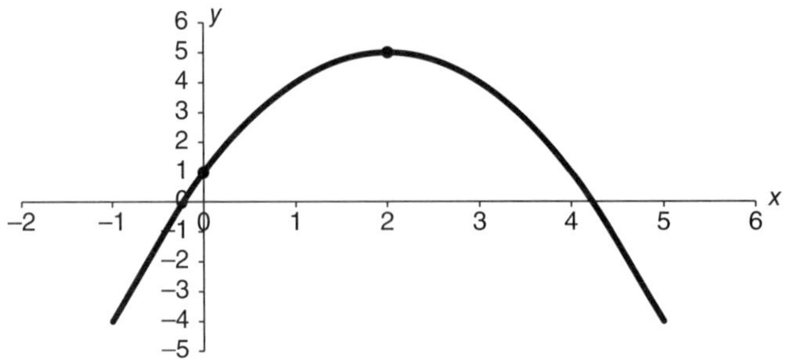

7. One possibility is depicted in the following graph.

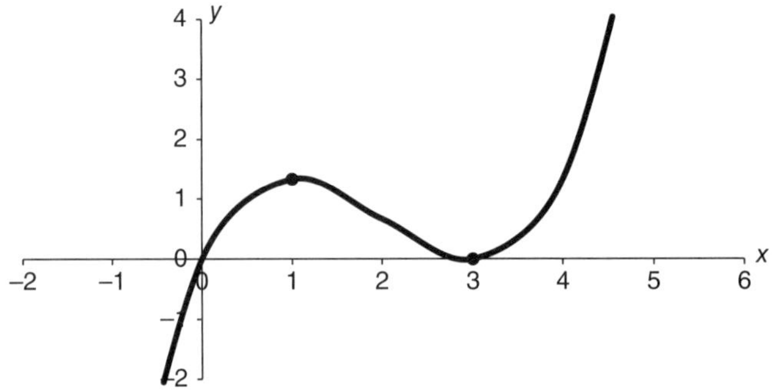

9. The table is filled in as follows

	f	f'	f''
A	+	−	+
B	−	0	+
C	−	−	0
D	0	+	+

11. $f'(5)$ is positive as the graph is shown above the axis here. A positive first derivative indicates an increasing function.

13. $x = 1$ 15. $y + (2/3) = -6(x - 2)$.

EXERCISES 3.3

1. There is a local minimum of −9 at $x = 2$.
3. Setting $3x^2 - 6x + 3 = 3(x - 1)^2 = 0$ or $x = 1$ as the only possible extremum. The first derivative is always positive, so the function has no local extrema.

5.

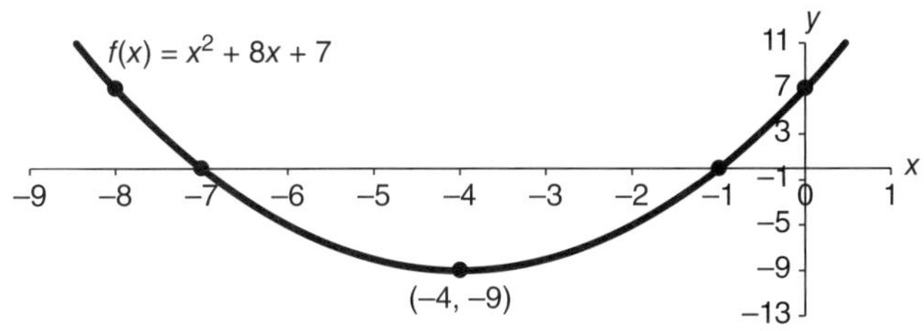

$f(x) = x^2 + 8x + 7$
y
11
7
3
−1
−5
−9
−13
x
−9 −8 −7 −6 −5 −4 −3 −2 −1 0 1
(−4, −9)

7.

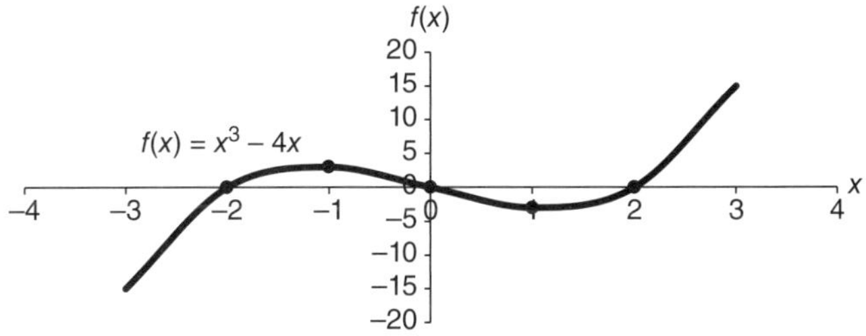

f(x)
20
15
10
5
0
−5
−10
−15
−20
$f(x) = x^3 - 4x$
x
−4 −3 −2 −1 0 1 2 3 4

9.

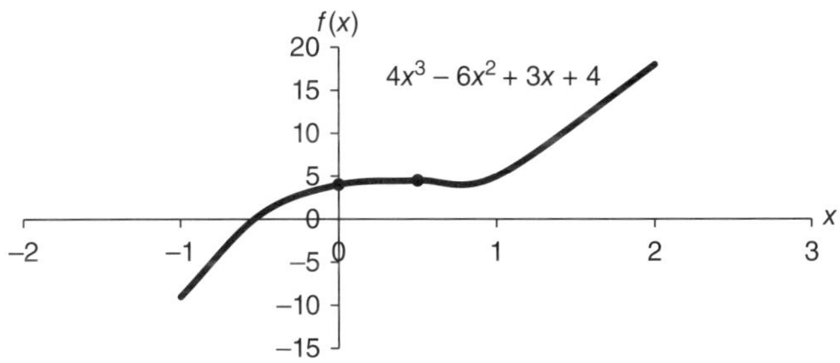

f(x)
20
15
10
5
0
−5
−10
−15
$4x^3 - 6x^2 + 3x + 4$
x
−2 −1 0 1 2 3

11.

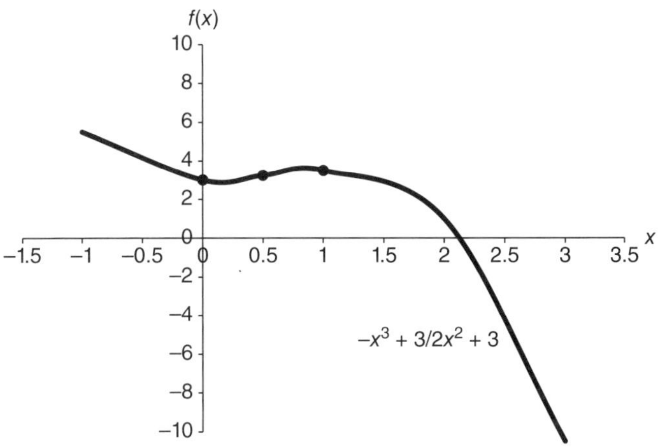

f(x)
10
8
6
4
2
0
−2
−4
−6
−8
−10
x
−1.5 −1 −0.5 0 0.5 1 1.5 2 2.5 3 3.5
$-x^3 + 3/2x^2 + 3$

13.

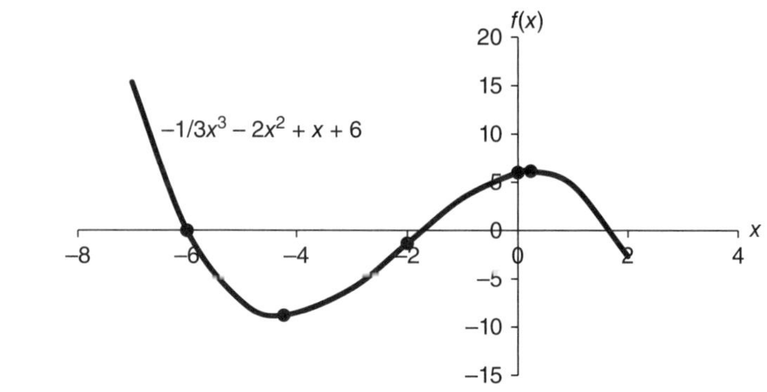

15.

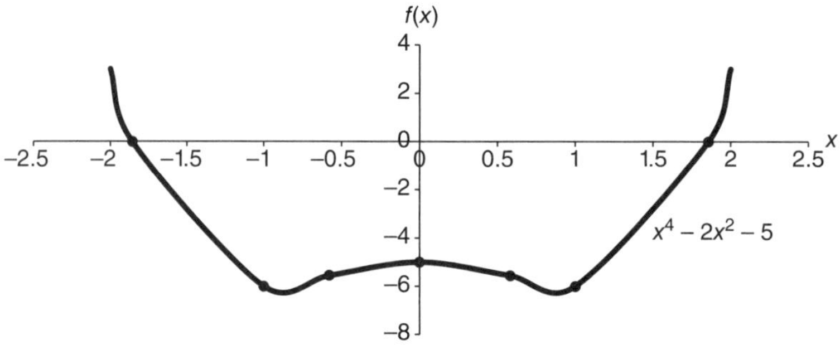

17.

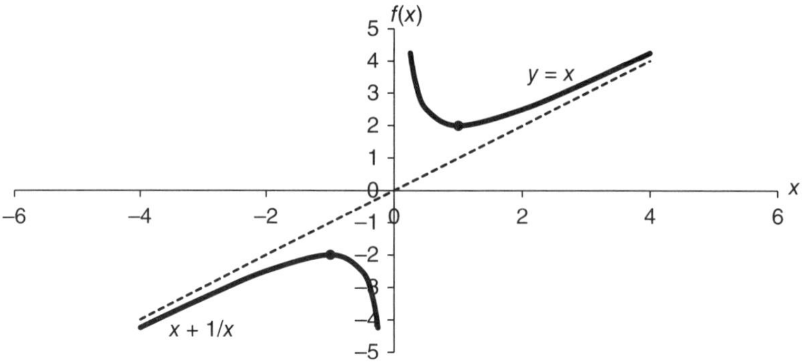

EXERCISES 3.4

1. $f'(x) = 2x - 6 = 0$, so a local minimum occurs when $x = 3$.
3. $f'(x) = -4x^3 + 6x$, and there are local maximums at $x = \dfrac{\pm\sqrt{6}}{2}$. The max is 9/4.
5. $f'(x) = 16 - 2x$, and there is a local maximum at $x = 8$. The maximum value is 64.
7. $P(x) = x(8 - x) = 8x - x^2$ and $P'(x) = 8 - 2x$, indicating a maximum at $x = 4$. The maximum of the product is 16.
9. $S(x) = x^2 + (20 - x)^2 = 2x^2 - 40x + 400$ and $S'(x) = 4x - 40$, indicating a minimum at $x = 10$. The minimum sum of squares is 200.

11. $f(x) = x - x^2$ and $f'(x) = 1 - 2x$, indicating 1/2 is the number that exceeds its square by the largest amount.
13. $A(x) = x(30 - x) = 30x - x^2$ and $A'(x) = 30 - 2x$, indicating a maximum when $x = y = 15$. (Alternatively, the second derivative being negative also indicates a maximum at the critical value $x = 15$).
The maximum area is 225 square feet and occurs when the rectangle is a square.
15. $A(x) = \frac{1}{2}(x)(10 - x) = \frac{-1}{2}x^2 + 5x$ and $A'(x) = -x + 5$ indicating a maximum when $x = 5$. If $x = 5$, then y is also 5, so the largest area for the right triangle is 25/2. (Alternatively, the second derivative being negative also indicates a maximum at the critical value $x = 5$).
17. The objective function is $S(x) = x^2 + (8 - x)^2 = 2x^2 - 16x + 64$ and $S'(x) = 4x - 16$, indicating a minimum when $x = 4$. The minimum value for the sum of squares is 32. (Alternatively, the second derivative being positive at the critical value $x = 4$ indicates a minimum there).
19. $A(x) = x\left(12 - \frac{6}{5}x\right) = 12x - \frac{6}{5}x^2$ and $A'(x) = 12 - \frac{12}{5}x$, indicating a local maximum when $x = 5$. The maximum area is 30 square feet.
21. $A(x) = x\left(120 - \frac{3}{4}x\right) = 120x - \frac{3}{4}x^2$ and $A'(x) = 120 - \frac{3}{2}x$, indicating a maximum when $x = 80$ feet and $y = 60$ feet.
23. $V(x) = x^2\left(\frac{15}{x} - \frac{1}{2}x\right) = 15x - \frac{1}{2}x^3$ and $V'(x) = 15 - \frac{3}{2}x^2$, indicating a local maximum when $x = h = \sqrt{10}$ feet. Therefore, the box is actually a cube with sides $\sqrt{10} \times \sqrt{10} \times \sqrt{10}$.
25. $S(r) = 2\pi r\left(\frac{V}{\pi r^2}\right) + 8r^2 = \frac{2V}{r} + 8r^2$ and $S'(r) = \frac{-2V}{r^2} + 16r$, indicating a minimum when $r = \frac{V^{1/3}}{2}$ and $h = \frac{4V^{1/3}}{\pi}$. The optimal height to diameter ratio is $\frac{h}{2r} = \frac{4V^{1/3}}{\pi(V^{1/3})} = \frac{4}{\pi}$.
This relationship holds regardless of the volume, V, and indicates that the can is slightly taller in this situation.
27. $V = \frac{\pi d^2}{4}(h)$. Solving for h yields $h = \frac{4V}{\pi d^2}$, so the cost as a function of d is $C(d) = 2\left(\frac{\pi d^2}{4}\right)(2c) + \pi d\left(\frac{4V}{\pi d^2}\right)(c) = \pi d^2 c + \frac{4cV}{d}$ and the first derivative is $C'(d) = 2\pi cd - \frac{4cV}{d^2}$, indicating a minimum when $d = \sqrt[3]{\frac{2V}{\pi}}$ and $h = \sqrt[3]{\frac{16V}{\pi}}$, so the best height to diameter ratio $\frac{h}{d} = 2$.

EXERCISES 3.5

1. $C'(x) = 6x^2 + 9$ 3. $C'(x) = 7x^6 - 15x^2 + 20$ so $C'(1) = 7(1)^6 - 15(1)^2 + 20 = \$12$
5. $MC = 0$ at $x = 1$ or 5 7. Minimum MC is \$12 at $x = 4$ 9. $R'(x) = -2x + 30$

11. $R'(x) = \frac{-x}{2} + 12$

13. The maximum profit is \$1500 when 10 units are produced at \$180 each.

CHAPTER 3 SUPPLEMENTARY EXERCISES

1. c and e are increasing for all x. 3. a and c are concave up for all x.

5. $y' = \dfrac{-500}{(x+1)^2}$, which is always negative indicating the function is decreasing for $x \geq 0$.

7. $f'(x) = 4x - 1$. The function decreases on $(-\infty, 1/4)$ and increases on $(1/4, \infty)$ and minimum value is $-49/8$ when x = 1/4.
$f''(x) = 4$ indicates the graph is concave up everywhere and no points of inflection.
y-intercept of -6 and x-intercepts of 2 and $-3/2$. The graph is

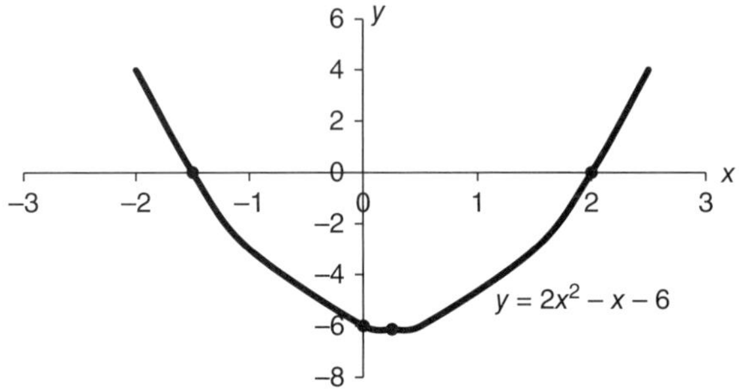

9. $f'(x) = 4x^3 - 18x$, decreasing on $\left(-\infty, \frac{-3\sqrt{2}}{2}\right) \cup \left(0, \frac{3\sqrt{2}}{2}\right)$ and increasing on $\left(-\frac{3\sqrt{2}}{2}, 0\right) \cup \left(\frac{3\sqrt{2}}{2}, \infty\right)$, local minimums at $\left(\frac{\pm 3\sqrt{2}}{2}, \frac{-81}{4}\right)$ and a local maximum at (0,0). $f''(x) = 12x^2 - 18$, concave down on $\left(\frac{-\sqrt{6}}{2}, \frac{\sqrt{6}}{2}\right)$ and concave up on $\left(-\infty, \frac{-\sqrt{6}}{2}\right) \cup \left(\frac{\sqrt{6}}{2}, \infty\right)$ with inflection points at $\left(\frac{\pm\sqrt{6}}{2}, \frac{-45}{4}\right)$. The function has a y-intercept at the origin, x-intercepts at $x = -3$, 0, and 3. The graph is

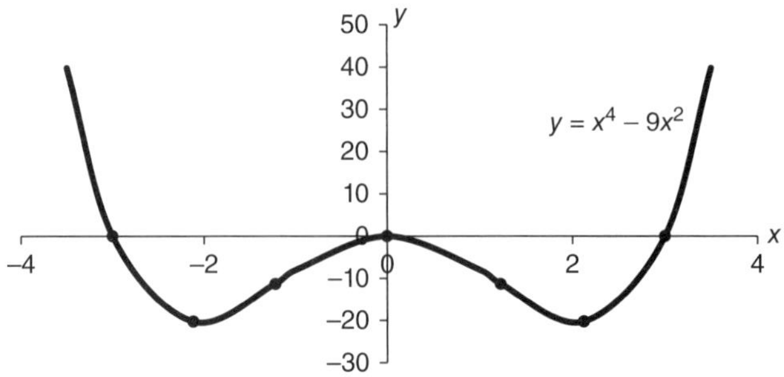

11. Here, $f'(x) = 12x^3 - 12x^2$ and $f''(x) = 36x^2 - 24x = 12x(3x - 2)$ and inflection points at (0, 0) and $\left(\frac{2}{3}, \frac{-16}{27}\right)$.

13. $A(x) = x\left(16 - \frac{2}{3}x\right) = 16x - \frac{2}{3}x^2$ and $A'(x) = 16 - \frac{4}{3}x$. The maximum area is 96 square yards when $x = 12$ yards and $y = 8$ yards.

15. a) $S(x) = 2x^2 + 4x\left(\frac{64}{x^2}\right) = 2x^2 + \frac{256}{x}$ and $S'(x) = 4x - \frac{256}{x^2}$ indicating a local maximum when $x = h = 4$ inches. The box is actually a cube with dimensions 4 in × 4 in × 4in.

b) $S(x) = x^2 + 4x\left(\frac{64}{x^2}\right) = x^2 + \frac{256}{x}$ and $S'(x) = 2x - \frac{256}{x^2}$. The derivative indicating a local maximum when $x = 4\sqrt[3]{2}$ inches and $h = 2\sqrt[3]{2}$ inches. The box is now half as high as the square base.

17. $V(x) = \frac{c_1^2}{x_1} + \frac{c_2^2}{x_2}$ subject to $c_1x_1 + c_2x_2 = B$. $x_2 = \frac{B - c_1x_1}{c_2}$, so we rewrite as

$$V(x) = \frac{c_1^2}{x_1} + \frac{c_2^2}{\frac{B - c_1x_1}{c_2}} = \frac{c_1^2}{x_1} + \frac{c_2^3}{B - c_1x_1}.$$

Differentiating yields $V'(x) = \frac{-c_1^2}{x_1^2} + \frac{c_2^3c_1}{(B - c_1x_1)^2}$.

Setting the derivative to zero yields the following $(c_1^3 - c_2^3)x_1^2 - 2c_1^2Bx_1 + c_1B^2 = 0$. There is a possible optimum when

$$x_1 = \frac{c_1^2B \pm c_2B\sqrt{c_1c_2}}{c_1^3 - c_2^3}$$

19. Here, $R'(x) - 4000 - 4x$ indicates the maximum revenue is \$2,000,000.

21. $R(x) = (x)(240 - 6x) = 240x - 6x^2$.

$\Pr(x) = (240x - 6x^2) - (x^3 - 21x^2 - 360x + 8350) = -x^3 + 15x^2 + 600x - 8,350$.

The maximum profit is \$1650 when 20 units are produced at \$120 each.

For complete solutions to these Exercises see the companion "Student Solutions Manual" by Morris and Stark.

EXERCISES 4.1

1. a) 2^{9x} b) 3^{6x} c) 2^{20x} 3. a) $(2)^{12x}$ b) $(3)^{-12x}$ c) $(3)^{6x}$ 5. a) 2^{5x} b) 2^{2x} c) $(3)^{-3x}$

7. $\frac{7x^6}{y^6}$ 9. x^2y^7 11. $\frac{2^{5x+3}(2^2)^{x+1}}{2^3(2^{3x-1})} = 2^{4x+3}$ 13. $x = 5$ 15. $x = 2$ 17. $x = 4$

19. $x = -2$ or $x = -3$. 21. $2^{3+h} = 2^h(2^3)$ 23. $7^{x+5} - 7^{2x} = 7^{2x}(7^{-x+5} - 1)$

25. $(7^h)^3 - 8 = (7^h - 2)[7^{2h} + 2(7^h) + 4]$

EXERCISES 4.2

1. Rewrite as $3^y = x$ to determine ordered pairs and graph as

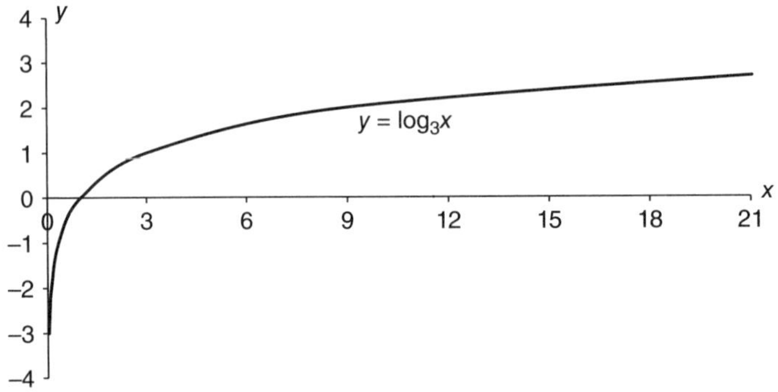

3. 6 5. 6 7. −5 9. 3 11. 7.65 13. 0 15. 3/2 17. 5/2 19. 7 21. 3

23. 5 25. 7 27. $e/4$ 29. $2\log_4(x+1) + 6\log_4(x-3) - 3\log 4(3x+5)$

31. $\ln \dfrac{x^2(z+1)^4}{(y+1)^3}$ 33. $\dfrac{1+\log_{10}105}{2}$ 35. $1 + \ln(4/3)$ 37. about 500 times as strong

EXERCISES 4.3

1. e^{3x}, e^{-5x} 3. e^{7x}, e^{5x} 5. 0 7. $2e^{2x}$ 9. $30e^{6x}$ 11. $125e^{25x}$ 13. $f'(x) = 432x^8e^{8x^9}$

15. $f'(x) = e^{2x^3+5x+1}[6x^2+5]$ 17. $f'(x) = 6e^{8x^7+9x^4}[56x^6 + 36x^3]$ 19. $y - 2 = 8(x - 0)$

EXERCISES 4.4

1. $\dfrac{1}{x}$ 3. $3x^2 + 10x + \dfrac{1}{x}$ 5. $\dfrac{4}{4x+7}$ 7. $\dfrac{15x^2+12x+9}{5x^3+6x^2+9x+2}$

9. $\dfrac{45x^4+24x^3+8}{9x^5+6x^4+8x}$ 11. $\dfrac{(1/2)(5x+1)^{-1/2}(5)}{\sqrt{5x+1}} = \dfrac{5}{2(5x+1)}$

13. $f'(x) = 8(17x^2 + 15x + \ln 2x)^7\left(34x + 15 + \dfrac{1}{x}\right)$ 15. $\dfrac{3e^{3x}}{e^{3x}+2}$

17. $y - 1 = (1/e)(x - e)$ 19. $\text{MC} = C'(x) = \dfrac{300}{3x+1}$.

21. Rewriting as $5\ln(3x+5) + 2\ln(4x^3 - 7x + 1)$ first yields

$$f'(x) = 5\left(\frac{3}{3x+5}\right) + 2\left(\frac{12x^2-7}{4x^3-7x+1}\right).$$

EXERCISES 4.5

1. $50{,}000e^{0.3(5)} \approx 58{,}092$ 3. $500(2)^7 = 64{,}000$ 5. about 12 days 7. 3.54 grams
9. 81.3 days 11. 0.7457 g

EXERCISES 4.6

1. \$68,639.29 3. \$76,0980.78 5. \$250,000 is worth more. 7. 34.66 years

9. 6.02 weeks 11. 58.82% 13. 40% 15. −20% 17. $p > 4$

CHAPTER 4 SUPPLEMENTARY EXERCISES

1. a) 2^{3x} b) 5^{5x-1} c) 5^{4x+4} 3. a) $x = -2$ b) $x = 2$ 5. $(3^x - 5)(3x + 5)$ 7. $36e^{9x}$
9. $f'(x) = 12x^2(e^{4x^3})$ 11. $y - 2 = 6\ (x - 0)$ or $y = 6x + 2$ 13. 6 15. 3/2 17. 7.23
19. $4\ln(2x + 3) + 2\ln(x + 1) + 7\ln(4x + 5)$ 21. about 25 times as strong

23. $f'(x) = \left[\dfrac{15x^2 + 4x + 8}{5x^3 + 2x^2 + 8x + 7}\right]$ 25. 1690 27. 97,200.

29. 3.125% when $t = 0$ and 7.53% when $t = 2$

For complete solutions to these Exercises see the companion "Student Solutions Manual" by Morris and Stark.

EXERCISES 5.1

1. $f'(x) = (2x^2 + 3x + 8)[5] + (5x - 1)[4x + 3]$

3. $f'(x) = (x^2 + 5x + 3)^4[3(3x^2 - 2x + 6)^2\,(6x - 2)] + (3x^2 - 2x + 6)^3[4(x^2 + 5x + 3)^3(2x + 5)]$

5. $f'(x) = 5x\left[\dfrac{2x}{x^2 + 5}\right] + \ln(x^2 + 5)[5]$

7. $f'(x) = x^3[5e^{5x}] + e^{5x}[3x^2] + \dfrac{1}{x} + \dfrac{10}{7}x^{-2/7}$

9. $f'(x) = -24(x^5 + 7x - 3)^{-9}[5x^4 + 7]$ or $\dfrac{-24[5x^4 + 7]}{(x^5 + 7x - 3)^9}$

11. $f'(x) = \dfrac{(5x^2 + 7x + 3)[3] - (3x + 2)[10x + 7]}{(5x^2 + 7x + 3)^2}$

13. $f'(x) = \dfrac{(5x^2 + 7x + 3)[12x^2 + 2] - (4x^3 + 2x + 1)[10x + 7]}{(5x^2 + 7x + 3)^2}$

15. $f'(x) = \dfrac{(7x^3 + 5x + 3)^4[11(5x^6 + 3x - 4)^{10}(30x^5 + 3)]}{(7x^3 + 5x + 3)^8} - \dfrac{(5x^6 + 3x - 4)^{11}[4(7x^3 + 5x + 3)^3(21x^2 + 5)]}{(7x^3 + 5x + 3)^8}$

⟶

17. $f'(x) = 100\left(\dfrac{30}{\left(x^7 - 5x^4 + 3\right)^3}\right)^{99}\left[\dfrac{-90\left(7x^6 - 20x^3\right)}{(x^7 - 5x^4 + 3)^4}\right]$

19. $f'(x) = \dfrac{(2x-5)\left[\frac{1}{x}\right] - \ln 7x[2]}{(2x-5)^2}$

21. $f'(x) = 25(x^3 + 7x + \ln 8x)^{24}\left[3x^2 + 7 + \dfrac{1}{x}\right]$

23. $f'(x) = \dfrac{(2x+7)^3\lfloor x^3(5e^{5x}) + e^{5x}(3x^2)\rfloor - (x^3e^{5x})\lfloor 3(2x+7)^2(2)\rfloor}{(2x+7)^6}$

25. $f'(x) = 4\left(\dfrac{\sqrt{3x-7}}{x^4+5x+3}\right)^3\left[\dfrac{(x^4+5x+3)\left[\frac{1}{2}(3x-7)^{-1/2}(3)\right] - \sqrt{3x-7}[4x^3+5]}{(x^4+5x+3)^2}\right]$

27. $f'(x) = (3x+1)^4\left[10\left(\sqrt[5]{x^3} + 2x\right)^9\left(\dfrac{3}{5}x^{-2/5} + 2\right)\right] + \left(\sqrt[5]{x^3} + 2x\right)^{10}[4(3x+1)^3(3)]$

29. $f'(x) = (6x^3 - \sqrt{x})^8[3x^2e^{x^3}] + e^{x^3}\left[8\left(6x^3 - \sqrt{x}\right)^7\left(18x^2 - \dfrac{1}{2}x^{-1/2}\right)\right]$

31. $f'(x) = \dfrac{(2x+1)^5[3e^{3x}] - e^{3x}[5(2x+1)^4(2)]}{(2x+1)^{10}}$

33. $f'(x) = \dfrac{(e^x+1)[4e^{4x}] - e^{4x}[e^x]}{(e^x+1)^2}$

35. $f'(x) = (5x-4)(3) + (3x+7)(5)$ and $f'(1) = 1(3) + 10(5) = 53 = m$.
The equation is $y - 10 = 53(x-1)$.

37. $f'(x) = (x-1)^4[2] + (2x+3)[4(x-1)^3(1)]$ and $f'(2) = (1)(2) + 7(4) = 30$.
The equation is $y - 7 = 30(x-2)$.

39. $f'(x) = (4x-3)^{3/2}[1] + (x+1)[3/2(4x-3)^{1/2}(4)]$ and
$f'(3) = (27)(1) + 4(3/2)(3)(4) = 99$. The equation is
$y - 108 = 99(x-3)$.

41. Setting the first derivative to zero yields $0 = \dfrac{(2x-1)[3x^2] - x^3[2]}{(2x-1)^2}$.
There is a horizontal tangent at (0, 0) and $\left(\dfrac{3}{4}, \dfrac{27}{32}\right)$.

EXERCISES 5.2

1. $f(g(x)) = (3x-1)^2 + 2(3x-1) + 4$

3. $f(g(x)) = (9x^2 + 5x + 7)^4 + 2(9x^2 + 5x + 7)^3 + 5(9x^2 + 5x + 7)$

5. $f(x) = x^4$ and $g(x) = 5x + 1$ so $f(g(x)) = h(x) = (5x+1)^4$

7. $f(x) = x^7 + \dfrac{2}{x^3}$ and $g(x) = 2x + 1$ so $f(g(x)) = h(x) = (2x+1)^7 + \dfrac{2}{(2x+1)^3}$

9. It is a composite of x^{19} and $x^2 + 5x + 3$.
The chain rule yields $f'(x) = 19(x^2 + 5x + 3)^{18}(2x + 5)$.

11. It is a composite of x^5 and $5x^7 - 3x^4 + 3$.
The chain rule yields $f'(x) = 5(5x^7 - 3x^4 + 3)^4\,(35x^6 - 12x^3)$.

13. It is a composite of $x^{1/5}$ and $2x^4 + 3x^2 + 3$.
The chain rule yields $f'(x) = \frac{1}{5}(2x^4 + 3x^2 + 3)^{-4/5}(8x^3 + 6x)$.

15. Here, $\dfrac{dy}{du} = \dfrac{5}{2}u^{3/2}$ and $\dfrac{du}{dx} = 8$.
Multiplying yields $\dfrac{dy}{du}\cdot\dfrac{du}{dx} = \dfrac{5}{2}u^{3/2}(8) = 20u^{3/2}$.
Substituting for u yields $\dfrac{dy}{dx} = 20(8x - 3)^{3/2}$.

17. Here, $\dfrac{dy}{du} = u[5(2u+3)^4(2)] + (2u+3)^5[1] = 3(2u+3)^4(4u+1)$ and
$\dfrac{du}{dx} = 6x^2 + 7$. Multiplying yields $\dfrac{dy}{du}\cdot\dfrac{du}{dx} = 3(2u+3)^4(4u+1)(6x^2+7)$
Substituting for u yields $\dfrac{dy}{dx} = 3(4x^3 + 14x + 5)^4(8x^3 + 28x + 5)(6x^2 + 7)$.

19. Here, $\dfrac{dy}{du} = \dfrac{(u+1)[2] - (2u-1)[1]}{(u+1)^2} = \dfrac{3}{(u+1)^2}$ and $\dfrac{du}{dx} = 16x$.
Multiplying yields $\dfrac{dy}{du}\cdot\dfrac{du}{dx} = \dfrac{3}{(u+1)^2}(16x) = \dfrac{48x}{(u+1)^2}$.
Substituting for u yields $\dfrac{dy}{dx} = \dfrac{48x}{(8x^2+4)^2}$.

EXERCISES 5.3

1. $\dfrac{dy}{dx} = \dfrac{-8x}{18y} = \dfrac{-4x}{9y}$ 3. $\dfrac{dy}{dx} = \dfrac{3 - 6x^2}{28y^3}$ 5. $\dfrac{dy}{dx} = \dfrac{2(x-1)}{3(1-y^2)}$

7. $\dfrac{dy}{dx} = \left(14 + \dfrac{2}{x^3}\right)\left(\dfrac{-y^5}{12}\right)$ 9. $\dfrac{dy}{dx} = \dfrac{18 - 2xy^2}{2x^2y} = \dfrac{9 - xy^2}{x^2y}$

11. $\dfrac{dy}{dx} = \dfrac{12x - 2 - 3x^2y^5}{5x^3y^4}$ 13. $\dfrac{dy}{dx} = \dfrac{-4y^3 + 2xy - 3x^2 + 9}{12xy^2 - x^2}$.

15. $x\left[\dfrac{dy}{dx}\right] + y[1] = 0$ and $\dfrac{dy}{dx} = \dfrac{-y}{x}$. At $(-3,\ 5)$ $\dfrac{dy}{dx} = \dfrac{-5}{-3} = \dfrac{5}{3}$.

17. $\dfrac{dy}{dx} = \dfrac{2 + 9x^2}{4y^3 + 2}$. At $(-1,\ 1)$ $\dfrac{dy}{dx} = \dfrac{11}{6}$. 19. $\dfrac{dy}{dx} = \dfrac{-y}{x + 3y^2}$. At $(5, 2)$ $\dfrac{dy}{dx} = \dfrac{-2}{17}$.

21. $\dfrac{dy}{dx} = \dfrac{-y^4}{4xy^3} = \dfrac{-y}{4x}$. At $(3, 2)$ $\dfrac{dy}{dx} = \dfrac{-2}{12} = \dfrac{-1}{6}$,

⟶

so the equation of the tangent is $y - 2 = \frac{-1}{6}(x - 3)$.

23. $\frac{dy}{dt} = \frac{-x^3}{y^3}\left(\frac{dx}{dt}\right)$ 25. $\frac{dy}{dt} = \left(\frac{2x + 4y - 7}{2y - 4x}\right)\frac{dx}{dt}$

27. $\frac{dr}{dt} = \frac{3}{\pi}\ ft/\text{min}.$ 29. $\frac{dI}{dR} = -4$ 31. $\frac{dy}{dx} = \frac{-27}{16}$

EXERCISES 5.4

1. $\Delta f(x) = (x + 1)^3 - x^3 = x^3 + 3x^2 + 3x + 1 - x^3 = 3x^2 + 3x + 1$

$\Delta^2 f(x) = f(x + 1) - \Delta f(x) = 3(x + 1)^2 + 3(x + 1) + 1 - (3x^2 + 3x + 1) = 6x + 6.$

3. Firstly, add the functions as $f(x) + g(x) = x^2 + 3x + 5 + 3^x$.
Next, differencing

$$\begin{aligned}
&[(x + 1)^2 + 3(x + 1) + 5] + 3^{x+1} - (x^2 + 3x + 5 + 3^x)\\
&\quad = x^2 + 2x + 1 + 3x + 3 + 5 + 3^{x+1} - x^2 - 3x - 5 - 3^x = 2x + 4 + 3^{x+1} - 3^x\\
&\quad = 2x + 4 + 2(3^x) \text{ (in agreement with the example)}
\end{aligned}$$

Firstly, subtract the functions as $f(x) - g(x) = x^2 + 3x + 5 - 3^x$.
Next, differencing

$$\begin{aligned}
&[(x + 1)^2 + 3(x + 1) + 5] - 3^{x+1} - (x^2 + 3x + 5 - 3^x)\\
&\quad = x^2 + 2x + 1 + 3x + 3 + 5 - 3^{x+1} - x^2 - 3x - 5 + 3^x = 2x + 4 - 3^{x+1} + 3^x\\
&\quad = 2x + 4 - 2(3^x) \text{ (in agreement with the example)}
\end{aligned}$$

Firstly, multiply the functions as $f(x)\,g(x) = x^2(3^x) + 3x(3^x) + 5(3^x)$
Next, differencing

$$\begin{aligned}
&(x + 1)^2(3^{x+1}) + 3(x + 1)(3^{x+1}) + 5(3^{x+1}) - [x^2(3^x) + 3x(3^x) + 5(3^x)]\\
&\quad = (x^2 + 2x + 1 + 3x + 3 + 5)(3^{x+1}) - [(x^2 + 3x + 5)(3^x)]\\
&\quad = (x^2 + 5x + 9)(3)(3^x) - [(x^2 + 3x + 5)(3^x)]\\
&\quad = (3x^2 + 15x + 27 - x^2 - 3x - 5)(3^x)\\
&\quad = (2x^2 + 12x + 22)(3^x)\\
&\quad = 2(x^2 + 6x + 11)(3^x) \text{ (in agreement with the example)}
\end{aligned}$$

Firstly, divide the functions as $\frac{f(x)}{g(x)} = \frac{x^2 + 3x + 5}{3^x}$

⟶

Next, differencing

$$\frac{(x+1)^2+3(x+1)+5}{3^{x+1}} - \frac{x^2+3x+5}{3^x}$$
$$= \frac{(x+1)^2+3(x+1)+5}{3^{x+1}} - \frac{3(x^2+3x+5)}{3^{x+1}}$$
$$= \frac{x^2+2x+1+3x+3+5-3x^2-9x-15}{3^{x+1}}$$
$$= \frac{-2x^2-4x-6}{3^{x+1}} = \frac{-2(x^2+2x+3)}{3^{x+1}} \text{ (in agreement with the example)}$$

5. First, ${}_nP_r = \dfrac{n!}{(n-r)!} = n(n-1)\cdots(n-r+1)$

$$\Delta_n P_r = (n+1)(n)(n-1)\cdots((n+1)-r+1) - n(n-1)\cdots(n-r+1)$$
$$\Delta_n P_r = (n)(n-1)\cdots((n+1)-r+1)[(n+1)-(n-r+1)]$$
$$= (n)(n-1)\cdots((n+1)-(r-1))[r]$$
$$\Delta_n P_r = (r)[{}_nP_{r-1}]$$

7. At a maximum $f(x^*)$ must be larger than any other value of the function, so $f(x^*) \geq f(x^*+1)$ and $f(x^*) \geq f(x^*-1)$. If $f(x^*) \geq f(x^*+1)$, subtract $f(x^*)$ from each side of the inequality $0 \geq f(x^*+1) - f(x^*) = \Delta f(x^*)$. Therefore, one condition for a maximum is that $\Delta f(x^*) \leq 0$. Another condition is $f(x^*) \geq f(x^*-1)$ and by subtracting $f(x^*-1)$ from each side of the inequality $f(x^*) - f(x^*-1) \geq 0$. As $f(x^*) - f(x^*-1) = \Delta f(x^*-1)$, we have another condition for a maximum is $\Delta f(x^*-1) \geq 0$. Combining both conditions yields $\Delta f(x^*) \leq 0 \leq \Delta f(x^*-1)$
9. When $a = 1$ and $b = 3$, then

$$\sum_{x=1}^{3} \Delta f(x)g(x) = \Delta[f(1)g(1)] + \Delta[f(2)g(2)] + \Delta[f(3)g(3)]$$
$$= [f(2)g(2) - f(1)g(1)] + [f(3)g(3) - f(2)g(2)] + [f(4)g(4) - f(3)g(3)]$$
$$= f(4)g(4) - f(1)g(1) = f(3+1)g(3+1) - f(1)g(1)$$
$$= f(b+1)g(b+1) - f(a)g(a)$$

11. Use antidifferences and summations to evaluate
   a) Using the technique of Example 5.4.8, we have $\sum_{x=1}^{n} x^2 = \sum_{x=1}^{n} \Delta\left(\frac{1}{3}x^3 - \frac{1}{2}x^2 + \frac{1}{6}x\right)$, which yields

$$\left[\frac{1}{3}(n+1)^3 - \frac{1}{2}(n+1)^2 + \frac{1}{6}(n+1)\right] - \left[\frac{1}{3} - \frac{1}{2} + \frac{1}{6}\right] = \frac{1}{3}n^3 + \frac{1}{2}n^2 + \frac{1}{6}n.$$

⟶

b) Using the earlier result that $a^x = \Delta\left(\frac{a^x}{a-1}\right)$ leads to

$$\sum_{x=1}^{n} a^x = \sum_{x=1}^{n} \Delta\left(\frac{a^x}{a-1}\right) = \frac{a^{n+1}}{a-1} - \frac{a^1}{a-1} = \frac{a^{n+1}-a}{a-1}$$

and, after simplifying,

$$\frac{a(a^n-1)}{a-1}.$$

c) Using the technique of Example 5.4.8, we have

$$\sum_{x=1}^{n} x^3 = \sum_{x=1}^{n} \Delta\left(\frac{1}{4}x^4 - \frac{1}{2}x^3 + \frac{1}{4}x^2\right), \text{ which yields}$$

$$\left[\frac{1}{4}(n+1)^4 - \frac{1}{2}(n+1)^3 + \frac{1}{4}(n+1)^2\right] - \left[\frac{1}{4} - \frac{1}{2} + \frac{1}{4}\right] = \frac{1}{4}n^4 + \frac{1}{2}n^3 + \frac{1}{4}n^2.$$

CHAPTER 5 SUPPLEMENTARY EXERCISES

1. $f'(x) = (5x^3 - 2x + 1)^4[2] + (2x+3)[4(5x^3 - 2x + 1)^3(15x^2 - 2)]$

3. $f'(x) = (2x^4 - 3x^2 + e^{3x})^2[3(2x+1)^2(2)]$
$\qquad + (2x+1)^3[2(2x^4 - 3x^2 + e^{3x})(8x^3 - 6x + 3e^{3x})]$

5. $f'(x) = 10\left(\frac{5x+1}{3x-5}\right)^9 \left[\frac{(3x-5)[5] - (5x+1)[3]}{(3x-5)^2}\right]$

7. $$f'(x) = \frac{(25x+3)^5\left[8(2x+3+\ln 4x)^7\left(2+\frac{1}{x}\right)\right] - (2x+3+\ln 4x)^8[5(25x+3)^4(25)]}{(25x+3)^{10}}$$

9. $$f'(x) = \frac{(4x^2-11x-9)^6\left[(x^3+5x+1)^3\left(\frac{1}{x}\right) + (\ln 3x)3(x^3+5x+1)^2(3x^2+5)\right]}{(4x^2-11x-9)^{12}} - \frac{(x^3+5x+1)^3\ln 3x[6(4x^2-11x-9)^5(8x-11)]}{(4x^2-11x-9)^{12}}$$

11. $f'(x) = (3x+1)^4\left[\frac{15x^2-18x+2}{5x^3-9x^2+2x+1}\right] + \ln(5x^3 - 9x^2 + 2x + 1)[4(3x+1)^3(3)]$

13. When $x = 2$, $y = 11$ and $f'(x) = (3x-5)^7[5] + (5x+1)[7(3x-5)^6(3)]$ evaluated when $x = 2$ yields 236. The line through (2, 11) with slope 236. The tangent line is $y - 11 = 236(x-2)$.

15. When $x = 0$, $y = 1$ and $f'(x) = (x+1)^5[3e^{3x}] + (e^{3x})[5(x+1)^4]$ evaluated when $x = 0$ yields 8. The line through (0, 1) with slope 8. The tangent line is $y - 1 = 8(x-0)$ or $y = 8x + 1$.

17. $\dfrac{dy}{du} = 14u^6$ and $\dfrac{dy}{dx} = 15x^4 - 18x + 5$ so $\dfrac{dy}{du} \cdot \dfrac{du}{dx} = 14u^6(15x^4 - 18x + 5)$.

Writing, in terms of x, yields

$$\frac{dy}{dx} = 14(3x^5 - 9x^2 + 5x + 10)^6(15x^4 - 18x + 5).$$

19. a) $f(g(x)) = f(x^7 + 2x^5 + x^2 + 1) = (x^7 + 2x^5 + x^2 + 1)^{10}$
b) Using the chain rule the derivative of x^{10} is $10x^9$. The chain rule begins with $10(x^7 + 2x^5 + x^2 + 1)^9$ this term is then multiplied by $(7x^6 + 10x^4 + 2x)$ to yield $10(x^7 + 2x^5 + x^2 + 1)^9(7x^6 + 10x^4 + 2x)$
c) The derivative of $f(g(x))$ is $10(x^7 + 2x^5 + x^2 + 1)^9(7x^6 + 10x^4 + 2x)$, in agreement with part b.

21. $3x^4\left[3y^2\dfrac{dy}{dx}\right] + y^3[12x^3] = 10y\left[\dfrac{dy}{dx}\right] + 12x$ and rearranging terms yields,
$(9x^4y^2 - 10y)\dfrac{dy}{dx} = 12x - 12x^3y^3$. Therefore, $\dfrac{dy}{dx} = \dfrac{12x - 12x^3y^3}{9x^4y^2 - 10y}$.

23. Here, $p\left(\dfrac{dv}{dt}\right) + v\left(\dfrac{dp}{dt}\right) = 0$ and rearranging terms,
$v\left(\dfrac{dp}{dt}\right) = -p\left(\dfrac{dv}{dt}\right)$ or that $\left(\dfrac{dp}{dt}\right) = -\dfrac{p}{v}\left(\dfrac{dv}{dt}\right)$. We are given $v = 75,\ p = 30$ and that $\dfrac{dv}{dt} = 5$.
Substituting yields $\dfrac{dp}{dt} = -2$.
The pressure is therefore decreasing at $\dfrac{2\text{lbs}}{\text{in}^2\text{min}}$.

25.
$$\begin{aligned}\Delta f(x) &= f(x+1) - f(x) = (x+1)^4 - x^4 = x^4 + 4x^3 + 6x^2 + 4x + 1 - x^4\\ &= 4x^3 + 6x^2 + 4x + 1\\ \Delta^2 f(x) &= \Delta(\Delta f(x))\\ &= 4(x+1)^3 + 6(x+1)^2 + 4(x+1) + 1 - [4x^3 + 6x^2 + 4x + 1]\\ &= 12x^2 + 24x + 14\end{aligned}$$

For complete solutions to these Exercises see the companion "Student Solutions Manual" by Morris and Stark.

EXERCISES 6.1

1. $\displaystyle\int 7dx = 7x + C$

3. $\displaystyle\int 5xdx = \frac{5x^2}{2} + C$

5. $\displaystyle\int 3x^{-5}dx = -\frac{3x^{-4}}{4} + C$

7. $\displaystyle\int 2tdt = t^2 + C$

9. Firstly, rewrite as $\displaystyle\int x^{1/2}dx$ to yield $\displaystyle\int x^{1/2}dx = \frac{2}{3}x^{3/2} + C$

11. Firstly, rewrite as $\int x^{2/3}dx$ to yield $\int x^{2/3}dx = \frac{3}{5}x^{5/3} + C$

13. Firstly, rewrite as $\int \frac{1}{2}x^{-3}dx$ to yield $\int \frac{1}{2}x^{-3}dx = -\frac{1}{4}x^{-2} + C$

15. Integrating yields $\int (x^{3/5} + x^{-2/3})dx = \frac{5}{8}x^{8/5} + 3x^{1/3} + C$

17. $\int e^{7t}dt = \frac{e^{7t}}{7} + C$

19. Integrating yields $\int (4x^3 + 3x^2 + 2x + 9)dx = x^4 + x^3 + x^2 + 9x + C$

21. Integrating yields $\int (e^{4x} + 1)dx = \frac{e^{4x}}{4} + x + C$

23. Integrating yields $\int \left(4 - \frac{2}{x}\right) dx = 4x - 2\ln|x| + C$

25. Rewrite $\int \left(\frac{3}{\sqrt{t}} - 2\sqrt{t}\right) dt$ as $\int (3t^{-1/2} - 2t^{1/2})dt$. Integrating yields

$$\int (3t^{-1/2} - 2t^{1/2})dt = 6t^{1/2} - \frac{4}{3}t^{3/2} + C$$

27. Firstly, integration yields $f(x) = x^3 + C$. Next, the initial condition indicates $6 = (-1)^3 + C$ and that $C = 7$ so $f(x) = x^3 + 7$.

29. Firstly, integration yields $f(x) = x^3 - x^2 + 4x + C$. Next, the initial condition indicates $10 = (2)^3 - (2)^2 + 4(2) + C$ and that $C = -2$ so $f(x) = x^3 - x^2 + 4x - 2$.

31. Differentiating part a yields $\frac{5}{2}[2xe^{x^2}] = 5xe^{x^2}$ and part b yields $5x[2xe^{x^2}] + e^{x^2}[5]$. Therefore, part a indicates that $\int 5xe^{x^2}dx = \frac{5}{2}e^{x^2} + C$

EXERCISES 6.2

1. $\Delta x = \frac{4-0}{4} = 1$ the four subintervals are [0, 1], [1, 2], [2, 3], and [3, 4]. The four left endpoints are 0, 1, 2, and 3.

3. $\Delta x = \frac{3-0}{4} = \frac{3}{4}$ the four subintervals are [0, 0.75], [0.75, 1.5], [1.5, 2.25], and [2.25, 3]. The four left endpoints are 0, 0.75, 1.5, and 2.25.

5. $\Delta x = \frac{15-1}{7} = 2$ the seven subintervals are [1, 3], [3, 5], [5, 7], [7, 9], [9, 11], [11, 13], and [13, 15]. The seven right endpoints are 3, 5, 7, 9, 11, 13, and 15.

7. $\Delta x = \frac{27-3}{6} = 4$ the six subintervals are [3, 7], [7, 11], [11, 15], [15, 19], [19, 23] and [23, 27]. The six right endpoints are 7, 11, 15, 19, 23, and 27.

9. $\Delta x = \frac{4-0}{8} = \frac{1}{2}$ the eight subintervals are [0, 0.5], [0.5, 1.0], [10, 1.5], [1.5, 2.0], [2.0, 2.5], [2.5, 3.0], [3.0, 3.5], and [3.5, 4.0]. The eight midpoints are 0.25, 0.75, 1.25, 1.75, 2.25, 2.75, 3.25 and 3.75.

11. $\Delta x = \frac{21-5}{4} = 4$ the four subintervals are [5, 9], [9, 13], [13, 17], and [17, 21]. The four midpoints are 7, 11, 15, and 19.

13. $\Delta x = \frac{9-1}{4} = 2$ the four subintervals are [1, 3], [3, 5], [5, 7], and [7, 9]. The four right endpoints are 3, 5, 7, and 9. Next, use $\frac{b-a}{n}\sum_{i=1}^{4} f(x_i)$ to yield $2[f(3)+f(5)+f(7)+f(9)] = 2[10+26+50+82] = 336.$

15. $\Delta x = \frac{20-0}{5} = 4$ the five subintervals are [0, 4], [4, 8], [8, 12], [12, 16], and [16, 20]. The five right endpoints are 4, 8, 12, 16, and 20 and $\frac{b-a}{n}\sum_{i=1}^{5} f(x_i)$ yields

$$4[f(4)+f(8)+f(12)+f(16)+f(20)]$$
$$= 4[64+512+1728+4096+8000] = 57{,}600.$$

17. $\Delta x = \frac{23-3}{4} = 5$ the four subintervals are [3, 8], [8, 13], [13, 18], and [18, 23]. The four left endpoints are 3, 8, 13, and 18 and $\frac{b-a}{n}\sum_{i=1}^{4} f(x_i)$ yields

$$5[f(3)+f(8)+f(13)+f(18)] = 5[19+44+69+94] = 1130.$$

19. $\Delta x = \frac{9-1}{4} = 2$ the four subintervals are [1, 3], [3, 5], [5, 7], and [7, 9]. The four left endpoints are 1, 3, 5, and 7. Next, use $\frac{b-a}{n}\sum_{i=1}^{4} f(x_i)$ to yield $2[f(1)+f(3)+f(5)+f(7)] = \frac{352}{105} \approx 3.3523.$

21. $\Delta x = \frac{9-1}{4} = 2$ the four subintervals are [1, 3], [3, 5], [5, 7], and [7, 9]. The four midpoints are 2, 4, 6, and 8. Next, use $\frac{b-a}{n}\sum_{i=1}^{4} f(x_i)$ to yield

$$2[f(2)+f(4)+f(6)+f(8)] = 248.$$

23. $\Delta x = \frac{3-1}{4} = \frac{1}{2}$ the four subintervals are [1, 1.5], [1.5, 2], [2, 2.5], and [2.5, 3]. The four midpoints are 1.25, 1.75, 2.25, and 2.75. Next, use $\frac{b-a}{n}\sum_{i=1}^{4} f(x_i)$ to yield

$$\frac{1}{2}[f(1.25)+f(1.75)+f(2.25)+f(2.75)] = \frac{1}{2}[e^5+e^7+e^9+e^{11}] \approx 34{,}611.13598.$$

25. $\Delta x = \dfrac{7-1}{6} = 1$ the six subintervals are [1, 2], [2, 3], [3, 4], [4, 5], [5, 6], and [6, 7]. The six right endpoints are 2, 3, 4, 5, 6, and 7. Next, use $\dfrac{b-a}{n}\sum_{i=1}^{6} f(x_i)$ to yield $1[f(2)+f(3)+f(4)+f(5)+f(6)+f(7)]$. Using the graph yields

$$1[6+10+7+4+7+10] = 44.$$

27. $\Delta x = \dfrac{10-2}{8} = 1$ the eight subintervals are [2, 3], [3, 4], [4, 5], [5, 6], [6, 7], [7, 8], [8, 9], and [9, 10]. The eight left endpoints are
2, 3, 4, 5, 6, 7, 8, and 9.
Next, use $\dfrac{b-a}{n}\sum_{i=1}^{8} f(x_i)$ to yield
$1[f(2)+f(3)+f(4)+f(5)+f(6)+f(7)+f(8)+f(9)]$ and using the graph yields
$1[6+10+7+4+7+10+14+12] = 70$.

29. $\Delta x = \dfrac{9-1}{4} = 2$ the four subintervals are [1, 3], [3, 5], [5, 7], and [7, 9]. The four midpoints are 2, 4, 6, and 8. Next, use $\dfrac{b-a}{n}\sum_{i=1}^{4} f(x_i)$ to yield
$2[f(2)+f(4)+f(6)+f(8)]$ and using the graph yields
$2[6+7+7+14] = 68$.

31. a) Using a rectangle and a right triangle the area is determined geometrically to be $(1\times 4)+(1/2)(4\times 16) = 36$ square units.
    b) Using a Riemann Sum, $\dfrac{b-a}{n}$ is unity and the four right endpoints are 1, 2, 3, and 4. The approximate area is
    $1[f(1)+f(2)+f(3)+f(4)] = 1[5+9+13+17] = 44$.
    c) Integration yields the true area $\displaystyle\int_0^4 (4x+1)dx = (2x^2+x)|_0^4 = 36$

EXERCISES 6.3

1. $\displaystyle\int_4^9 dx = x|_4^9 = 9-4 = 5$

3. $\displaystyle\int_1^4 3dx = 3x|_1^4 = 12-3 = 9$

5. $\displaystyle\int_2^5 (2x+3)dx = (x^2+3x)|_2^5 = 30$

7. $\displaystyle\int_1^2 (4x^3+2x+5)dx = x^4+x^2+5x|_1^2 = 23$

9. $\int_0^{15} (e^{3p})dp = \left(\frac{e^{3p}}{3}\right)\Big|_0^{15} = \frac{1}{3}(e^{45} - 1)$

11. $\int_3^6 \frac{1}{t}dt = \ln\ t|_3^6 = \ln\ 6 - \ln 3 = \ln 2$

13. $\int_1^3 \left(\frac{4}{t^2}\right) dt = \left(-\frac{4}{t}\right)\Big|_1^3 = \frac{8}{3}$

15. $\int_{-1}^4 (8x^3 + 5x + 2)dx = \left(2x^4 + \frac{5x^2}{2} + 2x\right)\Big|_{-1}^4 = \frac{1115}{2}$

17. $\int_1^9 \sqrt{t}dt = \left(\frac{2}{3}t^{3/2}\right)\Big|_1^9 = \frac{52}{3}$

19. $\int_1^8 \sqrt[3]{x^2}dx = \frac{3}{5}x^{5/3}\Big|_1^8 = \frac{93}{5}$

21. $\int_1^8 \left(x^3 + 3x^2 + \sqrt[3]{x}\right) dx = \left(\frac{x^4}{4} + x^3 + \frac{3}{4}x^{4/3}\right)\Big|_1^8 = 1546.$

23. $\int_0^3 \left(4 - \frac{x^2}{4}\right) dx = \left(4x - \frac{x^3}{12}\right)\Big|_0^3 = \left(12 - \frac{27}{12}\right) - (0 - 0) = \frac{39}{4}.$

25. Geometrically, the area is the rectangle from $x = 2$ to $x = 7$ with a height of 6, so the area is $6(5) = 30$. Using integration yields

$$\int_2^7 6\,dx = (6x)|_2^7 = 42 - 12 = 30.$$

27. Geometrically, the area can be found using a rectangle and a square with total area $(2 \times 3) + (1/2)(2 \times 6) = 12$. Using integration,

$$\int_1^3 3x\,dx = \left(\frac{3x^2}{2}\right)\Big|_1^3 = \frac{27}{2} - \frac{3}{2} = 12.$$

EXERCISES 6.4

1. $\int_{-1}^3 [-x^2 + 2x + 3]dx = \left(\frac{-x^3}{3} + x^2 + 3x\right)\Big|_{-1}^3 = \frac{32}{3}.$

3. $\int_0^1 (\sqrt{x} - x^2)dx = \left(\frac{2x^{3/2}}{3} - \frac{x^3}{3}\right)\Big|_0^1 = \frac{1}{3}.$

5. $\int_0^4 (4x - x^2)dx = \left(2x^2 - \frac{x^3}{3}\right)\Big|_0^4 = \frac{32}{3}.$

7. $\int_{-1}^3 [-x^2 + 2x + 3]dx = \left(\frac{-x^3}{3} + x^2 + 3x\right)\Big|_{-1}^3 = \frac{32}{3}.$

9. $\int_0^{5/2} [-2x^2 + 5x]dx = \left(\frac{-2x^3}{3} + \frac{5x^2}{2}\right)\Big|_0^{5/2} = \frac{125}{24}.$

11. $\int_{-2}^3 (x^2)dx = \left(\frac{x^3}{3}\right)\Big|_{-2}^3 = \frac{35}{3}.$

13. $\int_{-1}^0 [(x^3 - x^2 - 2x)]dx + \int_0^2 [(-x^3 + x^2 + 2x)]dx$

$= \left(\frac{x^4}{4} - \frac{x^3}{3} - x^2\right)\Big|_{-1}^0 + \left(\frac{-x^4}{4} + \frac{x^3}{3} + x^2\right)\Big|_0^2 = \frac{37}{12}.$

15. $\frac{1}{4-1}\int_1^4 (4x + 3)dx = \frac{1}{3}(2x^2 + 3x)|_1^4 = 13.$

17. $\frac{1}{6-2}\int_2^6 \left(\frac{1}{x}\right) dx = \frac{1}{4}(\ln x)|_2^6 = \frac{1}{4}(\ln 6 - \ln 2) = \frac{1}{4}\ln 3.$

19. $\frac{1}{8-0}\int_0^8 (x^{1/3})dx = \frac{3}{32}x^{4/3}|_0^8 = \frac{3}{2}.$

21. $\frac{1}{20-6}\int_6^{20} \left(\frac{1}{2}x^2 + x + 100\right) dx = \frac{1}{14}\left(\frac{x^3}{6} + \frac{x^2}{2} + 100x\right)\Big|_6^{20} \approx 205.67.$

23. $\int_0^5 [(-x^2 + 34) - (9)dx] = \int_0^5 (25 - x^2)dx = \left(25x - \frac{x^3}{3}\right)\Big|_0^5 = \frac{250}{3}.$

CHAPTER 6 SUPPLEMENTARY EXERCISES

1. $\int 4t^3 dt = \frac{t^4}{4} + C$. The derivative of $\frac{t^4}{4} + C$ is t^3, in agreement with the integrand so the integration checks.

3. $\int (e^{3x} + 6x + 5)dx = \frac{e^{3x}}{3} + 3x^2 + 5x + C.$

5. $\Delta x = \frac{21-3}{6} = 3$ and the right endpoints are 6, 9, 12, 15, 18, and 21.

7. $\Delta x = \frac{11-3}{4} = 2$ and the left endpoints are 3, 5, 7, and 9.
The fourth Riemann Sum is $2[f(3) + f(5) + f(7) + f(9)] = 328.$

9. $\displaystyle\int_1^4 (5 + 2e^{3x})dx = \left(5x + \frac{2e^{3x}}{3}\right)\Bigg|_1^4 = 15 + \frac{2}{3}(e^{12} - e^3).$

11. $\displaystyle\int_1^4 x^{3/2}\,dx = \left(\frac{2}{5}x^{5/2}\right)\Bigg|_1^4 = \frac{62}{5}.$

13. $\displaystyle\int_{-1}^0 [(x^3 - x)]dx + \int_0^1 [(-x^3 + x)]dx = \left(\frac{x^4}{4} - \frac{x^2}{2}\right)\Bigg|_{-1}^0 + \left(\frac{-x^4}{4} + \frac{x^2}{2}\right)\Bigg|_0^1 = \frac{1}{2}.$

15. $\displaystyle\int_{-2}^3 (-2x^2 + 2x + 12)dx = \left(\frac{-2x^3}{3} + x^2 + 12x\right)\Bigg|_{-2}^3 = \frac{125}{3}.$

17. $\displaystyle\frac{1}{2-0}\int_0^2 (4 - x^2)dx = \frac{1}{2}\left(4x - \frac{x^3}{3}\right)\Bigg|_0^2 = \frac{8}{3}.$

For complete solutions to these Exercises see the companion "Student Solutions Manual" by Morris and Stark.

EXERCISES 7.1

1. $u = 5x + 3,\ \frac{1}{5}du = dx,\ \displaystyle\int (5x + 3)^{-3/4}dx = \frac{4}{5}(5x + 3)^{1/4} + C.$

3. $u = x^2 + 5,\ \frac{1}{2}du = xdx,\ \displaystyle\int x(x^2 + 5)^4 dx = \frac{1}{10}(x^2 + 5)^5 + C.$

5. $u = x^4 + x^2 + 11,\ 2du = (8x^3 + 4x)dx,$
$\displaystyle\int (8x^3 + 4x)(x^4 + x^2 + 11)^9\ dx = \frac{1}{5}(x^4 + x^2 + 11)^{10} + C.$

7. $u = x^2,\ 4du = 8xdx,\ \displaystyle\int 8xe^{x^2}dx = 4e^{x^2} + C.$

9. $u = x^2 + 3x + 5,\ du = (2x + 3)dx,\ \displaystyle\int \frac{2x + 3}{x^2 + 3x + 5} = \ln|x^2 + 3x + 5| + C.$

11. $u = x^3 - 3x^2 + 1,\ \frac{1}{3}du = (x^2 - 2x)dx,\ \displaystyle\int \frac{x^2 - 2x}{x^3 + 3x^2 + 1}dx = \frac{1}{3}\ln|x^3 - 3x^2 + 1| + C.$

13. $u = x^5 + 1,\ \frac{1}{5}du = (x^4)dx,\ \displaystyle\int \frac{x^4}{x^5 + 1}dx = \frac{1}{5}\ln|x^5 + 1| + C.$

15. $u = x^2 + 2,\ \frac{1}{2}du = xdx,\ \displaystyle\int \left(\frac{x}{\sqrt{x^2 + 2}}\right) dx = \sqrt{x^2 + 2} + C.$

17. $u = x^2 + 9,\ 3du = 6xdx,\ \displaystyle\int 6x\sqrt{x^2 + 9}\ dx = 2(x^2 + 9)^{3/2} + C.$

19. $u = 3x^4 + 5x^2 + 8,\ \frac{1}{2}du = (6x^3 + 5x)dx,$
$\displaystyle\int (6x^3 + 5x)(3x^4 + 5x^2 + 8)^{10}dx = \frac{1}{22}(3x^4 + 5x^2 + 8)^{11} + C.$

EXERCISES 7.2

1. $\int xe^{9x}dx = x\left(\frac{e^{9x}}{9}\right) - \int \left(\frac{e^{9x}}{9}\right)(1dx) = \frac{xe^{9x}}{9} - \frac{e^{9x}}{81} + C.$

3. $\int xe^{-x}dx = -xe^{-x} + \int e^{-x}dx = -xe^{-x} - e^{-x} + C.$

5. $\int (xe^{7x} + 4x \quad + 3)dx = \int xe^{7x}dx + \int (4x + 3)dx = \frac{xe^{7x}}{7} - \frac{e^{7x}}{49} + 2x^2 + 3x + C.$

7. $\int (x^3 \ln 5x)dx = \frac{x^4}{4} \ln 5x - \int \frac{x^3}{4}dx = \frac{x^4}{4} \ln 5x - \frac{x^4}{16} + C.$

9. $\int (6x^5 \ln 9x)dx = x^6 \ln 9x - \int x^5 dx = x^6 \ln 9x - \frac{x^6}{6} + C.$

11. $\int (x^8 \ln 3x + e^{2x} + 6)dx = \int x^8 \ln 3x dx + \int (e^{2x} + 6)dx$
$= \frac{x^9}{9} \ln 3x - \frac{x^9}{81} + \frac{e^{2x}}{2} + 6x + C.$

13. $\int \frac{5x}{\sqrt{x+2}}dx = 10x(x+2)^{1/2} - \frac{20}{3}(x+2)^{3/2} + C.$

15. $\int x(x+4)^{-2}dx = \frac{-x}{x+4} + \int (x+4)^{-1}dx = \frac{-x}{x+4} + \ln|x+4| + C.$

17. $\int \ln x^3 dx = x \ln x^3 - \int 3dx = x \ln x^3 - 3x + C.$

19. $\int (3xe^x + 6xe^{x^2} + e^{5x})dx = \left[3xe^x - \int 3e^x dx\right] + \left[\int e^u (3dx)\right] + \int e^{5x}dx$
$= 3xe^x - 3e^x + 3e^{x^2} + \frac{e^{5x}}{5} + C.$

EXERCISES 7.3

1. Using substitution, $\int_1^4 3x^2 e^{x^3} dx = \int_1^{64} e^u du = e^{64} - e$

3. Using integration by parts,

$$\int_2^5 \left(\frac{x}{\sqrt{x-1}}\right) dx = 2x(x-1)^{1/2}\Big|_2^5 - \frac{4}{3}(x-1)^{3/2}\Big|_2^5 = \frac{20}{3}.$$

5. Using substitution,

$$\int_1^2 8x^3 e^{x^4} dx = \int_1^{16} e^u (2du) = 2e^u\Big|_1^{16} = 2e^{16} - 2e.$$

7. Using substitution,

$$\int_0^5 8x\sqrt{x^2+144}\ dx = \int_{144}^{169} u^{1/2}\ (4du) = \frac{8}{3}u^{3/2}\bigg|_{144}^{169} = \frac{3,752}{3}.$$

9. Using basic and substitution,

$$\int_1^2 (2xe^{x^2} + 4x + 3)dx = \int_1^4 e^u\, du + \int_1^2 (4x+3)dx$$
$$= e^u|_1^4 + (2x^2+3x)|_1^2 = e^4 - e + 9.$$

11. Using substitution,

$$\int_0^1 (9x^2+12x)(x^3+2x^2+1)^3 dx = \int_1^4 u^3\ (3du) = \frac{3u^4}{4}\bigg|_1^4 = \frac{765}{4}.$$

13. Using integration by parts,

$$\int_1^e x^4 \ln x\, dx = \frac{x^5}{5}\ln x\bigg|_1^e - \int_1^e \frac{x^4}{5}\, dx = \frac{x^5}{5}\ln x\bigg|_1^e - \frac{x^5}{25}\bigg|_1^e = \frac{4e^5}{25} + \frac{1}{25}.$$

15. Using integration by parts,

$$\int_1^5 6x(x+3)^{-3}\ dx = -3x(x+3)^{-2}\Big|_1^5 + \int_1^5 3(x+3)^{-2}dx$$
$$= -3x(x+3)^{-2}\Big|_1^5 - \frac{3}{(x+3)}\bigg|_1^5 = \frac{21}{64}.$$

EXERCISES 7.4

1. $\int \left(\frac{2}{x^2-1}\right) dx = \int \left(\frac{1}{x-1} + \frac{1}{x+1}\right) dx = \ln|x-1| - \ln|x+1| + C.$

3. $\int \left(\frac{x^3+3x^2-2}{x^2-1}\right) dx = \int \left(x+3+\frac{1}{x-1}\right) dx = \frac{x^2}{2} + 3x + \ln|x-1| + C.$

5. $\int \frac{dx}{x^2-9} = \int \left[\frac{1}{6}\left(\frac{1}{x-3}\right) - \frac{1}{6}\left(\frac{1}{x+3}\right)\right]dx = \frac{1}{6}\ln|x-3| - \frac{1}{6}\ln|x+3| + C.$

7. $\int \left(\frac{x^2+2}{x^3-3x^2+2x}\right) dx = \int \left(\frac{1}{x} - \frac{3}{x-1} + \frac{3}{x-2}\right)dx$

$$= \ln|x| - 3\ln|x-1| + 3\ln|x-2| + C.$$

9. $\int \left(\frac{6x^2+7x-4}{x^3+x^2-2x}\right) dx = \int \left(\frac{2}{x} + \frac{1}{x+2} + \frac{3}{x-1}\right)dx$

$$= 2\ln|x| + \ln|x+2| + 3\ln|x-1| + C.$$

EXERCISES 7.5

1. $\Delta x = \frac{5-2}{6} = \frac{1}{2}$. The six midpoints are 2.25, 2.75, 3.25, 3.75, 4.25, and 4.75.

3. $\Delta x = \frac{5-2}{6} = \frac{1}{2}$. The endpoints are 2, 2.5, 3, 3.5, 4, 4.5, and 5.

5. $\Delta x = \frac{4-0}{4} = 1$. The four midpoints are 0.5, 1.5, 2.5, and 3.5. Using the midpoint rule, $\int_0^4 (x^2+5)dx \approx 1[f(0.5)+f(1.5)+f(2.5)+f(3.5)] = 41$.

7. $\Delta x = \frac{9-1}{8} = 1$. The eight midpoints are 1.5, 2.5, 3.5, 4.5, 5.5, 6.5, 7.5, and 8.5. Using the midpoint rule,

$$\int_1^9 (x^2+9x+8)dx \approx 1[f(1.5)+f(2.5)+f(3.5)+f(4.5)+f(5.5)+f(6.5)$$
$$+f(7.5)+f(8.5)] = 666.$$

9. $\Delta x = \frac{3-0}{6} = \frac{1}{2}$. Using the trapezoidal rule,

$$\int_0^3 (x^3+5x+4)dx \approx \frac{1}{4}\left[f(0)+2f\left(\frac{1}{2}\right)+2f(1)+2f\left(\frac{3}{2}\right)+2f(2)+2f\left(\frac{5}{2}\right)+f(3)\right]$$
$$= 55.3125$$

11. $\Delta x = \frac{4-0}{4} = 1$. Using the Simpson's rule,

$$\int_0^4 (x^2+5)dx \approx \frac{1}{6}[f(0)+4f(0.5)+2f(1)+4f(1.5)+2f(2)+4f(2.5)+2f(3)$$
$$+4f(3.5)+f(4)] = 41.3$$

13. $\Delta x = \frac{9-1}{8} = 1$. Using Simpson's rule,

$$\int_1^9 (x^2+9x+8)dx \approx$$
$$\frac{1}{6}\left[\begin{array}{l} f(1)+4f(1.5)+2f(2)+4f(2.5)+2f(3)+4f(3.5)+2f(4)+4f(4.5)+2f(5) \\ \quad +4f(5.5)+2f(6)+4f(6.5)+2f(7)+4f(7.5)+2f(8)+4f(8.5)+f(9) \end{array}\right]$$
$$= \frac{2000}{3}.$$

15. $\Delta x = \frac{6-2}{4} = 1$. Therefore, using the trapezoidal rule,

$$\int_2^6 f(x)dx \approx \frac{1}{2}[f(2)+2f(3)+2f(4)+2f(5)+f(6)] = 23.55$$

EXERCISES 7.6

1. $\int_5^{\infty} \frac{1}{x-3} dx = \lim_{t\to\infty} \int_5^t \frac{1}{x-3} dx = \lim_{t\to\infty} [\ln|t-3| - \ln 2] = \infty$ (diverges)

3. $\int_1^{\infty} \frac{2}{x^4} dx = \lim_{t\to\infty} \int_1^t 2x^{-4} dx = \lim_{t\to\infty} \left[\frac{-2}{3t^3} + \frac{2}{3}\right] = \frac{2}{3}.$

5. $\int_{-\infty}^{4} e^{4x} dx = \lim_{t\to\infty} \int_1^4 e^{4x} dx = \lim_{t\to-\infty} \left[\frac{e^{16}}{4} - \frac{e^{4t}}{4}\right] = \frac{e^{16}}{4}.$

7. $\int_1^{\infty} e^{3x+1} dx = \lim_{t\to\infty} \int_1^t e^{3x+1} dx = \lim_{t\to\infty} \left[\frac{e^{3t+1}}{3} - \frac{e^4}{3}\right] = \infty.$ (diverges)

9. $\int_1^{\infty} 2xe^{-x^2} dx = \lim_{t\to\infty} \int_1^t 2xe^{-x^2} dx = \lim_{t\to\infty} [-e^{-t^2} + e^{-1}] = e^{-1}.$

11. $\int_2^{\infty} \frac{1}{x \ \ln x} dx = \lim_{t\to\infty} \int_2^t \frac{1}{x \ \ln x} dx = \lim_{t\to\infty} [\ln|\ln t| - \ln|\ln 2|] = \infty.$ (diverges)

13. $\int_2^{\infty} \frac{x^2}{\sqrt{x^3-4}} dx = \lim_{t\to\infty} \int_2^{\infty} \frac{(x^2)}{\sqrt{x^3-4}} dx = \lim_{t\to\infty} \left[\frac{2}{3}\left(t^3-4\right)^{1/2} - \frac{2}{3}(4)^{1/2}\right]$
$= \infty.$ (diverges)

15. $\int_{-\infty}^{2} \frac{2}{(4-x)^3} dx = \lim_{t\to-\infty} \int_t^2 \frac{2}{(4-x)^3} dx = \lim_{t\to-\infty} \left[\frac{1}{(4-2)^2} - \frac{1}{(4-t)^2}\right] = \frac{1}{4}.$

CHAPTER 7 SUPPLEMENTARY EXERCISES

1. $\int 8x^2\sqrt{x^3+9}\,dx = \int u^{1/2}\left(\frac{8}{3}du\right) = \frac{16}{9}u^{3/2} + C = \frac{16}{9}(x^3+9)^{3/2} + C.$

3. $\int \frac{4x^3+2x}{x^4+x^2+5} dx = \int \frac{1}{u} du = \ln|u| + C = \ln|x^4+x^2+5| + C.$

5. $\int \frac{5x}{\sqrt[3]{5x+4}} dx \ = \frac{3}{2}x(5x+4)^{2/3} - \int \left(\frac{3}{2}\right)(5x+4)^{2/3} dx$
$= \frac{3}{2}x(5x+4)^{2/3} - \frac{9}{50}(5x+4)^{5/3} + C.$

7. $\int (5x+1)e^{3x} dx = (5x+1)\frac{e^{3x}}{3} - \int \frac{5}{3}e^{3x} dx = (5x+1)\frac{e^{3x}}{3} - \frac{5e^{3x}}{9} + C.$

9. $\int_1^2 20xe^{5x^2} dx = \int_5^{20} 2e^u du = 2e^u|_5^{20} = 2e^{20} - 2e^5.$

11. $\displaystyle\int \frac{2x^2-25x-33}{(x+1)^2(x-5)}dx = \int \left[\frac{5}{x+1} + \frac{1}{(x+1)^2} - \frac{3}{x-5}\right]dx$
$\displaystyle = 5\ln|x+1| - \frac{1}{x+1} - 3\ln|x-5| + C.$

13. The midpoint rule yields $1[f(2.5)+f(3.5)+f(4.5)+f(5.5)] = e^{2.5}+e^{3.5}+e^{4.5}+e^{5.5}$.

Using the trapezoidal rule yields $\frac{1}{2}e^2 + e^3 + e^4 + e^5 + \frac{1}{2}e^6$.

Simpson's Rule yields $\frac{1}{6}[e^2+4e^{2.5}+2e^3+4e^{3.5}+2e^4+4e^{4.5}+2e^5+4e^{5.5}+e^6]$

15. $\displaystyle\int_{-\infty}^{3} e^{2x+1}dx = \lim_{t\to-\infty}\int_t^3 e^{2x+1}dx = \lim_{t\to-\infty}\left[\frac{e^7}{2} - \frac{e^{2t+1}}{2}\right] = \frac{e^7}{2}.$

For complete solutions to these Exercises see the companion "Student Solutions Manual" by Morris and Stark.

EXERCISES 8.1

1. $f(2, 5) = 2(2) + 3(5) = 19$, $f(3, -1) = 2(3) + 3(-1) = 3$ and
$f(4, -3) = 2(4) + 3(-3) = -1$.

3. $f(1, 0) = 4(1) + 3(0)^2 = 4$, $f(2, -1) = 4(2) + 3(-1)^2 = 11$, and
$f(2, 1) = 4(2) + 3(1)^2 = 11$.

5. $f(1, 2) = (1)^2 + 3(1) + (2)^3 + 2(2) + 5 = 21$,
$f(0, 1) = (0)^2 + 3(0) + (1)^3 + 2(1) + 5 = 8$, and
$f(-1, -2) = (-1)^2 + 3(-1) + (-2)^3 + 2(-2) + 5 = -9$.

7. $f(2, 0) = (2) + 2(0)^3 + e^0 = 3$, $f(0, 1) = (0) + 2(1)^3 + e^1 = 2 + e$, and
$f(3, 0) = (3) + 2(0)^3 + e^0 = 4$.

9. $f(0, 1, 2) = (0)^2 + 3(1) + (2)^3 = 11$, $f(0.5, 1, 1) = (0.5)^2 + 3(1) + (1)^3 = 4.25$,
and $f(-1, 0, 2) = (-1)^2 + 3(0) + (2)^3 = 9$.

11. $f(3+h, 4) - f(3, 4) = [(3+h)^2 + 4^2] - (3^2 + 4^2)$
$= (9 + 6h + h^2 + 16] - (25) = h^2 + 6h$

13. $f(2a, 2b) = 7(2a)^{2/3}(2b)^{1/3} = 7(2)^{2/3}(2)^{1/3}a^{2/3}b^{1/3} = 2(7a^{2/3}b^{1/3})$
$= 2f(a, b)$.

15. The level curves for $f(x, y) = x + y$ and $c = 1$, 4, and 9 are shown as follows

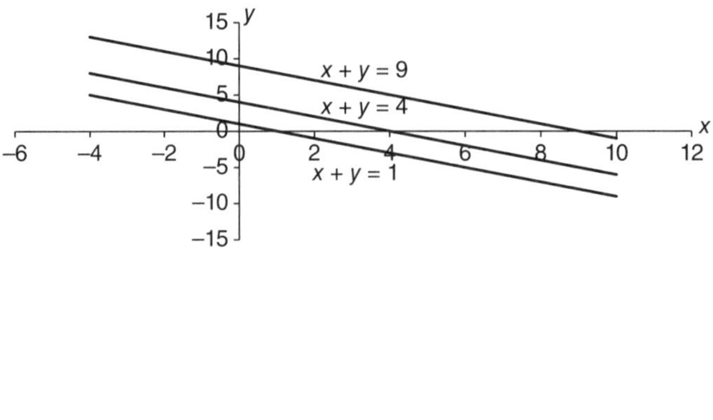

17. The level curves for $f(x, y) = xy$ and $c = 1, 4$, and 9 are shown as follows

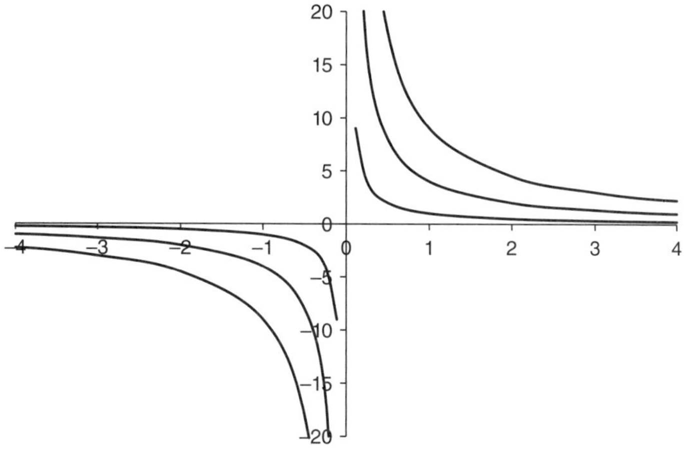

EXERCISES 8.2

1. $f_x = 20x^3 - 6$ and $f_y = -6y + 2$

3. $f_x = 25(5x^7 + 4y^5 + 7y + 3)^{24}(35x^6)$ and $f_y = 25(5x^7 + 4y^5 + 7y + 3)^{24}(20y^4 + 7)$

5. $f_x = 3x^2y^5(e^{x^3y^5}) + 36x^3$ and $f_y = 5x^3y^4(e^{x^3y^5}) + 3$

7. $f_x = \dfrac{-y^5}{x^2}$ and $f_y = \dfrac{5y^4}{x}$

9. $f_x = \dfrac{1}{3}(x^2y)^{-2/3}(2xy) = \dfrac{2}{3}x^{-1/3}y^{1/3}$ and $f_y = \dfrac{1}{3}(x^2y)^{-2/3}(x^2) = \dfrac{1}{3}x^{2/3}y^{-2/3}$

11. $f_x = (x^3e^x + 3x^2e^x)y^8$ and $f_y = x^3e^x(8y^7)$

13. $f_x = x^2\left[\dfrac{1}{3}(y-x)^{-2/3}(-1)\right] + (y-x)^{1/3}[2x]$ and $f_y = \left[\dfrac{1}{3}(y-x)^{-2/3}(1)\right](x^2)$

15. $f_x = 15x^2$, $f_y = 4y$, and $f_z = 12z^3$

17. $f_x = 3x^2y^2 + 4ze^{4xz}$, $f_y = 5 + 2x^3y + 3z^4$, and $f_z = 12yz^3 + 4xe^{4xz}$

19. $f_x = 6xy^3 + 2$ and $f(4, 1) = 6(4)(1)^3 + 2 = 26$
$f_y = 9x^2y^2 + 18y^5$ and $f(4, 1) = 9(4^2)(1)^2 + 18(1)^5 = 162$

21. The marginal productivities are $f_x = 3x^{-1/4}y^{1/4}$ and $f_y = x^{3/4}y^{-3/4}$.

EXERCISES 8.3

1. $f_x = 15x^2 - 4x$, $f_y = 4$, $f_{xx} = 30x - 4$, $f_{yy} = 0$, and $f_{xy} = f_{yx} = 0$.

3. $f_x = 10xy + 9y^5$, $f_y = 5x^2 + 45xy^4$, $f_{xx} = 10y$, $f_{yy} = 180xy^3$, and
$f_{xy} = f_{yx} = 10x + 45y^4$.

5. $f_x = 3x^2y + 9y^4$, $f_y = x^3 + 36xy^3$, $f_{xx} = 6xy, f_{yy} = 108xy^2$, and $f_{xy} = f_{yx} = 3x^2 + 36y^3$.

7. $f_x = \dfrac{-y^2}{x^2}$, $f_y = \dfrac{2y}{x}$, $f_{xx} = \dfrac{2y^2}{x^3}$, $f_{yy} = \dfrac{2}{x}$, and $f_{xy} = f_{yx} = \dfrac{-2y}{x^2}$.

9. $f_x = 3x^2y^4e^{x^3y^4} + 32x^7 + 28x^3y^5$, $f_y = 4x^3y^3e^{x^3y^4} + 35x^4y^4$,
$f_{xx} = 3x^2y^4[3x^2y^4(e^{x^3y^4})] + (e^{x^3y^4})[6xy^4] + 224x^6 + 84x^2y^5$,
$f_{yy} = 4x^3y^3[4x^3y^3(e^{x^3y^4})] + (e^{x^3y^4})[12x^3y^2] + 140x^4y^3$, and
$f_{xy} = f_{yx} = 3x^2y^4[4x^3y^3(e^{x^3y^4})] + (e^{x^3y^4})[12x^2y^3] + 140x^3y^4$.

11. $f_x = 2x$ and $f_y = -2y$, so the only possible extremum is at (0, 0).

13. $f_x = 2x + 8y$ and $f_y = 8x + 2y$, so the only possible extremum is at (0, 0).

15. $f_x = 2x + 2$, $f_y = 2y - 6$. Simultaneously setting these to zero yields (−1, 3) as a possible extremum. Using $f_{xx} = 2$, $f_{yy} = 2$, and $f_{xy} = f_{yx} = 0$, $D(-1, 3) = (2)(2) - (0)^2 > 0$. As $D(-1, 3)$ is positive and $f_{xx} > 0$, there is a local minimum $f(-1, 3) = -10$.

17. $f_x = 6x - 4y + 8$, $f_y = -4x + 6y - 17$. Simultaneously setting these to zero yields (1, 7/2) as a possible extremum. Using $f_{xx} = 6$, $f_{yy} = 6$, and $f_{xy} = f_{yx} = -4$, $D(1, 7/2) = (6)(6) - (-4)^2 > 0$. As D(1, 7/2) is positive and $f_{xx} > 0$, there is a local minimum $f(1, 7/2) = 17/4$.

19. $f_x = 40 - 2x - y$, $f_y = 50 - 2y - x$. Simultaneously setting these to zero yields (10, 20) as a possible extremum. Using $f_{xx} = -2$, $f_{yy} = -2$, and $f_{xy} = f_{yx} = -1$, $D(10,20) = (-2)(-2) - (-1)^2 > 0$. As $D(10, 20)$ is positive and $f_{xx} < 0$, there is a local maximum $f(10, 20) = 700$.

21. $f_x = 6x^2 - 6$, $f_y = -4y + 8$. Simultaneously setting these to zero yields (1, 2) and (−1, 2) as possible extrema. Using $f_{xx} = 12x$, $f_{yy} = -4$, and $f_{xy} = f_{yx} = 0$, $D(1, 2) = (12)(-4) - (0)^2 < 0$, indicating no extremum at (1, 2).
Next, as $D(-1, 2) = (-12)(-4) - (0)^2 > 0$ is positive and $f_{xx} < 0$, there is a local maximum $f(-1, 2) = 27$.

23. $f_x = 3x^2 - 2y$, $f_y = -2x + 12y^3$. Simultaneously setting these to zero yields $x = 0$ or $x = \sqrt[5]{\dfrac{4}{81}}$. The possible extrema are (0, 0) and (0.5479, 0.4503). Using $f_{xx} = 6x$, $f_{yy} = 36y^2$ and $f_{xy} = f_{yx} = -2$, $D(0, 0) = (6)(0) - (-2)^2 < 0$, indicating no extremum at (0, 0). Next, $D(0.5479, 0.4503) = (3.2874)(7.3) - (-2)^2 > 0$ and $f_{xx} > 0$, indicating a local minimum at $f(0.5479, 0.4503) = -0.2056$.

25. Use $z = 125/xy$ to minimize

$$2xy + 2y\left(\frac{125}{xy}\right) + 2x\left(\frac{125}{xy}\right) = 2xy + \frac{250}{x} + \frac{250}{y}.$$

The first partial derivatives are

$$f_x = 2y - \frac{250}{x^2} \text{ and } f_y = 2x - \frac{250}{y^2}.$$

Setting these to zero yields $x = 0$, which is not possible as no box is formed. The other possibility is that $x = 5$. The possible extrema occurs when x and y (and also z) = 5. The second partial derivatives are

$$f_{xx} = \frac{500}{x^3},\ f_{yy} = \frac{500}{y^3},\ \text{and } f_{xy} = f_{yx} = 2.\ D = (4)(4) - (2)^2 > 0 \text{ and } f_{xx} > 0,$$

so there is a minimum this case when the dimensions are 5ft by 5ft by 5ft or a cube using 150 square feet of insulation.

27. $f_x = 3 - y - 2x$ and $f_y = 4 - x - 4y$. Setting these to zero yields $\left(\frac{8}{7}, \frac{5}{7}\right)$ as a possible extremum. The second partial derivatives are

$$f_{xx} = -2,\ f_{yy} = -4 \text{ and } f_{xy} = f_{yx} = -1.\ D = (-2)(-4) - (-1)^2 > 0 \text{ and } f_{xx} < 0.$$

The maximum revenue is $f\left(\frac{8}{7}, \frac{5}{7}\right) = \frac{154}{49} = \frac{22}{7}$.

EXERCISES 8.4

1.
$$\begin{aligned} E &= (6 - B - 3A)^2 + (9 - B - 4A)^2 + (15 - B - 5A)^2 \\ E_A &= 2(6 - B - 3A)(-3) + 2(9 - B - 4A)(-4) + 2(15 - B - 5A)(-5) \\ &= 100A + 24B - 258 \\ E_B &= 2(6 - B - 3A)(-1) + 2(9 - B - 4A)(-1) + 2(15 - B - 5A)(-1) \\ &= 24A + 6B - 60 \end{aligned}$$

and the least squares line is $\hat{y} = \frac{9}{2}x - 8$.

3.
$$\begin{aligned} E &= (6 - B - 4A)^2 + (8 - B - 5A)^2 + (4 - B - 6A)^2 \\ E_A &= 2(6 - B - 4A)(-4) + 2(8 - B - 5A)(-5) + 2(4 - B - 6A)(-6) \\ &= 154A + 30B - 176 \\ E_B &= 2(6 - B - 4A)(-1) + 2(8 - B - 5A)(-1) + 2(4 - B - 6A)(-1) \\ &= 30A + 6B - 36 \end{aligned}$$

and the least squares line is $\hat{y} = -x + 11$.

5.
$$\begin{aligned} E &= (5 - B)^2 + (11 - B - 4A)^2 + (18 - B - 8A)^2. \\ E_A &= 2(11 - B - 4A)(-4) + 2(18 - B - 8A)(-8) = 160A + 24B - 376 \\ E_B &= 2(5 - B)(-1) + 2(11 - B - 4A)(-1) + 2(15 - B - 8A)(-1) \\ &= 24A + 6B - 68 \end{aligned}$$

and the least squares line is $\hat{y} = \frac{13}{8}x + \frac{29}{6}$.

7.
$$\begin{aligned} E &= (12 - B - A)^2 + (11 - B - 2A)^2 + (9 - B - 3A)^2 + (6 - B - 4A)^2. \\ E_A &= 2(12 - B - A)(-1) + 2(11 - B - 2A)(-2) + 2(9 - B - 3A)(-3) \\ &\quad +2(6 - B - 4A)(-4) = 60A + 20B - 170 \\ E_B &= 2(12 - B - A)(-1) + 2(11 - B - 2A)(-1) + 2(9 - B - 3A)(-1) \\ &\quad +2(6 - B - 4A)(-1) = 20A + 8B - 76 \end{aligned}$$

and the least squares line is $\hat{y} = -2x + \frac{29}{2}$.

9. $E = (10 - B - 3A)^2 + (27 - B - 4A)^2 + (52 - B - 5A)^2 + (102 - B - 6A)^2.$

$E_A = 2(10 - B - 3A)(-3) + 2(27 - B - 4A)(-4) + 2(52 - B - 5A)(-5)$
$\quad +2(102 - B - 6A)(-6) = 172A + 36B - 2020$

$E_B = 2(10 - B - 3A)(-1) + 2(27 - B - 4A)(-1) + 2(52 - B - 5A)(-1)$
$\quad +2(102 - B - 6A)(-1) = 36A + 8B - 382$

and the least squares line is $\hat{y} = \frac{301}{10}x - \frac{877}{10}$.

11. $E = (9 - B - 2A)^2 + (7 - B - 3A)^2 + (6 - B - 4A)^2 + (10 - B - 5A)^2.$

$E_A = 2(9 - B - 2A)(-2) + 2(7 - B - 3A)(-3) + 2(6 - B - 4A)(-4)$
$\quad +2(10 - B - 5A)(-5) = 108A + 28B - 226$

$E_B = 2(9 - B - 2A)(-1) + 2(7 - B - 3A)(-1) + 2(6 - B - 4A)(-1)$
$\quad +2(10 - B - 5A)(-1) = 28A + 8B - 64$

and the least squares line is $\hat{y} = \frac{1}{5}x + \frac{73}{10}$.

13. $E = (5 - B - A)^2 + (9 - B - 3A)^2 + (11 - B - 4A)^2.$

$E_A = 2(5 - B - A)(-1) + 2(9 - B - 3A)(-3) + 2(11 - B - 4A)(-4)$
$\quad = 52A + 16B - 152$

$E_B = 2(5 - B - A)(-1) + 2(9 - B - 3A)(-1) + 2(11 - B - 4A)(-1)$
$\quad = 16A + 6B - 50$

and the least squares line is $\hat{y} = 2x + 3$.

The least squares error is $E = (2+3-5)^2 + (3(2)+3-9)^2 + (4(2)+3-11)^2 = 0.$

15. $E = (25 - B - A)^2 + (30 - B - 4A)^2 + (32 - B - 5A)^2.$

The first partial derivatives are

$$E_A = 2(25 - B - A)(-1) + 2(30 - B - 4A)(-4) + 2(32 - B - 5A)(-5)$$
$$= 84A + 20B - 610$$
$$E_B = 2(25 - B - A)(-1) + 2(30 - B - 4A)(-1) + 2(32 - B - 5A)(-1)$$
$$= 20A + 6B - 174$$

and the least squares line is $\hat{y} = \frac{45}{26}x + \frac{302}{13}$. The price 2 months after t is \$26.69.

EXERCISES 8.5

1. $F(x, y, \lambda) = x^2 - y^2 + \lambda(3 - 2x - y)$, $F_x = 2x - 2\lambda$, $F_y = -2y - \lambda$
Therefore, $x = 2$ and $y = -1$ and the maximum of $x^2 - y^2$ is $(2)^2 - (-1)^2 = 3$.

3. $F(x, y, \lambda) = xy + \lambda(16 - x - y)$, $F_x = y - \lambda$, $F_y = x - \lambda$
Therefore, $x = 8$ and $y = 8$, so the maximum of xy is $(8)(8) = 64$.

5. $F(x, y, \lambda) = (x - 1)^2 + y^2 + \lambda(4x - y^2)$, $F_x = 2(x - 1) + 4\lambda$, $F_y = 2y - 2y\lambda$
Therefore, $y = 0$ and then $x = 0$. The point $(0, 0)$ is the point on the parabola $y^2 = 4x$ that is closest to $(1, 0)$.

7. $F(x, y, \lambda) = x^2y + \lambda(9 - x - y)$, $F_x = 2xy - \lambda$, $F_y = x^2 - \lambda$
   Therefore, $x = 6$ and $y = 3$ and that the maximum value of x^2y is 108.
9. $F(x, y, \lambda) = xy + \lambda(280 - 16x - 10y)$, $F_x = y - 16\lambda$, $F_y = x - \lambda$
   Therefore, $x = 8.75$ feet and $y = 14$ feet.
11. $F(x, y, \lambda) = 2x^2 + 4xy + \lambda(64{,}000 - x^2y)$, $F_x = 4x + 4y - 2xy\lambda$, $F_y = 4x - x^2\lambda$
   Therefore, $x = y = 40$ inches. A cube uses the least amount of material.

EXERCISES 8.6

1. $$\int_0^3 \left[\int_{-1}^2 (5x - 2y)\,dy\right] dx = \int_0^3 (15x - 3)dx = \left(\frac{15x^2}{2} - 3x\right)\Big|_0^3 = \frac{117}{2}.$$

3. $$\int_1^3 \left[\int_{-2}^3 \left(6xy^2 - x^4\right) dy\right] dx = \int_1^3 (70x - 5x^4)dx = (35x^2 - x^5)|_1^3 = 38.$$

5. $$\int_1^4 \left[\int_0^2 x^4y^5 dy\right] dx = \int_1^4 \left(\frac{32x^4}{3}\right) dx = \left(\frac{32x^5}{15}\right)\Big|_1^4 = \frac{10{,}912}{5}.$$

7. $$\int_0^2 \left[\int_0^x (x + y + 2)\,dy\right] dx = \int_0^2 \left(\frac{3x^2}{2} + 2x\right) dx = \left(\frac{x^3}{2} + x^2\right)\Big|_0^2 = 8.$$

9. $$\int_0^3 \left[\int_0^{x/2} \left(e^{2y-x}\right) dy\right] dx = \int_0^3 \left(\frac{1}{2} - \frac{e^{-x}}{2}\right) dx = \left(\frac{1}{2}x + \frac{e^{-x}}{2}\right)\Big|_0^3 = 1 + \frac{e^{-3}}{2}.$$

11. $$\int_0^1 \left[\int_{y^2}^{y-1} (5x - 2y)\,dx\right] dy = \int_0^1 \left(-\frac{5y^4}{2} + 2y^3 + \frac{y^2}{2} - 3y + \frac{5}{2}\right) dy$$
$$= \left(-\frac{y^5}{2} + \frac{y^4}{2} + \frac{y^3}{6} - \frac{3y^2}{2} + \frac{5}{2}y\right)\Big|_0^1 = \frac{7}{6}.$$

CHAPTER 8 SUPPLEMENTARY EXERCISES

1. The first partial derivatives are $f_x = 7x^6y^5e^{x^7y^5} + 36x^3y^3 + 32$ and
   $f_y = 5x^7y^4e^{x^7y^5} + 27x^4y^2$
3. $f(mx, my) = k(mx)^\alpha(my)^\beta = km^{\alpha+\beta}x^\alpha y^\beta = m^{\alpha+\beta}\lfloor kx^\alpha y^\beta\rfloor = m^1\lfloor kx^\alpha y^\beta\rfloor$
   $= m(f(x, y))$.
5. $f_x = 10(2x^3 - 3x + y^4)^9[6x^2 - 3]$ and $f_y = 10(2x^3 - 3x + y^4)^9[4y^3]$
7. The first partial derivatives are $f_x = 3x^2 - 2xy + y^2$ and $f_y = -x^2 + 2xy - 3y^2$.
   The second partial derivatives are $f_{xx} = 6x - 2y$, $f_{yy} = 2x - 6y$, $f_{xy} = -2x + 2y$, and
   $f_{yx} = -2x + 2y$. The mixed derivatives f_{xy} and f_{yx} are equal.

9. First $E = (7 - B - 3A)^2 + (15 - B - 4A)^2 + (23 - B - 7A)^2$. The first partial derivatives are

$$\begin{aligned} E_A &= 2(7 - B - 3A)(-3) + 2(15 - B - 4A)(-4) + 2(23 - B - 7A)(-7) \\ &= 148A + 28B - 484 \\ E_B &= 2(7 - B - 3A)(-1) + 2(15 - B - 4A)(-1) + 2(23 - B - 7A)(-1) \\ &= 28A + 6B - 90 \end{aligned}$$

Solving the system

$$\begin{aligned} 148A + 28B &= 484 \\ 28A + 6B &= 90 \end{aligned}$$

yields $A = 48/13$ and $B = -29/13$ so the least squares line is $\hat{y} = \frac{48}{13}x - \frac{29}{13}$.

11. Using $E = \sum_{i=1}^{n} (y_i - \beta_0 - \beta_1 x_i)^2$ the first partial derivatives are

$$\begin{aligned} E_{\beta_1} &= 2(y_1 - \beta_0 - \beta_1 x_1)(-x_1) + 2(y_2 - \beta_0 - \beta_1 x_2)(-x_2) \\ &\quad + \cdots + 2(y_n - \beta_0 - \beta_1 x_n)(-x_n) \\ E_{\beta_0} &= 2(y_1 - \beta_0 - \beta_1 x_1)(-1) + 2(y_2 - \beta_0 - \beta_1 x_2)(-1) \\ &\quad + \cdots + 2(y_n - \beta_0 - \beta_1 x_n)(-1) \end{aligned}$$

rearranging terms yields

$$2\beta_1(x_1^2 + x_2^2 + \cdots x_n^2) + 2\beta_0(x_1 + x_2 + \cdots + x_n) = 2(x_1 y_1 + x_2 y_2 + \cdots + x_n y_n)$$

and

$$2\beta_1(x_1 + x_2 + \cdots + x_n) + 2n\beta_0 = 2(y_1 + y_2 + \cdots + y_n)$$

Solving the system of equations as

$$\begin{aligned} n &\left[2\beta_1 \sum x^2 + 2\beta_0 \sum x = 2\sum xy\right] \\ \left(-\sum x\right) &\left[2\beta_1 \sum x + 2n\beta_0 = 2\sum y\right] \end{aligned}$$

yields $n\beta_1 \sum x^2 - \beta_1 \left(\sum x\right)^2 = n\sum xy - \sum x \sum y$ and therefore,

$$\beta_1 = \frac{n\sum xy - \sum x \sum y}{n\sum x^2 - \left(\sum x\right)^2} = \frac{\sum xy - \dfrac{\sum x \sum y}{n}}{\sum x^2 - \dfrac{\left(\sum x\right)^2}{n}} = \frac{SS_{xy}}{SS_{xx}}$$

and $\beta_0 = \dfrac{\sum y}{n} - \beta_1 \dfrac{\sum x}{n} = \bar{y} - \beta_1 \bar{x}$

13. Minimize $F(x, y) = x^2 + y^2$ subject to $x + y = f$. The Lagrangian is $F(x, y, \lambda) = x^2 + y^2 + \lambda(f - x - y)$. The partial derivatives are $F_x = 2x - \lambda$ and $F_y = 2y - \lambda$, indicating $\lambda = 2x = 2y$ or that $x = y$. To minimize the sum of squares each number is equal to f/2.

15. $$\int_1^2 \left[\int_0^1 e^{x+y} dx\right] dy = \int_1^2 e^{1+y} - e^y) dy$$
$$= (e^{1+y} - e^y)|_1^2 = (e^3 - e^2) - (e^2 - e) = e^3 - 2e^2 + e.$$

For complete solutions to these Exercises see the companion "Student Solutions Manual" by Morris and Stark.

EXERCISES 9.1

1. $a = 1,\ r = \frac{1}{2}$, and $S = \dfrac{1}{1 - \frac{1}{2}} = 2.$

3. $a = \frac{2}{3},\ r = \frac{1}{3}$, and $S = \dfrac{\frac{2}{3}}{1 - \frac{1}{3}} = 1.$

5. $a = \frac{4}{5},\ r = \frac{-1}{5}$, and $S = \dfrac{\frac{4}{5}}{1 - \left(\frac{-1}{5}\right)} = \frac{2}{3}.$

7. $a = \frac{5}{2},\ r = \frac{1}{2}$, and $S = \dfrac{\frac{5}{2}}{1 - \frac{1}{2}} = 5.$

9. $a = 5,\ r = \frac{-1}{4}$, and $S = \dfrac{5}{1 - \left(\frac{-1}{4}\right)} = 4.$

11. Rewritten as $\frac{1}{10} + \frac{1}{10^2} + \frac{1}{10^3} + \cdots$ yields $a = \frac{1}{10},\ r = \frac{1}{10}$, and $S = \dfrac{\frac{1}{10}}{1 - \frac{1}{10}} = \frac{1}{9}.$

13. Rewritten as $\frac{16}{10^2} + \frac{16}{10^4} + \frac{16}{10^6} + \cdots$ yields $a = \frac{16}{100},\ r = \frac{1}{100}$, and
$$S = \frac{\frac{16}{100}}{1 - \frac{1}{100}} = \frac{16}{99}.$$

15. Rewritten as $\frac{135}{10^3} + \frac{135}{10^6} + \frac{135}{10^9} + \cdots$ yields $a = \frac{135}{1,000}$, $r = \frac{1}{1,000}$ and

$$S = \frac{\frac{135}{1,000}}{1 - \frac{1}{1,000}} = \frac{135}{999} = \frac{5}{37}.$$

17. Rewritten as $(1) + \frac{4}{10} + \frac{4}{10^2} + \frac{4}{10^3} + \cdots$ yields $a = \frac{4}{10}$, $r = \frac{1}{10}$ and

$$S = (1) + \frac{\frac{4}{10}}{1 - \frac{1}{10}} = (1) + \frac{4}{9} = \frac{13}{9}.$$

19. Rewritten as $(3) + \frac{43}{10^2} + \frac{43}{10^4} + \frac{43}{10^6} + \cdots$ yields $a = \frac{43}{100}$, $r = \frac{1}{100}$ and

$$S = (3) + \frac{\frac{43}{100}}{1 - \frac{1}{100}} = (3) + \frac{43}{99} = \frac{340}{99}.$$

21. $a = \frac{1}{27}$, $r = \frac{1}{3}$ and $S = \frac{\frac{1}{27}}{1 - \frac{1}{3}} = \frac{1}{18}$.

23. $a = \frac{3}{5}$, $r = \frac{3}{5}$ and $S = \frac{\frac{3}{5}}{1 - \frac{3}{5}} = \frac{3}{2}$.

25. We seek the sum of $15 + 15(0.80) + 15(0.80)^2 + 15(0.80)^3 + \cdots$ Therefore, $a = 15$ and $r = 0.80$ and $S = \frac{15}{1 - 0.80} = 75$. There is a 75-million-dollar multiplier effect for the state economy.

EXERCISES 9.2

1. $f(x) = e^x$, $f'(x) = e^x$, $f''(x) = e^x$, and $f'''(x) = e^x$. The third-degree Maclaurin Polynomial is

$$1 + x + \frac{x^2}{2} + \frac{x^3}{6}.$$

3. $f(x) = 2x^3 - 3x^2 + 1$, $f'(x) = 6x^2 - 6x$, $f''(x) = 12x - 6$ $f'''(x) = 12$. The third-degree Maclaurin Polynomial is

$$1 - 3x^2 + 2x^3.$$

5. $f(x) = \ln(x+1)$, $f'(x) = \dfrac{1}{x+1}$, $f''(x) = \dfrac{-1}{(x+1)^2}$, and $f'''(x) = \dfrac{2}{(x+1)^3}$.

The third-degree Maclaurin Polynomial is

$$x - \frac{x^2}{2} + \frac{x^3}{3}.$$

7. $f(x) = (x+1)^{3/2}$, $f'(x) = \dfrac{3}{2}(x+1)^{1/2}$, $f''(x) = \dfrac{3}{4}(x+1)^{-1/2}$, and

$f'''(x) = \dfrac{-3}{8}(x+1)^{-3/2}$. The Third-degree Maclaurin Polynomial is

$$1 + \frac{3}{2}x + \frac{3}{8}x^2 - \frac{1}{16}x^3.$$

9. $f(x) = \dfrac{1}{1-x}$, $f'(x) = \dfrac{1}{(1-x)^2}$, $f''(x) = \dfrac{2}{(1-x)^3}$, and $f'''(x) = \dfrac{6}{(1-x)^4}$

The third-degree Maclaurin Polynomial is

$$1 + x + x^2 + x^3.$$

11. $f(x) = \ln\, x$, $f'(x) = \dfrac{1}{x}$, $f''(x) = \dfrac{-1}{x^2}$, and $f'''(x) = \dfrac{2}{x^3}$.

The third-degree Taylor Polynomial is

$$(x-1) - \frac{1}{2}(x-1)^2 + \frac{1}{3}(x-1)^3.$$

13. $f(x) = \dfrac{1}{3-x}$, $f'(x) = \dfrac{1}{(3-x)^2}$, $f''(x) = \dfrac{2}{(3-x)^3}$, $f'''(x) = \dfrac{6}{(3-x)^4}$ and $f^{iv}(x) = \dfrac{24}{(3-x)^5}$. The fourth-degree Taylor Polynomial is

$$1 + (x-2) + (x-2)^2 + (x-2)^3 + (x-2)^4.$$

15. $f(x) = e^x$, $f'(x)e^x$, $f''(x) = e^x$, $f'''(x) = e^x$, $f^{iv}(x) = e^x$, $f^{v}(x) = e^x$ and $f^{vi}(x) = e^x$ The sixth-degree Taylor Polynomial is

$$e + e(x-1) + e\frac{(x-1)^2}{2} + e\frac{(x-1)^3}{6} + e\frac{(x-1)^4}{24} + e\frac{(x-1)^5}{120} + e\frac{(x-1)^6}{720}$$

17. $f(x) = e^{2x}$, $f'(x) = 2e^{2x}$, $f''(x) = 4e^{2x}$, $f'''(x) = 8e^{2x}$, and $f^{iv}(x) = 16e^{2x}$.
The fourth-degree Taylor Polynomial is

$$e^6 + 2e^6(x-3) + 2e^6(x-3)^2 + \frac{4e^6(x-3)^3}{3} + \frac{2e^6(x-3)^4}{3}.$$

19. $f(x) = x^{3/2}, f'(x) = \frac{3}{2}x^{1/2}, f''(x) = \frac{3}{4}x^{-1/2}$, and $f'''(x) = \frac{-3}{8}x^{-3/2}$.
The third-degree Taylor Polynomial is

$$8 + 3(x-4) + \frac{3}{16}(x-4)^2 - \frac{1}{128}(x-4)^3.$$

21. $f'(x) = \frac{1}{2}x^{-1/2}$, $f''(x) = \frac{-1}{4}x^{-3/2}$, and $f'''(x) = \frac{3}{8}x^{-5/2}$. The Taylor polynomial takes the form

$$f(x) = 1 + \frac{1}{2}(x-1) - \frac{1}{8}(x-1)^2 + \frac{1}{16}(x-1)^3 + \cdots \text{ and}$$

$$f(1.04) = 1 + \frac{1}{2}(0.04) - \frac{1}{8}(0.04)^2 \approx 1.020.$$

23. $f'(x) = \frac{1}{3}x^{-2/3}$, and $f''(x) = \frac{-2}{9}x^{-5/3}$. The Taylor polynomial takes the form

$$f(x) = 3 + \frac{1}{27}(x-27) - \frac{1}{2187}(x-27)^2 + \cdots \text{ and } f(26.98) = 3 + \frac{1}{27}(-0.02)$$
$$\approx 2.999.$$

25. $f(x) = e^x, f'(x) = e^x, f''(x) = e^x$, and $f'''(x) = e^x$. The third-degree Taylor (Maclaurin) Polynomial is $f(x) = 1 + x + \frac{x^2}{2} + \frac{x^3}{6}$ and
$f(0.1) = 1 + (0.1) + \frac{(0.1)^2}{2} + \frac{(0.1)^3}{6} \approx 1.105$

EXERCISES 9.3

1. $f(x) = \frac{1}{1+x}$, $f'(x) = -(1+x)^{-2}$, $f''(x) = 2(1+x)^{-3}$, $f'''(x) = -6(1+x)^{-4}$, and $f^{iv}(x) = 24(1+x)^{-5}$, which yields

$$\frac{1}{1+x} = 1 - x + x^2 - x^3 + x^4 + \cdots \quad + x^n + \cdots \quad = \sum_{n=0}^{\infty}(-x)^n \quad \text{for} \quad -1 < x < 1.$$

3. $f(x) = \frac{1}{1+2x}, f'(x) = -2(1+2x)^{-2}, f''(x) = 8(1+2x)^{-3}, f'''(x) = -48(1+2x)^{-4}$, and $f^{iv}(x) = 384(1+2x)^{-5}$. The expansion yields

$$\frac{1}{1+2x} = 1 + (-2x) + (-2x)^2 + (-2x)^3 + (-2x)^4 + \cdots \quad + (-2x)^n + \cdots$$
$$= \sum_{n=0}^{\infty}(-2x)^n.$$

5. $f(x) = \frac{1}{1+x^2}, f'(x) = -2x(1+x^2)^{-2}, f''(x) = \frac{6x^2-2}{(1+x^2)^3}$,

$$f'''(x) = \frac{24x - 24x^{-3}}{(1+x^2)^4}, \text{ and } f^{iv}(x) = \frac{120x^4 - 240x^2 + 24}{(1+x^2)^5}.$$

$\longrightarrow$

The expansion yields

$$\frac{1}{1+x^2} = 1 + (-x^2) + (-x^2)^2 + \cdots + (-x^2)^n + \cdots = \sum_{n=0}^{\infty} (-x^2)^n.$$

(Notice that it would be simpler to replace x by $-x^2$ in Exercise 1 for $-1 < x < 1$).

7. $f(x) = \ln(1+x)$, $f'(x) = \dfrac{1}{1+x}$, $f''(x) = -(1+x)^{-2}$, $f'''(x) = 2(1+x)^{-3}$, and $f^{\text{iv}}(x) = -6(1+x)^{-4}$. The expansion yields

$$\ln(1+x) = x - \frac{x^2}{2} + \frac{x^3}{3} - \frac{x^4}{4} + \cdots + (-1)^{n+1}\frac{x^n}{n} + \cdots = \sum_{n=1}^{\infty} (-1)^{n+1}\frac{x^n}{n}$$

or $\sum_{n=0}^{\infty} (-1)^n \dfrac{x^{n+1}}{n+1}$.

9. $f(x) = 4e^{x/2}$, $f'(x) = 2e^{x/2}$, $f''(x) = e^{x/2}$, $f'''(x) = \dfrac{1}{2}e^{x/2}$, and $f^{\text{iv}}(x) = \dfrac{1}{4}e^{x/2}$. The expansion yields

$$4e^{x/2} = 4 + 2x + \frac{x^2}{2} + \frac{x^3}{12} + \frac{x^4}{96} + \cdots + (4)\frac{\left(\frac{x}{2}\right)^n}{n!} + \cdots = 4\sum_{n=0}^{\infty} \frac{(1/2)^n x^n}{n!}.$$

(The series for e^x could be used by replacing x with x/2 and then multiplying 4).

11. $f(x) = xe^x - x$, $f'(x) = (x+1)e^x - 1$, $f''(x) = (x+2)e^x$, $f'''(x) = (x+3)e^x$, and $f^{\text{iv}}(x) = (x+4)e^x$. The expansion yields

$$0 + (0)(x) + 2\left(\frac{x^2}{2!}\right) + 3\left(\frac{x^3}{3!}\right) + 4\left(\frac{x^4}{4!}\right) + \cdots n\frac{(x)^n}{n!} + \cdots = \sum_{n=2}^{\infty} \frac{x^n}{(n-1)!}.$$

13. Making use of the expansion for e^x yields

$$\begin{aligned}\frac{1}{2}(e^x + e^{-x}) &= \frac{1}{2}\left[1 + x + \frac{x^2}{2!} + \frac{x^3}{3!} + \cdots + \frac{x^n}{n!} + \cdots\right] \\ &\quad + \frac{1}{2}\left[1 + (-x) + \frac{(-x)^2}{2!} + \frac{(-x)^3}{3!} + \cdots + \frac{(x)^n}{n!} + \cdots\right] \\ &= \frac{1}{2}\left[2 + 0\,(x) + 2\left(\frac{x^2}{2!}\right) + 0\left(\frac{x^3}{3!}\right) + 2\left(\frac{x^4}{4!}\right) + \cdots\right] \\ &= 1 + \frac{x^2}{2!} + \frac{x^4}{4!} + \frac{x^n}{n!} \qquad n = 0,2,4\ \ldots\end{aligned}$$

$$\text{or } \frac{1}{2}(e^x + e^{-x}) = \sum_{n=0}^{\infty} \frac{(x)^{2n}}{(2n)!}$$

15. $f\left(\frac{1}{2}\right) = \ln\left(\frac{1}{2}\right)$, $f'(x) = \frac{1}{x}$, $f'\left(\frac{1}{2}\right) = \frac{1}{2}$, $f''(x) = \frac{-1}{x^2}$, $f''\left(\frac{1}{2}\right) = -4$
$f'''(x) = \frac{2}{x^3}$, $f'''\left(\frac{1}{2}\right) = 16$, and $f^{\text{iv}}(x) = \frac{-6}{x^4}$, $f^{\text{iv}}\left(\frac{1}{2}\right) = -96$.
Substituting yields,

$$\ln\left(\frac{1}{2}\right) + 2\left(x - \frac{1}{2}\right) - 4\frac{\left(x - \frac{1}{2}\right)^2}{2!} + 16\frac{\left(x - \frac{1}{2}\right)^3}{3!} - 96\frac{\left(x - \frac{1}{2}\right)^4}{4!} + \cdots$$

$$\ln\left(\frac{1}{2}\right) + \sum_{n=1}^{\infty} \frac{(-1)^{n+1}(2)^n\left(x - \frac{1}{2}\right)^n}{n}.$$

17. Exercise 9.3.1 yielded (for $-1 < x < 1$)

$$\frac{1}{1+x} = 1 - x + x^2 - x^3 + x^4 + \cdots + x^n + \cdots = \sum_{n=0}^{\infty} (-x)^n$$

$$\int \frac{1}{1+x}dx = \int (1 - x + x^2 - x^3 + x^4 + \cdots + (-x)^n + \cdots)dx = \int \sum_{n=0}^{\infty} (-x)^n dx$$

$$\ln|x+1| + C = x - \frac{x^2}{2} + \frac{x^3}{3} - \frac{x^4}{4} + \frac{x^5}{5} - \cdots + \frac{(-1)^n x^{n+1}}{n+1} + \cdots$$

If $x = 0$, then $\ln 1 + C = 0$ and therefore, $C = 0$ and therefore,
$\ln|x+1| = \sum_{n=0}^{\infty} \frac{(-1)^n x^{n+1}}{n+1}$.

EXERCISES 9.4

1. $\displaystyle\int_3^{\infty} \frac{1}{x}dx = \lim_{t\to\infty} \int_3^{t} \frac{1}{x}dx = \lim_{t\to\infty}(\ln\ t - \ln\ 3)$

As the limit approaches infinity, the series is divergent.

3. $\displaystyle\int_1^{\infty} \frac{\ln x}{x}\,dx = \lim_{t\to\infty} \int_1^{t} \frac{\ln x}{x}dx.$

Using substitution to integrate yields $\displaystyle\lim_{t\to\infty}\left[\frac{(\ln\ t)^2}{2} - \frac{(\ln\ 1)^2}{2}\right]$.
As the limit approaches infinity, the series is divergent.

5. $\displaystyle\int_2^{\infty} \frac{1}{x(\ln x)^3}\,dx = \lim_{t\to\infty} \int_2^{t} \frac{1}{x(\ln\ x)^3}\,dx.$

Using substitution to integrate yields $\displaystyle\lim_{t\to\infty}\left[-\frac{1}{2(\ln\ t)^2} + \frac{1}{2(\ln\ 2)^2}\right]$.

The limit approaches $\frac{1}{2(\ln\ 2)^2}$, implying the series is convergent.

$\longrightarrow$

7. $\int_2^{\infty} 2xe^{-x^2}\,dx = \lim_{t\to\infty} \int_2^{t} 2xe^{-x^2}\,dx.$

Using substitution to integrate yields $\lim_{t\to\infty}[-e^{-t^2} + e^{-4}]$.

The limit approaches e^{-4}, implying the series is convergent.

9. Compare to $\sum_{n=1}^{\infty} \frac{1}{\sqrt{n^3}} = \sum \frac{1}{n^{3/2}}$, which is a p series that converges as $p > 1$ $(p = 3/2)$ and any series which term for term is less than a convergent series will converge.

11. Compare to $\sum_{n=1}^{\infty} \frac{1}{\sqrt{n^2}} = \sum \frac{1}{n}$, which is the divergent harmonic series $(p = 1)$. Any series which term for term is greater than a divergent series will also diverge.

13. Use $\sum_{n=1}^{\infty} \frac{n}{n^3+1} < \sum_{n=1}^{\infty} \frac{n}{n^3} = \sum_{n=1}^{\infty} \frac{1}{n^2}$ and as the latter series is a convergent p series $(p = 2)$, the series $\frac{1}{2} + \frac{2}{9} + \frac{3}{28} + \frac{4}{65} + \cdots + \frac{n}{n^3+1} + \cdots$ is convergent.

15. Use $\sum_{n=2}^{\infty} \frac{\ln n}{n^2} < \sum_{n=2}^{\infty} \frac{1}{n^{3/2}}$. The latter is a convergent p series with $p > 1$ $(p = 3/2)$. A series term for term less than a convergent series also converges so

$$\frac{\ln 2}{4} + \frac{\ln 3}{9} + \frac{\ln 4}{16} + \cdots + \frac{\ln n}{n^2} + \cdots \text{ is convergent.}$$

17. Using an integral test with substitution yields $\lim_{t\to\infty}\left(\frac{t}{(1-p)\,t^p} - \frac{\ln 2}{(1-p)(\ln 2)^p}\right)$ so when $p < 1$ the limit is infinite so the integral is divergent.

When $p > 1$ $\int_2^{\infty} \frac{1}{x(\ln x)^p}dx = \frac{1}{1-p}\frac{\ln 2}{(\ln 2)^p}$ and is convergent.

19. Using $\lim_{n\to\infty}\left|\frac{a_{n+1}}{a_n}\right| = \lim_{n\to\infty}\left|\frac{\frac{3^{n+1}}{(n+1)!}}{\frac{3^n}{n!}}\right| = \lim_{n\to\infty}\left|\frac{3^{n+1}}{(n+1)!}\cdot\frac{n!}{3^n}\right| = \lim_{n\to\infty}\left|\frac{3}{n+1}\right| = 0 < 1.$

The series $\sum_{n=0}^{\infty} \frac{3^n}{n!} = \frac{3}{1} + \frac{9}{2} + \frac{27}{6} + \cdots$ is convergent.

21. Using

$$\lim_{n\to\infty}\left|\frac{a_{n+1}}{a_n}\right| = \lim_{n\to\infty}\left|\frac{\frac{4^{n+1}}{(n+1)^3}}{\frac{4^n}{n^3}}\right| = \lim_{n\to\infty}\left|\frac{4^{n+1}}{(n+1)^3}\cdot\frac{n^3}{4^n}\right| = \lim_{n\to\infty}\left|4\frac{n^3}{(n+1)^3}\right| = 4 > 1.$$

The series $\sum_{n=1}^{\infty} \frac{4^n}{n^3} = \frac{4}{1} + \frac{16}{8} + \frac{64}{27} + \cdots$ is divergent.

23. Using

$$\lim_{n\to\infty}\left|\frac{a_{n+1}}{a_n}\right| = \lim_{n\to\infty}\left|\frac{\dfrac{(-1)^{n+1}6^{n+1}}{(n+1)!}}{\dfrac{(-1)^n 6^n}{n!}}\right| = \lim_{n\to\infty}\left|\frac{(-1)(6)n!}{(n+1)!}\right| = \lim_{n\to\infty}\left|\frac{-6}{n+1}\right| = 0 < 1.$$

The series $\sum_{n=0}^{\infty}\frac{(-1)^n 6^n}{n!} = \frac{1}{1} - \frac{6}{1} + \frac{36}{2} + \cdots$ is convergent.

EXERCISES 9.5

1. Here, $n = 14$, $a = 1$, and $d = 3$, so the sum is $14\left(1 + \frac{13}{2}(3)\right) = 287.$

3. Here, $n = 7$, $a = 11$, and $d = 4$, so the sum is $7\left(1 + \frac{6}{2}(4)\right) = 161.$

5. Here, $a = 3/2$, $r = 3/2$ and $n = 5$, so the sum is $\frac{3}{2}\left(\frac{1-\left(\frac{3}{2}\right)^5}{1-\frac{3}{2}}\right) = \frac{633}{32}.$

7. Here, $a = 1/5$, $r = 1/5$ and $n = 12$, so the sum is $\frac{1}{5}\left(\frac{1-\left(\frac{1}{5}\right)^5}{1-\frac{1}{2}}\right) = \frac{61{,}035{,}156}{244{,}140{,}625}.$

9. Here, $a = 5/2$, $r = -5/8$ and $n = 12$, so the sum is $\frac{5}{2}\left(\frac{1-\left(-\frac{5}{8}\right)^{12}}{1-\left(-\frac{5}{8}\right)}\right) \approx 1.533.$

11. Here, $a = 10$, $r = 1.2$ and $n = 6$, so the sum is $10\left(\frac{1-(1.2)^6}{1-(1.2)}\right) = \frac{62{,}062}{625}.$

13. Here, $a = 3$, $r = 5$ and $n = 6$, so the sum is $3\left(\frac{1-(5)^6}{1-(5)}\right) = 11{,}718.$

15. Here, $a = 3$, $r = -2$ and $n = 10$, so the sum is $3\left(\frac{1-(-2)^{10}}{1-(-2)}\right) = -1{,}023.$

17. Here $a = 1$, $r = 0.51$, so we seek n such that $1\left(\frac{1-(0.51)^n}{1-(0.51)}\right) < 2$. Therefore,

$$2.0408163\lfloor 1-(0.51)^n\rfloor < 2$$

$$\lfloor 1-(0.51)^n\rfloor < 0.98$$

$$(0\ .51)^n > 0\ .02$$

$$n < \frac{\ln(0.02)}{\ln(0.51)} = 5.80984$$

After the fifth day, a dose should be withheld to avoid a double dose.

CHAPTER 9 SUPPLEMENTARY EXERCISES

1. $a = \frac{2}{9}$ and $r = \frac{1}{3^3} = \frac{1}{27}$. Therefore, $S = \dfrac{\frac{2}{9}}{1 - \frac{1}{27}} = \frac{3}{13}$.

3. $a = \frac{36}{100}$ and $r = \frac{1}{100}$ and $S = \dfrac{\frac{36}{100}}{1 - \frac{1}{100}} = \frac{36}{99} = \frac{4}{11}$.

5. $a = \frac{45}{1000}$ and $r = \frac{1}{100}$ and $S = 3 + \dfrac{\frac{45}{1000}}{1 - \frac{1}{100}} = 3 + \frac{5}{110} = \frac{335}{110} = \frac{67}{22}$.

7. $\sum_{x=1}^{\infty} \left(\frac{3}{5}\right)^x$ Here, $a = \frac{3}{5}$ and $r = \frac{3}{5}$ and $S = \dfrac{\frac{3}{5}}{1 - \frac{3}{5}} = \frac{3}{2}$.

9. $2 + 2(x-3) + 4\left(\frac{(x-3)^2}{2!}\right) + 12\left(\frac{(x-3)^3}{3!}\right)$
$= 2 + 2(x-3) + 2(x-3)^2 + 2(x-3)^3$

11. $2 + \left(\frac{2}{5}\right)(x) + \left(\frac{2}{25}\right)\left(\frac{x^2}{2!}\right) + \left(\frac{2}{125}\right)\left(\frac{x^3}{3!}\right) + \cdots \left(\frac{2}{5^n}\right)\left(\frac{x^n}{n!}\right) + \cdots$
$= 2\sum_{n=0}^{\infty} \frac{(1/5)^n x^n}{n!}$.

13. The integral test yields

$$\int_1^{\infty} \frac{x^2}{x^3+2}dx = \lim_{t\to\infty} \int_1^t \frac{x^2}{x^3+2}dx.$$

Using substitution to integrate yields $\lim_{t\to\infty} \left[\frac{1}{3} \ln |t^3 + 2| - \frac{1}{3} \ln 3\right]$.

As the limit approaches infinity, the series is divergent.

15. Using

$$\lim_{n\to\infty}\left|\frac{a_{n+1}}{a_n}\right| = \lim_{n\to\infty}\left|\frac{\dfrac{(n+2)(-1)^{n+1}}{(n+1)!}}{\dfrac{(n+1)(-1)^n}{n!}}\right| = \lim_{n\to\infty}\left|\frac{(n+2)(-1)^{n+1}}{(n+1)!}\cdot\frac{n!}{(n+1)(-1)^n}\right|$$

$$= \lim_{n\to\infty}\left|\frac{(-1)(n+2)}{(n+1)^2}\right| = 0 < 1.$$

The series $\sum_{n=1}^{\infty}\dfrac{(n+1)(-1)^n}{n!}$ is convergent.

17. $5+9+13+\cdots\ 45 = 11\left(5+\frac{10}{2}(4)\right) = 275.$

19. $5\left(\dfrac{1-(3)^5}{1-3}\right) = 605.$

21. $9\left(\dfrac{1-(4)^{10}}{1-4}\right) = 3{,}145{,}725.$

For complete solutions to these Exercises see the companion "Student Solutions Manual" by Morris and Stark.

EXERCISES 10.1

1. a) continuous b) discrete c) discrete d) continuous e) continuous
3. a) $0.14+0.16+0.30+0.50 = 1.1$, which exceeds unity, so it is not a probability distribution.
   b) $0.4+0.3+0.2+0.1 = 1$, each probability is non-negative, so it is a probability distribution.
   c) $0.12+0.18+0.14+0.16+0.20+0.25+0.05 = 1.1$, which exceeds unity, so it is not a valid probability distribution.
5. (a) $P(x \geq)53 = 0.65$ (b) $P(x > 55) = 0.60$ (c) $P(x \leq 58) = 0.80$
   (d) $P(52 \leq x \leq 60) = 0.60$ (e) $P(x \leq 57) = 0.70$ (f) $P(x = 59) = 0$
7. The function $f(x)$ is at least 0 on the interval [1, 8] and

$$\int_1^8 \frac{1}{7}dx = \frac{1}{7}x\Big|_1^8 = \frac{8}{7} - \frac{1}{7} = 1.$$

9. The function $f(x)$ is at least 0 on the interval [0, 10] and

$$\int_0^{10} \frac{1}{50}xdx = \frac{x^2}{100}\Big|_0^{10} = \frac{100}{100} - \frac{0}{100} = 1.$$

11. The function $f(x)$ is at least 0 on the interval [0, 1] and

$$\int_0^1 3x^2dx = x^3|_0^1 = 1 - 0 = 1.$$

13. The function $f(x)$ is at least 0 on the interval [0, 1] and

$$\int_0^1 4x^3 dx = x^4|_0^1 = 1 - 0 = 1.$$

15. The function f(x) is at least 0 on the interval $[0, \infty)$ and

$$\int_0^\infty 3e^{-3x} = \lim_{x\to\infty} \int_0^t 3e^{-3x}dx = \lim_{t\to\infty}(-e^{-3t} + 1) = 1.$$

17. $\int_2^7 \frac{1}{7}dx = \frac{1}{7}x\Big|_2^7 = \frac{7}{7} - \frac{2}{7} = \frac{5}{7}.$

19. $\int_3^7 \frac{1}{50}xdx = \frac{1}{100}x^2\Big|_3^7 = \frac{49}{100} - \frac{9}{100} = \frac{40}{100} = \frac{2}{5}.$

21. $\int_{0.5}^1 3x^2dx = x^3|_{0.5}^1 = 1 - \frac{1}{8} = \frac{7}{8}.$

23. $\int_{0.1}^{0.8} 4x^3dx = x^4|_{0.1}^{0.8} = 0.4096 - 0.001 = 0.4095.$

25. $\int_{1/3}^5 3e^{-3x}dx = -e^{-3x}|_{1/3}^5 = -e^{-15} + e^{-1} \approx 0.3679.$

EXERCISES 10.2

1. $E(x) = 50(0.20) + 100(0.10) + 150(0.30) + 200(0.40) = 145$

$$\sigma^2 = (50 - 145)^2(0.20) + (100 - 145)^2(0.10) + (150 - 145)^2(0.30) + (200 - 145)^2(0.40) = 3225$$

$$\sigma = \sqrt{3225} = 56.789$$

3. $E(x) = 1(0.15) + 4(0.15) + 7(0.25) + 10(0.20) + 12(0.25) = 7.5$

$$\sigma^2 = (1 - 7.5)^2(0.15) + (4 - 7.5)^2(0.15) + (7 - 7.5)^2(0.25) + (10 - 7.5)^2(0.20) + (12 - 7.5)^2(0.25) = 14.55$$

$$\sigma = \sqrt{14.55} = 3.184$$

5. The mean $E(x)$, the variance σ^2, and the standard deviation σ are, respectively,

$$E(x) = \int_1^8 \frac{1}{7}xdx = \frac{x^2}{14}\Big|_1^8 = \frac{64}{14} - \frac{1}{14} = \frac{63}{14} = 4.5.$$

$$\sigma^2 = \left(\int_1^8 \frac{1}{7}x^2dx\right) - (4.5)^2 = \frac{x^3}{21}\Big|_1^8 - (4.5)^2 = \frac{343}{84} = \frac{49}{12},$$

and $\sigma = \sqrt{\frac{49}{12}} = 2.0207.$

7. The mean $E(x)$, the variance σ^2, and the standard deviation σ are, respectively,

$$E(x) = \int_0^{10} \frac{1}{50}x^2 dx = \left.\frac{x^3}{150}\right|_0^{10} = \frac{1,000}{150} - \frac{0}{150} = \frac{20}{3}$$

$$\sigma^2 = \left(\int_0^{10} \frac{1}{50}x^3 dx\right) - \left(\frac{20}{3}\right)^2 = \left.\frac{x^4}{200}\right|_0^{10} - \left(\frac{400}{9}\right) = \left(\frac{50}{9}\right),$$

and $\sigma = \sqrt{\dfrac{50}{9}} = 2.357.$

9. The mean $E(x)$, the variance σ^2, and the standard deviation σ are, respectively,

$$E(x) = \int_0^{1} 3x^2 dx = \left.\frac{3x^4}{4}\right|_0^{1} = \frac{3}{4} - 0 = \frac{3}{4}.$$

$$\sigma^2 = \left(\int_0^{1} 3x^4 dx\right) - \left(\frac{3}{4}\right)^2 = \left.\frac{3x^5}{5}\right|_0^{1} - \left(\frac{9}{16}\right) = \frac{3}{5} - \frac{9}{16} = \frac{3}{80},$$

and $\sigma = \sqrt{\dfrac{3}{80}} = 0.1936.$

11. The mean $E(x)$, the variance σ^2, and the standard deviation σ are, respectively,

$$E(x) = \int_0^{1} 4x^4 dx = \left.\frac{4x^5}{5}\right|_0^{1} = \frac{4}{5},$$

$$\sigma^2 = \left(\int_0^{1} 4x^5 dx\right) - \left(\frac{4}{5}\right)^2 = \left.\frac{2x^6}{3}\right|_0^{1} - \left(\frac{16}{25}\right) = \frac{2}{75},$$

and $\sigma = \sqrt{\dfrac{2}{75}} = 0.1633.$

13. The mean $E(x)$, the variance σ^2, and the standard deviation σ are, respectively,

$$E(x) = \int_0^{\infty} 3xe^{-3x} dx = \lim_{t\to\infty} \int_0^{t} 3xe^{-3x} dx = \lim_{t\to\infty} \left[\left.\left(-xe^{-3x} - \frac{e^{-3x}}{3}\right)\right|_0^{t}\right] = \frac{1}{3},$$

$$\sigma^2 = \left(\int_0^{\infty} 3x^2e^{-3x} dx\right) - \left(\frac{1}{3}\right)^2 = \lim_{t\to\infty} \left(\int_0^{t} 3x^2e^{-3x} dx\right) - \left(\frac{1}{9}\right)$$

$$= \lim_{t\to\infty} \left[\left.\left(-x^2e^{-3x} - \frac{2xe^{-3x}}{3} - \frac{2e^{-3x}}{9}\right)\right|_0^{t} - \left(\frac{1}{9}\right)\right] = \frac{1}{9}, \text{ and } \sigma = \frac{1}{3}.$$

EXERCISES 10.3

1. A bell curve sketch is useful. Here, the probabilities are obtained directly from a Normal table.

   a) $P(0 \leq z \leq 1.47) = 0.4292$
   b) $P(0 \leq z \leq 0.97) = 0.3340$
   c) $P(-2.36 < z < 0) = 0.4909$
   d) $P(-1.24 < z < 0) = 0.3925$
   e) $P(-2.13 \leq z \leq 0) = 0.4834$
   (f) $P(-0.19 \leq z \leq 0) = 0.0753$

3. a) This includes the entire upper half of the distribution. Sum 0.5000 and $P(-1.55 < z < 0) = 0.4394$ to yield 0.9394.

   b) This is in the upper tail so $0.5000 - 0.4686 = 0.0314$

   c) This is the lower tail so $0.5000 - 0.4292 = 0.0708$

   d) This is the entire lower half of the curve plus $P(0 < Z < 1.30)$. Therefore, $0.5000 + 0.4032 = 0.9032$.

5. a) $$P(460 \leq x \leq 640) = P\left(\frac{460-550}{100} \leq x \leq \frac{640-550}{100}\right)$$
   $$= P(-0.90 \leq z \leq 0.90) = 0.3159 + 0.3159 = 0.6318.$$

   b) $$P(x \geq 730) = P\left(z \geq \frac{730-550}{100}\right) = P(z \geq 1.80) = 0.5000 + 0.4641$$
   $$= 0.9641.$$

   c) $$P(x \geq 410) = P\left(z \geq \frac{410-550}{100}\right) = P(z \geq -1.40)$$
   $$= 0.5000 + 0.4192 = 0.9192.$$

7. Here, $\mu = 128.4$
   We seek $P(x \leq 128) \leq 0.01$. The z score corresponding to this probability is −2.33. Therefore,
   $$-2.33 = \frac{128 - 128.4}{\sigma}$$

   Solving yields a value of 0.171674 for the standard deviation, σ.

CHAPTER 10 SUPPLEMENTARY EXERCISES

1. a) $0.13 + 0.17 + 0.35 + 0.45 = 1.1$, which exceeds unity so it is not a probability distribution.

   b) $0.33 + 0.27 + 0.22 + 0.18 = 1$, each probability is non-negative, so it is a probability distribution.

3. a) $P(15) + P(16) + P(18) + P(20) = 0.30 + 0.05 + 0.20 + 0.10 = 0.65$

   b) $P(18) + P(20) = 0.20 + 0.10 = 0.30$

   c) $P(12) + P(15) = 0.20 + 0.30 = 0.50$

   d) $P(x \leq 18) = 1 - P(20) = 1 - 0.10 = 0.90$

5. Firstly, $f(x)$ is non-negative on the interval [0, 1]. Secondly,

$$\int_0^1 5x^4 dx = x^5|_0^1 = 1 - 0 = 1.$$

Therefore, the function satisfies both criteria to make it a valid probability density function.

7. The mean $E(x)$, the variance σ^2, and the standard deviation σ are, respectively,

$$E(x) = \int_0^1 5x^5 dx = \left.\frac{5x^6}{6}\right|_0^1 = \frac{5}{6},$$

$$\sigma^2 = \int_0^1 5x^6 dx - \left(\frac{5}{6}\right)^2 = \left.\frac{5x^7}{7}\right|_0^1 - \left(\frac{5}{6}\right)^2 = \frac{5}{7} - \frac{25}{36} = \frac{5}{252},$$

and $\sigma = \sqrt{\frac{5}{252}} = 0.141.$

INDEX

Fundamentals of Calculus, First Edition. Carla C. Morris and Robert M. Stark.

Companion Website: http://www.wiley.com/go/morris/calculus